INVADING NATURE -
SPRINGER SERIES IN INVASION ECOLOGY

Volume 5

Series Editor: JAMES A. DRAKE
University of Tennessee,
Knoxville, TN, U.S.A.

For other titles published in this series, go to
www.springer.com/series/7228

Management of Invasive Weeds

Inderjit
Editor

Management of Invasive Weeds

Springer

Editor
Inderjit
Centre for Environmental Management of Degraded Ecosystems (CEMDE)
University of Delhi
Delhi 110007
India

ISBN 978-1-4020-9201-5 e-ISBN 978-1-4020-9202-2

Library of Congress Control Number: 2008935887

Printed on acid-free paper

springer.com

Preface

Biological invasions are one of the major threats to our native biodiversity. The magnitude of biodiversity losses, land degradation and productivity losses of managed and natural ecosystems due to invasive species is enormous. It has an adverse impact on our efforts to maintain biodiversity and on our conservation programs, and thus could create societal instability. The ecological and environmental aspects of nonnative invasive plants are of great importance to (1) understand ecological principles involved in the management of invasives, (2) design management strategies, (3) find effective management solutions for some of the worst invaders, and (4) frame policies and regulations.

The aim of this book is to provide up-to-date insights into the management of invasives by discussing (1) ecological approaches needed to design effective management strategies, (2) recent progress in management methods and tools, (3) success and failure of management efforts for some of the worst invaders, and (4) restoration and conservation of invaded land. In an effort to achieve these objectives, contributing authors provided up-to-date reviews and discussions on the management of invasives. In the introductory chapter, the role of invasive species in species extinction and economic losses due to exotic invaders is discussed. Chapters 2–7 show the importance of understanding ecology in relation to management. Chapters 8 and 9 discuss the problem of plant invasion in reference to agriculture and horticulture. Chapter 10 outlines the biological control of forest weeds through microbial agents. Chapters 11–14 discuss invasiveness and management aspects of certain specific exotic plants. The management of invasives in aquatic ecosystems is discussed in Chaps. 15 and 16. The concluding chapter elaborates on science-based invasive plant management. Together, these chapters highlight the complexity of invasive species management and suggest that management of certain invasives will be a difficult struggle.

I am grateful to the contributors for submitting their work on time, and for their patience with manuscript revisions. I am indebted to the following referees for reviewing various chapters: Scott Steinmaus, Timothy Seastedt, Sarah Brunel, Rob Colautii, Curt Daehler, Carla D'Antonio, M. Germino, Ruth Hufbauer, David Knochel, Catherine Jarnevich, Meyerson, Mathew L. Brooks, Lockwood, Richard Mack, Travis Belote, James O. Luken, Helen Murphy, Stefan Nehring, Linda Walters, Gritta Schrader, Dean Pearson, Heinz Müller-Schärer, K. Neil Harker,

Jessica Gurevitch, Piero Genovesi, Jodie Holt, and David Chapman. I appreciate the help of Suzanne Mekking and Martine van Bezooijen, acquisition editors, Springer. It is my hope that the book will be useful to graduate students, researchers, managers, and policy makers involved in the management of exotic invasives.

Delhi, India Inderjit

Contents

Chapter 1
Invasive Plants: Their Role in Species Extinctions and Economic Losses to Agriculture in the USA

David Pimentel

Abstract The more than 50,000 species of plants, animals, and microbes introduced into the United States cause more extinction of native species than most any other threats and cause more than $120 billion in damages and control costs each year. An assessment of the invasive plants that have been introduced and their control and damage costs will be estimated.

Keywords Economic losses European purple loosestrife *Lythrum salicaria* Bog turtle Yellow star thistle *Centaurea solstitialis* European cheatgrass *Bromus tectorum* Exotic aquatic weeds *Hydrilla verticillata Pistia stratiotes* Eurasian watermilfoil *Myriophyllum spicatum* Yellow rocket *Barberia vulgaris* Canada thistle *Cirsium arvense* US Crop losses

1.1 Introduction

There are approximately 50,000 nonnative species in the United States, including plants, animals, and microbes (Pimentel et al. 2000). Some of these species are beneficial and include our introduced food crops and livestock species, and these species make up about 99% of agriculture. The value of US agriculture is more than $800 billion per year (USCB 2007).

However, there are many species of plants, animals, and microbes that have caused major economic and environmental damages to agriculture and other aspects of the US ecosystem. We have reported about $120 billion per year in environmental and public health damages in the USA (Pimentel et al. 2007). Estimating the full damage and control costs of invasive species is extremely difficult because

---

D. Pimentel
College of Agriculture and Life Sciences, Cornell University, Ithaca, NY
dp18@cornell.edu

Inderjit (ed.), *Management of Invasive Weeds,*

the actual species that have been introduced and their impact on agriculture and other aspects of the US managed and unmanaged natural systems are not fully understood. In this article, an assessment of the invasive plants that have been introduced, and their control and damage costs will be estimated.

1.2 Native and Introduced Plants

Most alien plants introduced and established in the US were introduced for food, fiber, and ornamental purposes. An estimated 5,000 species of plants have been introduced and are present in the natural or wild ecosystem (Morse et al. 1995; Audubon 2007). In addition, there are an estimated 17,000 species of native plants in US (Morin 1995). Florida has the largest number of alien plants that have been introduced. Most of these 25,000 introduced species of plants were introduced for ornamental and agricultural purposes (Florida Native Plant Society 2005). An estimated 900 species have escaped and have become established in the natural ecosystems (Refuge Net 2007). Also in California, about 3,000 species of plants have been introduced and have become established in unmanaged natural ecosystems (Dowell and Krass 1992; Pimentel et al. 2007).

Some of the nonindigenous plant species that have become established in the US have displaced several native species of plants (Pimentel et al. 2007). Some of these plant species are serious weed species and have invaded an estimated 700,000 ha of US wildlife habitat each year (Babbitt 1998). For example, the European purple loosestrife (*Lythrum salicaria*) that was introduced in the early nineteenth century as an ornamental plant has been spreading at a rate of 100,000 ha per year. Purple loosestrife is changing the wetland ecosytems that it has invaded (Costly Invaders 2006). The invading plant has reduced the biomass of 44 native plant species in various habitats and reduced the numbers of some animals, including the bog turtle (Glyptemys muhlenbergii) and several duck species (Costly Invaders 2006).

The invading purple loosestrife now exists in 48 states, and the annual control costs are estimated to be $45 million per year (Aquatic Invasives 2007). In addition, several species of biological control insects have been introduced and are providing partial control of purple loosestrife in the Northeast and Mid-west (University of Illinois 2007).

Several other species of introduced plants are having an impact on natural federal lands (Christen 2007). In the Great Smoky Mountains National Park, for example, 400 of the estimated 1,500 vascular plant species are exotic, about 26% of the flora, and ten of these are currently displacing and threatening several native plant species (Pimentel et al. 2007). The problem of introduced plants is especially significant in Hawaii, where 946 of the 2,690 plant species are nonindigenous, about 35% of the flora (Eldredge and Miller 1997). Moreover, Hawaii is particularly vulnerable because it is an island.

In some cases, one invasive plant species competitively overruns an entire ecosystem. For instance, in California, yellow star thistle (*Centaurea solstitialis*) now dominates more than 4 million hectares of northern California grassland, resulting in the total loss of this once productive grassland, valued at an estimated $200 million (Campbell 1994). Similarly, European cheatgrass (*Bromus tectorum*) is dramatically changing the vegetative flora of many natural ecosystems. This invasive annual grass has spread throughout the shrub-steppe habitat of the Great Basin in Idaho and Utah, predisposing the invaded habitat to fires (Kurdila 1995). Before the invasion of cheatgrass, fire burned once every 60–110 years, and the shrubs had a chance to become well established. Now, the occurrence of fires about every 5 years has led to a decline in shrubs and other vegetation and to the occurrence of competitive monocultures of cheatgrass on several million hectares in Idaho and Utah (University of Nevada 2007). The animals and microbes dependent on the shrubs and other indigenous vegetation have been reduced and/or exterminated.

An estimated 138 nonnative species of tree and shrub species have been introduced into native US forest and shrub ecosystems (Campbell 1998). Some of the introduced trees include salt cedar (*Tamarix ramosissima* Ledeb), eucalyptus (*Eucalyptus globulus* Labill), Brazilian pepper-tree (*Schinus terebinthifolius*), and Australian melaleuca tree (*Melaleuca alternifolia*) (Randall 1996). Some of these trees have displaced native trees, shrubs, and other vegetation types, and populations of some associated native animals and microbes. For example, the melaleuca tree is spreading at a rate of 11,000 ha per year throughout the forest and grassland ecosystems of the Florida Everglades (Campbell 1994), where it damages the native vegetation and wildlife.

Exotic aquatic weeds are also a significant problem in the United States. For example, in the Hudson River basin of New York, there are 53 exotic aquatic weed species (Mills et al. 1997). In Florida, exotic aquatic plants include hydrilla (*Hydrilla verticillata*), water hyacinth (*Eichhornia crassipes*), and water lettuce (*Pistia stratiotes*), and these invasives are altering the aquatic ecosystem for fish and animal species. These invasives are choking waterways, changing nutrient cycles, and reducing the recreational use of rivers and lakes. Active control measures are needed in the aquatic ecosystems. For example, Florida spends an estimated $14.5 million each year on just hydrilla control, mostly herbicides (Center et al. 1997). Despite this control expenditure, hydrilla infestations in just two Florida lakes cost the state an estimated $10 million per year in recreational losses, such as swimming and boating (Center et al. 1997).

In the United States as a whole, an estimated total of more than $800 million is spent on the damages and control costs of aquatic weed species (Pimentel 2005). This includes an estimated $400 million for Eurasian watermilfoil (*Myriophyllum spicatum*), a total of $229 million for purple loosestrife (*Lythrum salicaria*), and $200 million for water chestnut (*Trapa natans*) (Pimentel 2005).

In the Great Lakes Basin, there are an estimated 85 exotic plant and algae species (Pimentel 2005). Including exotic plant species and all introduced animal species, the total damage and control costs annually in the Great Lakes Basin is $5.7 billion per year (Pimentel 2005).

1.2.1 Agricultural and Forest Invasive Plants

Many weeds and plant pathogens (primarily fungi) are biological invaders and cause several billion dollars worth of losses to US crops, pastures, and forests each year. In addition, several billion dollars are spent controlling these plant pests.

1.2.1.1 Weeds

In crop ecosystems, including forage crops, an estimated 500 introduced plant species have become serious pests. These include Johnson grass (*Sorghum halepense*) and kudzu (*Pueraria lobata*), which were actually introduced as crops and became pests (Pimentel et al. 1989). Most other weeds were accidentally introduced with crop seeds, soil used as ballast, or various imported plant materials. Two of the most costly accidental introductions were yellow rocket (*Barberia vulgaris*) and Canada thistle (*Cirsium arvense*).

In US agriculture, weeds cause an overall reduction of 12% in crop yields, and this represents approximately $32 billion in lost crop production each year (USCB 2007). On the basis of the research that found that approximately 73% of weed species in the US are nonindigenous, this suggests that about $23 billion (of the $32 billion above) per year are losses from invasive weeds (Pimentel 1993). However, nonindigenous weeds are often more serious pests than native weeds. Thus, the $23 billion per year loss is a conservative estimate. In addition to the direct losses, approximately $4 billion is spent each year on herbicides used to control pest weeds. Thus, the total annual cost of introduced weeds to US agricultural economy is about $26 billion.

Please note that in making the calculation, I simply calculated the proportion of potential losses caused by nonindigenous weeds on the basis of the percentage of weed species that were nonindigenous. Clearly, if there were no nonindigenous weeds in crops, native weeds would replace them. One way to assess the impacts of nonindigenous weed introductions would be to assess their impacts relative to native weeds.

The literature confirms that nonindigenous weeds have a greater impact on crops than native weeds, but there is no estimate as to how much more severe are the nonindigenous weeds. Even though our approach does not take into account the fact that native weeds would partially substitute for exotic weeds, any potential overestimation of the impacts of exotic weeds would be cancelled out by the fact that the cost figure did not include other potential losses caused by nonindigenous weeds. For instance, I did not include the approximately $11 billion in environmental and public health impacts caused by the large quantities of herbicides and other pesticides used to control exotic weeds and other pests each year in the United States (Pimentel 2005).

Also not yet taken into account has been the effect of exotic weeds on food prices. For every 1% decrease in crop yield, on average there is a 4.5% increase in price value of the crop at the farm gate (Pimentel 1997). Consequently, because nonindigenous weeds cause more extensive crop losses than native weeds, they cause a greater increase in the cost of food.

Weeds, both native and exotic, are also a problem in pastures, where 45% of the weed species are nonindigenous (Pimentel 1993). US pastures provide approximately $10 billion in forage crops annually (USDA 2006), and the losses due to inedible weeds are estimated to be $2 billion per year. Forage loss due to nonindigenous weeds, therefore, amounts to about $1 billion per year.

Some introduced weeds, such as leafy spurge (*Euphorbia esula*), are toxic to cattle and other ungulates (Trammel and Butler 1995). In addition, several nonindigenous thistles have reduced native forage plant species in pastures, rangeland, and forests, thus reducing cattle grazing (Cotton Thistle 2007). According to Babbitt (1998), ranchers spend about $5 billion each year to control invasive nonindigenous weeds in pastures and rangeland; nevertheless, these weeds continue to spread.

Control of weeds in lawns, gardens, and golf courses makes up a significant proportion of the total management costs for lawns, gardens, and golf courses of about $36 billion per year (USCB 2007). In fact, Templeton et al. (1998) estimated that each year, about $1.3 billion of the $36 billion is spent on residential weed, insect, and disease pest control. Because a large proportion of the residential weeds, such as dandelions (*Taxaxcum officinale*), are exotics, the estimate is that $500 million is spent to control exotic weeds in residential areas and an additional $1 billion is spent to control nonindigenous weeds on golf courses. Weed trees also have an economic impact. For instance, $3 to $6 million per year is spent in efforts to control the melaleuca tree (*Melaleuca alternifolia*) in Florida (Pimentel et al. 2000). Valuable cropland may be devalued in the USA because too contaminated by Silverleaf Nightshade (*Solanum elaeagnifolium)* (Mekki 2007).

1.2.1.2 Plant Pathogens

There are an estimated 50,000 parasitic and nonparasitic diseases of plants in the United States, most of these are fungi (USDA 1960). In addition, more than 1,300 species of viruses are plant pests in the US (USDA 1960). Many of these plant microbes are nonnative and were introduced inadvertently with the seeds and other parts of host plants that were introduced. Including the introduced plant pathogens and other soil microbes, it is estimated that conservatively more than 20,000 species of microbes have invaded the United States.

US crop losses to all plant pathogens total about $33 billion per year (Pimentel 1997; USCB 2007); $21 billion each year of these losses are attributable to nonindigenous plant pathogens. In addition, growers spend $720 million each year on fungicides; about $500 million of that is used to combat nonindigenous plant pathogens specifically. The total damage and control costs of nonindigenous plant pathogens therefore amount to about $22 billion per year. In addition, on the basis of the fact that 65% of the plant pathogens are exotic, the estimated control costs of plant pathogens in lawns, gardens, and golf courses are at least $2 billion per year.

In addition, plant pathogens of forests cause a loss of about 9%, or about $7 billion of forest products each year (Hall and Moody 1994; USCB 2007). The proportion of introduced plant pathogens in forests is similar to that of introduced insects or about 30%. Thus, about $2.1 billion in forest products are lost each year to exotic plant pathogens in the United States. Again, damages from exotic pests appear to be more severe than those from native pests.

1.3 Conclusions

With more than 50,000 introduced species of plants, animals, and microbes in the United States, only a portion of these cause significant damage to agriculture, forestry, and natural ecosystems, and require costly control measures. Plants and plant pathogens are one group of invasive species that cause significant ecological damage. More research is needed for the prevention of these invasions and to improve management of pest species using environmentally safe methods.

References

Aquatic Invasives (2007) The impact of aquatic invasive species on the Great Lakes. House Subcommittee on Water Resources and the Environment. Retrieved July 25, 2007 from http://transportation.house.gov/hearings/hearingdetail.aspx?NewsID = 75

Audubon (2007) Remove exotic plant pests. Retrieved July 25, 2007 from http://www.audubon.org/bird/at_home/InvasivePests.html

Babbitt B (1998). Statement by Secretary of the Interior on invasive alien species. Proceedings, National Weed Syposium, BLM Weed Page. April 8–10, 1998

Campbell FT (1994). Killer pigs, vines, and fungi: alien species threaten native ecosystems. Endang Species Tech Bull 19:3–5

Campbell FT (1998). “Worst” invasive plant species in the conterminous United States. Report. Springfield, VA: Western Ancient Forest Campaign

Center TD, Frank JH, Dray FA (1997). Biological control. In: Simberloff D, Schmitz DC, Brown TC (eds) Strangers in paradise. Island Press, Washington, DC. pp. 245–266

Christen K (2007) Combating Alien Invaders. Retrieved July 26, 2007 from http://eerc.ra.utk.edu/sightline/V3N1/Alien.htm.

Costly Invaders (2006) Costly Invaders: The Economic Impact of Invasive Species. Retrieved July 26, 2007 from http://www.jjfnew.com/ViewNews.asp?NewsID = 42.

Cotton Thistle (2007) Cotton Thistle. Retrieved July 26, 2007 from http://en.wikipedia.org.wiki/Cotton_thistle.

Dowell RV, Krass CJ (1992) Exotic pests pose growing problem for California. California Agric 46:6–10

Eldredge LG, Miller SE (1997) Numbers of Hawaiian species: supplement 2, including a review of freshwater invertebrates. Bishop Museum Occasional Papers 48:3–32

Florida Native Plant Society (2005) Invasive exotic plants. Retrieved July 26, 2007 from http://www.fnps.org/pages/plants/invasives.php.

Hall JP, Moody B (1994) Forest depletions caused by insects and diseases in Canada 1982–1987. Forest Insect and Disease Survey Information Report ST-X-8, Ottawa, Canada: Forest Insect and Disease Survey, Canadian Forest Service, Natural Resources Canada

Kurdila J (1995) The introduction of exotic species into the United States: there goes the neighborhood. Environ Aff 16:95–118

Mekki M (2007) Biology, distribution and impacts of silverleaf nightshade (*Solanum elaeagnifolium* Cav.)* EPPO Bull 37 (1):114–118. Retrieved February 22, 2008, from, http://www.blackwell-synergy.com/doi/abs/10.1111/j.1365–2338.2007.01094.x

Mills EL, Scheuerell MD, Carlton JT, Strayer DL (1997) Biological invasions in the Hudson River Basin. New York State Museum Circular No. 57. The University of the State of New York, State Education Department.

Morin N (1995). Vascular plants of the United States. In: LaRoe ET, Farris GS, Puckett CE, Doran PD, Mac MD (eds) Our living resources: a report to the nation on the distribution, abundance, and health of U.S. plants, animals and ecosystems. U.S. Department of the Interior, National Biological Service, Washington, DC. Pp. 200–205

Morse LE, Kartesz JT, Kutner LS (1995) In: LaRoe ET, Farris GS, Puckett CE, Doran PD, Mac MD (eds) Our living resources: a report to the nation on the distribution, abundance, and health of U.S. plants, animals and ecosystems. U.S. Department of the Interior, National Biological Service, Washington, DC. Pp. 205–209

Pimentel D (1993) Habitat factors in new pest invasions. In: Kim KC, McPheron BA (eds) Evolution of insect pests — patterns of variation. Wiley, New York. pp. 165–181

Pimentel D (2005) Aquatic nuisance species in the New York State canal and Hudson River systems and the Great Lakes Basin: an economic and environmental assessment. Environ Manage 35:692–701

Pimentel D, Pimentel M, Wilson A (2007) Plant, animal and microbe invasive species in the United States and world. In: Nentwig W (ed) Biological invasions. Springer-Verlag, Berlin. pp. 315–330

Pimentel D, Hunter MS, LaGro JA, Efroymson RA, Landers JC, Mervis FT, McCarthy CA, Boyd AE (1989) Benefits and risks of genetic engineering in agriculture. BioScience 39:606–614

Pimentel D (1997). Pest management in agriculture. In: Pimentel D. (ed) Techniques for reducing pesticide use: environmental and economic benefits. Wiley, Chichester, UK. pp. 1–11

Pimentel D, Lach L, Zuniga R, Morrison D (2000) Environmental and economic costs of nonindigenous species in the United States. BioScience 50:53–65

Randall JM (1996) Weed control for the preservation of biological diversity. Weed Technol 10:370–381

Refuge Net (2007) Invasive species fact sheet. National Wildlife Refuge Association. Retrieved August 17, 2007 from http://www.refugenet.org/New-issues/invasives.html#toc01

Templeton SR, Zilberman D, Yoo SJ (1998) An economic perspective on outdoor residential pesticide use. Environ Sci Technol 32:416–423

Trammel MA, Butler JL (1995) Effects of exotic plants on native ungulate use of habitat. J Wildlife Manage 59:808–816

University of Illinois (2007) Biological control of purple loosestrife program. Retrieved July 29, 2007 from http://www.inhs.unuc.edu/cee/loosestrife/bepl.html.

University of Nevada (2007) Cheatgrass and fire. Retrieved July 29, 2007 from http://www.cabnr.unr.edu/CABNR/Newsletter/FullStory.aspx?StoryID = 41.

USCB (2007) Statistical abstracts of the United States. U.S. Census Bureau, Washington, DC

USDA (1960) Index of plant diseases in the United States. Crop Research Division, ARS. U.S. Department of Agriculture, Washington, DC

USDA (2006) Agricultural statistics. U.S. Department of Agriculture, Washington, DC

Chapter 2
Practical Considerations for Early Detection Monitoring of Plant Invasions

Matthew L. Brooks and Robert C. Klinger

Abstract Invasions by multiple nonnative species into wildland areas require that decisions be made on which species and sites to target for early detection monitoring efforts and ultimately management actions. Efficient allocation of resources to detect invasions from outside of a management unit, and to monitor their spread within a management unit, leaves more resources available for control efforts and other management priorities. In this chapter, we describe three types of monitoring plans that are possible given three typical scenarios of data availability within or adjacent to the management unit: (1) there are *no data* on invasive species, (2) there are *species lists* of invasives, and (3) there are *georeferenced abundance* data for invasive species. In the absence of invasive species data, monitoring must be guided based on the general principals of invasion biology related to propagule pressure and plant resource availability. With invasive species lists, prioritization processes can be applied to narrow the monitoring area. It is also helpful to develop separate prioritized lists for species that are currently colonizing, established but not spreading, and those that have begun to spread within a management unit, because management strategies differ for species at different phases of the invasion process. With georeferenced abundance data, predictive models can be developed for high priority species to further increase the efficiency of early detection monitoring. For the majority of invasive species management programs, we recommend a design based on integrating prioritization and predictive modeling into an optimized monitoring plan, but only if the required species information and resources to process them are available and the decision is based on well-defined management goals. Although the up-front costs of this approach appear to be high, its long-term benefits can ultimately make it more cost-effective than less systematic approaches that typify most early detection programs.

Keywords Modeling • Niche • Prioritization • Prediction • Species distribution models • Vegetation management

---

M.L. Brooks(✉) and R.C. Klinger
United States Geological Survey, Western Ecological Research Center, Yosemite Field Station, El Portal Office, El Portal, California 95318,
matt_brooks@usgs.gov

2.1 Introduction

Early detection monitoring forms the foundation of all invasive plant management programs, and is often coupled with rapid response to control incipient populations of undesirable invaders. Collectively, early detection and rapid response provide the first line of defense against plant colonizations. Compared with the spread and equilibrium phases of invasions, the colonization and, to a somewhat lesser degree, establishment phases are typically the only points at which eradication is possible (Rejmanek and Pitcairn 2002). Once invading species have established populations, or are in the process of subsequently spreading into new areas, eradication quickly becomes unfeasible. Thus, prevention of new invasions into a management unit is predicated primarily on an effective early detection and response program.

Invasive plants are managed within local project areas, preserves or agency units, counties, states, nations, and continents. Although priorities and challenges vary among these different types of management units, there are certain issues common to all which we emphasize in this chapter. One major issue is the daunting task of accounting for large numbers of potentially invading species within large areas. Resources will never be sufficient to monitor all invading species in all places. Guidelines are needed on how best to narrow search parameters for the types of species that are poised to invade and focus efforts on areas where they are most likely to invade and/or are most important to protect from invasion.

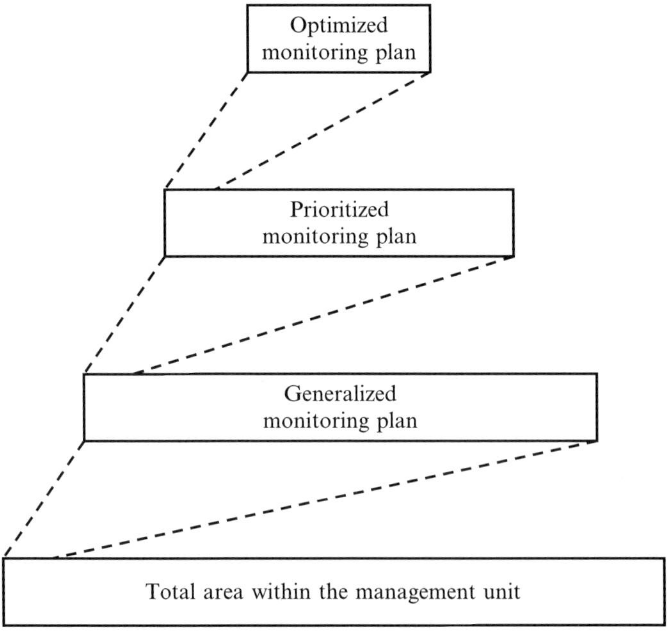

Fig. 2.1 A generalized example demonstrating how the relative proportion of sampling area can decline with each successive monitoring approach

An initial step in any early detection program involves compiling existing information on species and site characteristics to develop an efficient monitoring approach. This information is used to prioritize among species and sites that are most important to monitor, and develop predictive models to optimize monitoring efforts by narrowing their spatial and temporal scope. In this chapter, we discuss the issues associated with compiling and using information to develop and improve the efficiency of early detection monitoring plans. This chapter does not address monitoring tools (e.g., remote sensing) or specific monitoring methodologies (e.g., plot-based sampling), but rather describes a framework for narrowing the search area within which those other tools and methodologies can be applied. The framework we present is structured around three types of monitoring plans that successively reduce the size of the area within which early detection monitoring is conducted: the (1) generalized, (2) prioritized, and (3) optimized monitoring plans (Fig. 2.1) .

2.2 Evaluating Available Data

The information collection stage is perhaps more important than any other step in developing monitoring programs, because all future actions are based on analyses stemming from the information collected. Consequently, we feel it is important not to just supply a "cookbook" of what information to collect and what to do with it, but also to emphasize the importance on thinking about what types of information are most useful for different phases of the invasion process, including colonization, population establishment, and subsequent spread (Groves 1986; Cousins and Mortimer 1995; Rejmanek 2000; Richardson et al. 2000).

Before any information is compiled, the resources available for conducting an early detection program should be realistically evaluated. Time spent compiling vast amounts of information to develop an early detection plan is wasted if there is little hope of supporting the efforts needed to synthesize the information into an implementation plan or to implement the plan itself. Time and money are obvious limitations, but so too are institutional support and the personal commitment of staff. Turn-over rates of personnel can also be a hindrance, since extensive training is often required to develop effective early detection teams (M. Brooks pers. obs.).

Spatial and temporal scales are also very important to consider prior to compiling data. As mentioned above, early detection programs can be developed for areas as small as local projects to as large as continents. Clearly, the amounts and types of information needed vary among these spatial scales. For example, as geographic scale increases, so too do landscape variability, land-use variability, the range of potential sources of nonnative propagules, and many other factors influencing plant invasions, which should be considered when developing early detection programs.

In most cases there is information available on the site characteristics within a management unit. This includes vegetation maps and assessments that can be used to evaluate landscape invasibility, and information on the natural, cultural, recreational, and/or economic values at potential risk due to plant invasions. In contrast, there is a much wider range of information availability regarding invasive species data. In this chapter, we focus primarily on what to do with different amounts of invasive species data.

There are typically three scenarios relative to the availability of invasive species data within or adjacent to a management unit:

1. There are *no data* on invasive species.
2. There are species lists of invasives.
3. There are *georeferenced abundance* data for invasive species.

Because data limitations are a fundamental consideration in developing any monitoring plan, we organized this chapter around these three scenarios. As data quantity and quality increase so too do their range of potential applications for designing early monitoring plans. Accordingly, the sections of this chapter that deal with each of the three scenarios presented above become progressively longer and more detailed. We realize that the most common situation involves having no data or only having species lists. However, we devote significant attention to what can be done with georeferenced data because scientists advising land managers often emphasize the need for this type of data. We feel it necessary to explain just how resource intensive this process of generating and using georeferenced data is, so that those who may be considering this path can better determine whether the effort required is worth the potential improvement in monitoring efficiencies that may result. We hope that this approach will ultimately make it easier to translate the information we present into practice.

2.3 What can be Done in the Absence of Species Data?

It is becoming increasingly rare that there are absolutely no species data available within or near a land management unit, either because most have some sort of species inventory (e.g., plots used to validate vegetation maps) or land managers have access to regional lists of invasive plant species (e.g., invasive plant council lists). Even if species data are present, the resources may not be available to compile, synthesize, and evaluate the data. In the event that species data or resources to process the data do not exist, all efforts to develop efficient monitoring plans must rely on general invasion theory to develop a generalized monitoring plan (Fig. 2.2).

2.3.1 General Invasion Theory

Numerous interacting factors influence rates and extent of biological invasions, and their relative effects have been widely discussed and debated (Hobbs and Huenneke 1992; Lonsdale 1999; Williamson 1999; Davis et al. 2000; Rejmanek et al. 2005).

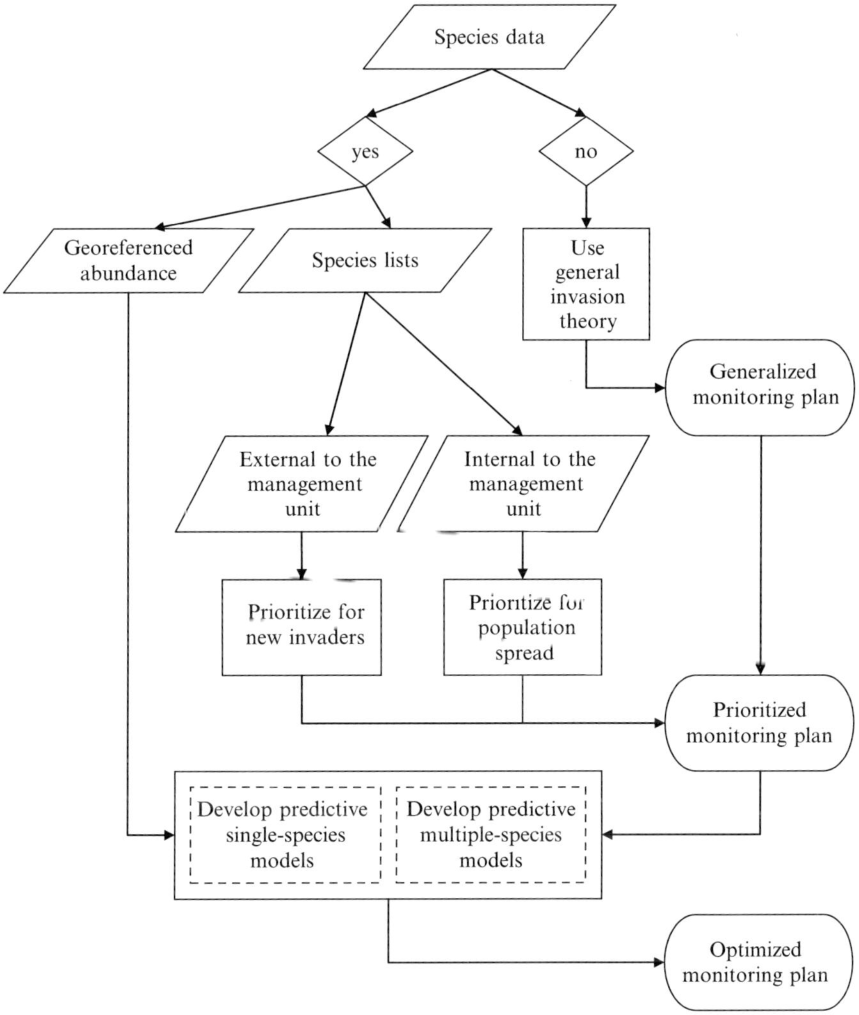

Fig. 2.2 Flowchart linking available species data with synthesis processes resulting in different hierarchical levels of final sampling plans

However, two factors appear particularly important: plant propagule pressure and plant resource availability (Davis et al. 2000; Brooks 2007). Collectively, these two factors can be used to develop a basic program for monitoring specific sites. Information collected during this basic monitoring program can then be used to evaluate and adjust monitoring as needed (Holling 1978).

Plant propagule pressure is related to the number of disseminules (e.g., seeds, rhizomes) introduced into an area per unit time and the species that they represent (Lockwood et al. 2005) Dispersal rates are positively associated with pathways such as roads and trails, vectors such as livestock, land use practices such as seeding burned areas, and the extent of area open to invasion (Forman et al. 2003; Brooks 2007). The species pool is the number of nonnative species in a region, and the larger that pool the greater the likelihood that at least one or several species will invade other areas within the region (Lockwood et al. 2005). Propagules can originate from populations outside of, or within, a management unit.

Plant resource availability is a function of the supply of light, water, and mineral nutrients, and the proportion of these resources that are unused by existing vegetation (Davis et al. 2000; Brooks 2007). Resource availability can increase due to direct additions (e.g., atmospheric nitrogen deposition), increased rates of production (e.g., nutrient cycling rates), or by reduced rates of uptake following declines in plant abundance after they are thinned or removed. Feedback processes from established populations of nonnative plants can also affect resource supply. This can occur by direct increases in nutrient supply (e.g., nitrifying plants) or indirect increases brought about by limiting the growth of other species through competition or inhibition. Areas of high resource availability are often disturbed sites. Fire, landslides, floods, and grazing not only increase the pool of available resources but may also reduce abundance of native species that would otherwise compete with invading species or, conversely, reduce invasion rates by consuming potential colonizers (Marty 2005).

2.3.2 *Generalized Monitoring Plan*

The role of disturbance in facilitating invasions is well established (Lonsdale 1999; Mack and D'Antonio 1998; Mack et al. 2000), probably because they often lead to increases in both propagule pressure and resource availability. Accordingly, disturbed areas are often high or very high priorities for early detection monitoring. However, disturbances are typically pulsed events that often cannot be predicted. Early detection monitoring plans must, therefore, include two parts: (1) a strategic baseline plan that should be updated periodically (e.g., 5 year intervals) on the basis of an assessment of propagule pressure and resource availability across the entire management unit; and (2) tactical incident plans for each major event that results in major landscape-scale pulses of propagules and/or resources (e.g., a large fire or construction project). Part 1 should be supported by a consistent and predictable source of funding, whereas part 2 should be supported as part of monitoring efforts associated with each major landscape-scale event.

Very high priority areas for early detection monitoring occur where both propagule pressure and resource availability are high (Fig. 2.3). If significant sources of invading species are present, and resources are readily available, then plant invasions have the greatest probability of occurring.

High priority areas for monitoring occur where propagule pressure is high, but resource availability is low (Fig. 2.3). Any time when propagule pressure is high there is a chance that invasive plants can establish following unanticipated

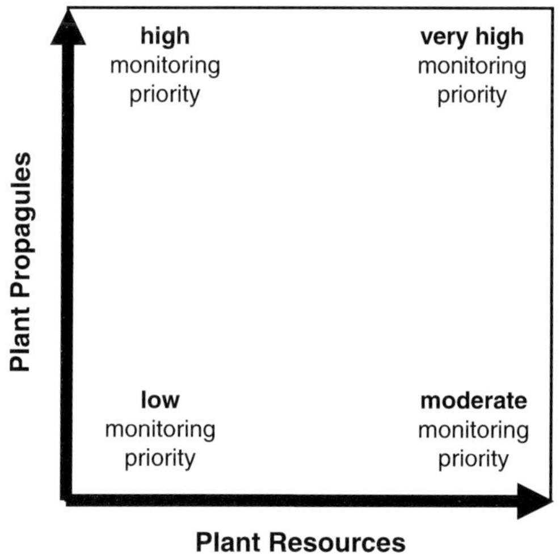

Fig. 2.3 Relative priorities for early detection monitoring relative to propagule pressure of invading plants and resource availability

surges in resource availability that would shift a site to a very high priority for monitoring. These changes can literally occur overnight, most commonly following a major disturbance such as fire, flood, or other agents of vegetation removal.

Moderate priority areas occur where there are few or no vectors and pathways to the site, and thus propagule pressure is low, but resource availability is high (Fig. 2.3). In this case long-distance dispersal is the primary means by which invasions might occur. These types of sites can quickly upgrade in priority following major influxes of propagules, which may occur following revegetation or soil stabilization projects (e.g., in seed mixes or straw mulches) or the establishment of temporary logistical support sites (e.g., fire camps).

Low priority areas occur where both propagule pressure and resource availability are low (Fig. 2.3). However, as mentioned above, these conditions can rapidly change causing a concomitant upgrade in monitoring priority.

The efficiency of generalized monitoring plans is relatively low compared with other approaches described later (Fig. 2.1), but so are the costs necessary to develop them (Fig. 2.4). However, one must remember that time and resources saved up front with generalized monitoring plans may be eclipsed by the time and resources lost due to the inefficiencies of the monitoring efforts that follow. For example, these generalized monitoring plans do not integrate information about the life history characteristics, specific habitat requirements, or potential impacts of invading species which could otherwise be used to further focus monitoring efforts.

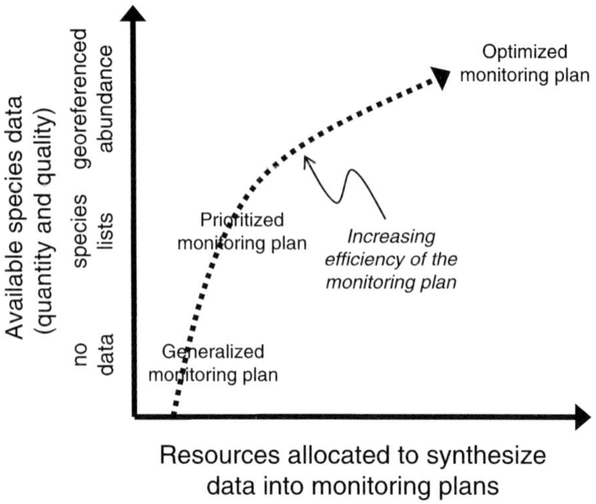

Fig. 2.4 Relationship between available species data and the resources applied to synthesize the data into monitoring plans. Both data and resource investment are required to improve the efficiency of early detection monitoring plans

2.4 What can be Done with Species Lists?

2.4.1 Types of Lists

Species lists provide the fundamental data upon which early detection programs should be based. Even programs designed to monitor sites (as opposed to searching for species; see later) benefit tremendously if species lists are used in the program design. Species lists vary in usefulness depending on their geographic scope, ancillary information, and the time that has passed since they were compiled.

Species lists have been developed for many states or multistate geographic regions within the United States. Examples from the western United States include lists for Arizona (AZ-WIPWG 2005), California (Cal-IPC 2006), and Oregon and Washington (Reichard et al. 1997). Other regions with state lists include Connecticut (Mehrhoff et al. 2003), Florida (Anonymous 1993; Florida Exotic Pest Plant Council Plant List Committee 2005), Illinois (Schwegman 1994), Rhode Island (Gould and Stuckey 1992), Tennessee (Bowen and Shea 1996), and Virginia (Virginia Department of Conservation and Recreation and Virginia Native Plant Society 2003; Heffernan et al. 2001).

Species lists can also be derived from coarse-scale regional surveys, or from finer-scale local studies. Regional lists are generally less useful than site-specific lists for programs focused on local scales, although combining the two can be particularly useful. For example, a site-specific list can be used to target management actions for species already occurring within a management unit, and a regional list can be used to design programs focused on detecting the initial establishment of species that currently occur elsewhere in the region.

Lists that are compiled to specifically document the status of nonnative plants are highly preferable over lists that are compiled for other purposes, such as general botanical surveys or validation of vegetation maps. Monitoring plans vary according to their intended purpose, and there is no single optimal plan for all applications. Consequently, the resulting species lists vary in level of specificity, accuracy, and scope. For example, surveys done to validate vegetation maps are often focused on plant associations, noting only dominant species and other species of interest. Rare occurrences (i.e., the primary targets for early detection) may be left off intentionally or simply overlooked. Accordingly, surveys that are not designed to specifically inventory nonnative plants will most likely underreport the actual number of nonnative species present in the monitoring area.

Numerous types of useful ancillary information can be included in species lists and are almost always useful in designing early detection programs. Estimates of distribution and abundance in the area of concern, even if they are qualitative (e.g., widely distributed but not abundant), are the most basic types of ancillary information that can be included. If the program goal is to monitor areas based on statistical models of the likelihood of a species colonizing a site, then geo-referenced data on environmental conditions where the species is known to occur are highly desirable (see later).

Although it may seem counter-intuitive, it is often useful to also have data on environmental conditions where species do not occur (i.e., absence data). If data on environmental variables are not available, then life history traits (e.g., perennial vs. annual, presence of rhizomatous roots, seed mass, etc.) should be included in the lists. If the program goal is to implement management based on a prioritized list of species, then data on life history characteristics, tendency to be invasive in other geographic regions, known ecological impacts, and feasibility of control are highly desirable. Older species lists (e.g., > 20–30 years) can be useful in documenting occurrence of a species in an area, but data on environmental conditions associated with them may be obsolete.

2.4.2 *The Prioritization Process*

If species lists exist and resources to evaluate them are available, then the suite of species that early-detection should most optimally focus upon can be developed using a process known as prioritization (Fig. 2.2). The prioritization process initially requires more of an obligation of time and resources than do generalized monitoring methods, but this investment results in monitoring plans of greater efficiency focused on smaller areas (Fig. 2.1) that can be more cost-effective in the long run (Fig. 2.3).

The prioritization process has been typically applied to reduce the number of species targeted for active management, but it can also be used to reduce the number of species targeted for early detection monitoring. In both cases, prioritization addresses the desire to focus management efforts, whether for control or early detection, on a reduced subset of the total species pool where they will be most effective.

2.4.2.1 Prioritization for Control of Nonnative Plants

Prioritization for control efforts has commonly been used to maximize the cost-effectiveness of efforts designed to manage species that are known to reside within a particular management unit. When faced with lists of tens to hundreds of invasive species, land managers need guidance on how best to allocate scarce resources to control them. Randall et al. (2008) recently reviewed 17 examples of systems used to help place nonnative plants into categories to facilitate their management, and compared them to a system that they developed themselves (Morse et al. 2004). Twelve of these systems were designed to prioritize management actions for nonnative species that are already established within a management unit. Two prioritized among invaded sites (Timmins and Owens 2001; Wainger and King 2001) and ten prioritized among invaded species within sites, states, or nations (Orr et al. 1993; Weiss and McLaren 1999; Thorp and Lynch 2000; Champion and Clayton 2001; Fox et al. 2001; Heffernan et al. 2001; Virtue et al. 2001; Hiebert and Stubbendieck 1993; Warner et al. 2003; Morse et al. 2004). Only two (Warner et al. 2003; Morse et al. 2004) focus heavily on species' impacts on biodiversity, whereas the rest focus mostly on feasibility of control, or potential effects on agricultural, horticultural, or other economic factors.

Prioritization decisions are typically made based on some combination of the following four factors:

1. The relative ecological and/or economic threats that the species pose
2. Their potential to spread and establish populations quickly (i.e., their "weediness")
3. Their potential geographic and/or ecological ranges
4. The feasibility in which they can be controlled (Timmins and Williams 1987; Hiebert and Stubbendieck 1993, Hiebert 1998, Weiss and McLaren 1999; Fox et al. 2000; Mehrhoff 2000; Warner et al. 2003; Morse et al. 2004)

The scoring systems for these prioritization efforts generally emphasize the threat potential and spread potential over the other two factors, with the weighted sum of the ranks for all four resulting in the net priority assessment.

Although the large number of systems may appear bewildering at first, many can be directly applied to a wide variety of areas and situations. Using an existing system will reduce the cost of developing a new system and provide managers with choices and flexibility. However, it is important to stress the necessity of selecting the system that is most appropriate for a given situation (Randall et al. 2008).

Prioritization is generally done for species that are known to be invasive, or for sites that have high conservation value but may be susceptible to invasion. In some instances, both species and sites can be prioritized for management actions (Timmins and Owens 2001), and if adequate resources and information are available this can be an extremely useful strategy. Prioritization is most often based on a synthesis of preexisting studies, expert opinion, or both (Randall et al. 2001; Hiebert and Stubbendieck 1993; Timmins and Owens 2001). Attributes are then scored on an ordinal scale.

2.4.2.2 Prioritization for Early-Detection of Nonnative Plants

Prioritization for early-detection monitoring has not resulted in the wide range of approaches that have been developed for the task of prioritizing for control efforts. However, the basic premise of both is the same, and there is no compelling reason that systems developed to inform control efforts could not be used (with minor modifications) to help inform early-detection monitoring efforts. They both rely on information related to threat potential, spread potential, range of potential geographic/ecological sites, and feasibility of control. The one primary difference is that species that have low feasibility of control should raise their priority level in terms of early-detection monitoring, but may lower its priority level in terms of control. Basically, species that are more difficult to control should have higher priority in situations where early-detection monitoring is used to identify new populations and keep them from establishing. In contrast, among species already established within a region, those that are more difficult to control may be prioritized lower for control efforts than those which are easier to control.

2.4.3 *Information Needed for Prioritization*

Relatively few life history characteristics have been found to be consistently good predictors of invasiveness (Kolar and Lodge 2001). Therefore, rather than spending inordinate amounts of time trying to collect as much information as possible on a very large number of species and site attributes (the "shotgun" approach), a more logical and focused approach will produce better (and more timely) results. When prioritizing species, careful attention needs to be given to what phase of the invasion process the rankings are meant to address. Management objectives will differ among the phases as will the relative importance of species attributes.

2.4.3.1 Information for Prioritizing Species

The management objective for species in the colonization phase of invasion is to prevent their introduction and establishment. Developing a list of species with the greatest potential for being introduced into the area of interest is a critical step in any effort to prevent such introductions. In most cases, this phase will be the most difficult to develop a prioritized list for because the pool of potential species will likely be quite large. Once a list of candidate species is developed, useful information for prioritizing includes: (1) invasiveness potential, (2) biogeographic range, (3) land cover types where typically invasive, and (4) potential impacts (Table 2.1).

After invading species have established localized populations, eradication becomes a priority (Rejmanek and Pitcairn 2002). If eradication is not feasible then control of populations (i.e., reducing abundance and/or dispersal pathways and vectors) within the boundaries of local infestations may be an alternative. However, it is important to recognize that even if eradication or control is successful, species could be reintroduced into an area. Clearly, high priority species in this stage would be those that tend to fit the definition of a "transformer species," which cause significant changes in community and ecosystem characteristics (Richardson et al. 2000) and have ecological and life-history characteristics associated with rapid spread potential. Therefore, the primary focus of prioritization at the establishment phase includes: (1) actual and potential impacts, (2) distribution and abundance, (3) life-history characteristics,| (4) biogeographic range, and (5) management feasibility (Table 2.1).

Species in the more advanced invasion stages of spread and equilibrium are widely distributed and are often relatively abundant. Eradication is unlikely (Rejmanek and Pitcairn 2002), so containment of existing populations or preventing them from becoming established in high priority sites (see next section) are

Table 2.1 Information needed to develop prioritized lists of species in different phases of the invasion process

A. Colonization phase	
1) Invasiveness potential	Tendency to be invasive elsewhere
2) Biogeographic range	Natural ("native") range Nonnative ("invasive") range
3) Land cover types where invasive	
4) Potential Impacts	
B. Establishment phase	
1) Actual and potential impacts	Ecosystems Structure Species composition
2) Distribution and abundance	Distribution in target sites Distribution in adjacent sites Abundance in adjacent sites
3) Life history characteristics	Dispersal Reproduction
4) Biogeographic range	Regional range
5) Management feasibility	Availability of control methods
C. Spread and equilibrium phases	
1) Management feasibility	Availability of control methods Size of infestation Accessibility to infestations
2) Distribution and abundance	Trend in target sites Distribution in target sites Abundance in target sites
3) Life history characteristics	Dispersal
4) Actual impacts	Ecosystems Structure Species composition

The categories within each phase are ranked in general order of importance probably the most reasonable management objectives. The likelihood of success for limiting further spread and reducing existing populations will depend on the availability and effectiveness of containment methods and size of existing populations. Data on trends in abundance and distribution and dispersal capability can help distinguish species that are spreading rapidly from those with slower spread rates. Although the general categories of information on species in the spread stage are the same as those in the establishment stage, the specific information that is of most use is generally different (Table 2.1). Information on management feasibility and distribution and abundance are more important than at other phases. Information on impacts can still be useful for prioritizing species in the spread phase, but it is more focused on actual impacts that have been observed than on the potential to cause future impacts. Biogeographic information is not particularly helpful at this phase because it should already be apparent which biogeographic regions (e.g. habitat types) are being invaded by the species.

2.4.3.2 Information for Prioritizing Sites

There are two main categories of information to collect when prioritizing sites: (1) susceptibility to invasion, and (2) the conservation value of the site (Table 2.3). Management feasibility is another consideration, but of lesser importance.

Predicting the susceptibility of vegetation communities to invasion has long been an active area of research (Rejmanek et al. 2005). Success of predictions for general patterns has been elusive, but predictions are often reliable only when done at local scales. Besides basic ecological information on nonnative species and land use within the area of interest (intrinsic factors; Table 2.2), landscape configuration and characteristics are also important (extrinsic factors; Table 2.2). This is because invasive species may initially spread from neighboring lands. Attributes at the landscape scale should also be considered when prioritizing sites, especially patchiness of vegetation communities (some communities are more prone to invasion caused by edge effects; e.g., grasslands) and corridors connecting vegetation types to particular sites. Conservation value includes information on local hotspots of native diversity, endemism, and threatened and endangered species, as well as other cultural or recreational site values.

2.4.4 Prioritized Monitoring Plan

Prioritization can help further reduce the area identified for monitoring in a generalized monitoring plan (Fig. 2.1), and thus increase monitoring efficiency (Fig. 2.4). The specific approach will depend on whether the prioritization was developed for colonizing species, species established in an area but not yet spreading, or species currently spreading through an area. Preventing colonization will require monitoring vectors and pathways to the site, as well as areas where the species is likely to

Table 2.2 Information needed to develop prioritized lists of sites to protect from invasion by nonnative species

A. Susceptibility to invasion		
1) Intrinsic (site-specific)	Nonnatives richness	
	Nonnative distribution	Spatial
	Nonnative abundance	Vegetation community
	Land use	Disturbance
		Historic
		Contemporary
2) Extrinsic (off-site)	Vectors and pathways	Roads
	Neighbor perimeter	Trails
	Neighbor area	Watercourses
	Land use	Disturbance
		Contemporary
3) Invasion Rates	Temporal trend in nonnative species accumulation	
B. Conservation value		
1) Hotspots		
2) Endemics		
3) T & E species		
4) Rare community types		
5) Sensitive areas of other value	e.g. cultural or recreational	
C. Management feasibility		
1) Management constraints	e.g. in wilderness	
2) Site accessibility		

The categories within each level are ranked in general order of importance

become established. Management of established species not yet spreading should be focused on eradication. Attention should be given not just to sites with larger infestations but satellite populations as well which often serve as propagule sources from which larger infestations can develop and spread. Species that are actively spreading are especially hard to deal with. A strategy with dual objectives of containing further spread and reducing density is recommended, but resources may not always allow this. If resources are limited, the decision to focus on containment vs. control will be determined by how rapidly the species is spreading.

2.5 What can be Done with Geo-Referenced Abundance Data?

Geo-referenced abundance data provide the opportunity to develop the most efficient types of early-detection monitoring plans possible. Specifically, these types of data can be used to develop predictive models to help focus monitoring efforts where species, or suites of species, are most likely to appear on the landscape. Because the development of these predictive models can be costly, prioritization is often employed first to narrow a large list of candidate species to a manageable number (Fig. 2.2). This is often done at relatively local scales such as parks and reserves.

The development of predictive models can increase search efficiency by focusing searches on areas that are most likely to be invaded. Predictive models can also be used to estimate the threat posed by specific species and thus can be integrated into the prioritization process. Regardless of scale, the goal of predictive models is to identify sites where invasive species are most likely to occur. Models can be developed for individual species as well as groups of species (Guisan et al. 1999; Underwood et al. 2004; Ferrier and Guisan 2006). Good predictive models substantially reduce the enormous amounts of resources required to detect populations before they become established or before nascent populations begin to expand (Rejmanek and Pitcairn 2002).

The uses of predictive models in wildlife management and other areas of conservation are extensive (Ejrnaes et al. 2002; Scott et al. 2002; Guisan and Thuiller 2005). In contrast, despite a plethora of research predicting what species are likely to be invasive and what communities are likely to be invaded (Rejmanek 1989; Reichard and Hamilton 1997; Daehler and Carino 2000; Kolar and Lodge 2001; Rejmanek et al. 2005, Krivanek and Pysek 2006), the modeling of invasive species distributions has been relatively limited until only recently (Peterson 2003; Rouget et al. 2004; Underwood et al. 2004; Thuiller et al. 2008).

2.5.1 Types of Predictive Modeling Approaches

There are two general approaches for predicting which species will likely become invasive in an area. One is based on decision trees, usually with binary answers (yes/no) to a series of questions on species biogeography, biology/ecology, and traits generally considered to be legitimate indicators of invasiveness (Daehler et al. 2004; Pheloung et al. 1999; Reichard and Hamilton 1997). The number of questions can range from a few (e.g., 7; Reichard and Hamilton 1997) to many (e.g., 50; Pheloung et al. 1999). In many ways, this approach resembles prioritization with the use of decision trees and ordinal scores. It is simple in concept and has proven effective in predicting species likely to colonize a large geographic area (e.g., a country or state) and become invasive (Krivanek and Pysek 2006).

The other approach is based on statistical models using geo-referenced environmental data at sites where a species is known to occur and, ideally, also where it does not occur. Standard environmental data are correlated with species distribution and abundance patterns including climate, topographic, soil, and land cover variables (Table 2.3). Some of these variables directly influence species distribution patterns (e.g., soil pH, light), while others indirectly influence patterns (e.g., elevation, aspect). In addition, invasive species biologists have identified other variables that

Table 2.3 Information needed to develop predictive models of invasive species in different phases of the invasion process

A. Pre-introduction and introduction phases		
1) Species data	Biogeographic	Native range
		Nonnative range
	Tendency to be invasive elsewhere	
2) Environmental data	Climate	Temperature
		Precipitation
	Productivity	
	Evapotranspiration	
B. Establishment and spread phases		
1) Species data	Distribution	
	Abundance	
2) Environmental data	Topography	Elevation
		Slope
		Aspect
	Soils	Structure
		Chemistry
	Land cover	Vegetation association
		Land use
3) "Invasion Theory" data	Disturbance	Grazing
		Fire
		Logging
		Roads & trails
	Species pool	
	Propagule pressure	Site-specific land use
		Off-site land use
	Neighboring land perimeters	
	Neighboring land area	
	Vectors (sources of transport)	

The categories within each level are ranked in general order of importance

are often correlated with invasive plant species (Mack and D'Antonio 1998; Lonsdale 1999). These include factors such as disturbance, propagule pressure, and the species pool of potential invaders.

2.5.2 *Preintroduction Prediction Models for Single Species*

Many studies have focused on predicting the likelihood of a species being introduced and becoming established in an area in which it does not yet occur. Until recently, there has been a great deal of pessimism regarding the success of these studies (Williamson 1999). However, important advances have been made in recent years, and there do appear to be traits that have some generality for predicting invasiveness (Kolar and Lodge 2001; Rejmanek et al. 2005), especially for particular taxa (Rejmanek and Richardson 1996; Grotkopp et al. 2002).

A number of models have been developed that attempt to predict the likelihood of different species becoming invasive if they are introduced in an area (e.g. Rejmanek and Richardson 1996; Reichard and Hamilton 1997; Pheloung et al. 1999; Daehler et al. 2004). Some of these models have good predictive ability even outside geographic areas in which they were developed (Krivanek and Pysek 2006; Pauchard et al. 2004). A potential limitation is that both the decision tree and statistical models require a large amount of detailed information which is not always available, such as species life-history characteristics or environmental conditions. On a more fundamental level, the models have often been applied at much larger scales (e.g., countries, bioregions, or even continents) than the effective scale of most early detection programs (i.e., local or designated management units). Although they may be useful for predicting *what* species might become invasive over a large geographic region, they generally do not predict *where* species are most likely to become established at a scale appropriate for most early detection programs.

Early detection programs are generally targeted at species early in the colonization phase of invasion and implemented at local or, perhaps, regional scales. However, in some instances, there may be a need to develop an early detection program for a large geographic area. In these cases, there is a group of predictive models known as climatic-envelope models (CEM) that form a bridge between the preintroduction models discussed above and postintroduction models. CEMs are based on general relationships between climate and species biogeographic patterns (Rouget et al. 2004), and require little if any detailed species life-history information or environmental characteristics. Predictions are for large geographic areas, but they have the flexibility to be applied to species in either preintroduction or postintroduction phases. Information needed for developing CEMs includes the native and nonnative ranges of the species, basic climatic data for where the species occurs, productivity (which rainfall can often be a surrogate for), and evapotranspiration (Table 2.3).

2.5.2.1 Postintroduction Prediction Models for Single Species

Postintroduction predictive models are often developed with preexisting data from plant surveys and GIS data. The fundamental ecological concept that is the foundation of most predictive modeling studies is the ecological niche (Grinnell 1917; Hutchinson 1957; MacArthur 1968). A species' fundamental niche is determined by a large number of abiotic, biotic, and behavioral factors. Where species actually occur is best conceptualized as its realized niche (e.g., Austin and Meyers 1996). Although a species could have greater ranges of distribution, biotic interactions (e.g., competition, predation, pathogens), the lack or limitation of important resources (e.g., moisture, light), and/or the inability to cross barriers restricts its actual distribution. Consequently, predictive models are based on data of a species' realized niche. Differentiation between the fundamental niche and the realized niche has important practical considerations for evaluating the scale to which model predictions can be extended and for the information collected for developing models (Thuiller et al. 2005). Because environments are dynamic and heterogeneous, factors that influence a species' realized niche can be expected to vary unpredictably, both spatially and temporally. Therefore, a good rule of thumb is to assemble data on species (e.g., distribution, abundance) and environmental variables (e.g., elevation, soils) from areas in close geographic proximity to where the early detection program will be applied. It is also very important that the environment has not substantially changed since the time when the data were collected.

2.5.3 Postestablishment Prediction Information for Single Species

Models of species in the spread and equilibrium phases are focused on local scales (e.g., a reserve, national park, or state forest). At this phase of invasion, nonnative species have a proven ability to establish themselves and survive regional climatic conditions. The objective of modeling efforts then becomes predicting where the species can reproduce, persist, and disperse.

For obvious reasons, developing statistical models for species that are in the equilibrium phase would not be a good investment of financial or human resources. Therefore, statistical models are most appropriate for species in the establishment and, to a lesser degree, the spread phase of invasion. Even then, the usefulness of these models may be limited. Data might be too sparse for developing models for species in the establishment phase, because populations are restricted in distribution and/or abundance. Although species known to be spreading are better suited for modeling, they may be beyond the point of practical control efforts.

Basic information to gather on species in the establishment and spread phases are estimates of distribution and abundance (Table 2.3). Predictive models are often based on presence–absence (incidence) of species in an area, but abundance data (e.g., cover, density) give a far more ecologically meaningful correlation of the species along environmental gradients (Austin 2002; Klinger et al. 2006). Although incidence-based models have utility, we strongly recommend the use of abundance data if they can be obtained. Models based on incidence data essentially give equal weight for species relationships along environmental gradients; a species that occurs at 10%, 30%, 50%, and 70% values for a given predictor variable provides the same amount of information at each value (it simply occurs there, but in what amount we do not know). A species with densities of 10, 40, 60, and 80 at 10%, 30%, 50%, and 70% values for the predictor variable provides much more ecological information and has greater predictive value.

Standard environmental data to correlate with species distribution and abundance patterns include topographic, soil, and land cover variables (Table 2.3). In addition to these standard environmental variables, invasive species biologists have identified other variables that are often correlated with the occurrence of invasive plant species (Mack and D'Antonio 1998; Lonsdale 1999). These factors include disturbance, propagule pressure, and the species pool of potential invaders (Rouget and Richardson 2003).

Invasions can be facilitated by biological interactions such as pollination and seed dispersal. Theoretically, incorporating these processes into predictive models could be very useful, but in most instances it would be extremely difficult to do in a meaningful way (Araujo and Luoto 2007). Lack of data on the processes, deciding what metric to use in the models, and matching the scale of the process to the scale where species and environmental data have been collected would be problematic. The issue of matching scales where predictor variables and species data are collected is a general issue that confronts even models found to have reasonable predictive value (Underwood et al. 2004).

In developing a useful predictive model, it is essential to only include predictor variables that are available in the management unit's database, especially in the case of spatial data. Although other predictors may be very important, if spatially explicit information is not available for the management unit the model cannot be used to predict areas of the unit that should be searched for invasives. It may be possible to include some important predictors, such as propagule pressure, through the use of available surrogates such as vectors and pathways.

2.5.4 *Predicting Risk of Occurrence Using Multispecies Models*

Information that can be used for modeling species assemblages is essentially the same as that for individual species. The main difference is the statistical methods used to develop the models, not the data themselves (Guisan et al. 1999; Underwood et al. 2004; Ferrier and Guisan 2006). Most landscapes have been invaded by multiple species, so an approach focused on assemblages may be very efficient (Underwood et al. 2004). Because of computerized databases, the time required to collect information on species assemblages is not much greater than for a single species. Nevertheless, care must be taken with assemblage-based models. Because species tend to respond individualistically to environmental gradients, predictions of distribution patterns could either be narrower or broader depending upon the shape of the species response curves (Austin 2002). In an early detection program, this could result in areas not being monitored where invasive species do occur, or spending time searching areas where few if any occur. An additional consideration is that within an assemblage only one or a few species are truly prone to be problematic. In these instances, it is more useful to predict where the problem species occur rather than the entire assemblage (Zimmerman and Kienast 1999; Ferrier and Guisan 2006).

Multiple species models assume that species within an assemblage respond similarly to environmental gradients. Numerous studies have shown this assumption is tenuous, so great care needs to be used when using these models. Careful analysis of species distribution data is needed before developing models to determine whether the assumption of similar niche responses among species is justified. Even if the assumption appears justified, the results need to be interpreted cautiously.

2.5.5 Predictive Models Applied to Multiple Sites

Predictive models can be applied to multiple sites. However, models developed at one site may have poor prediction success at other sites, because the relative importance of different realized niche dimensions can change between areas (see above). For this reason, multisite models should be based on information for species and predictor variables from each site. If this is not possible, then predictions of invasive species distributions in areas where the models could not be validated should be interpreted very cautiously. It is also a strong argument for the need to validate predictions in the field before full implementation of an early detection program.

2.5.6 Optimized Monitoring Plan

An optimized monitoring plan allows for a further reduction in the search area required for early detection monitoring (Fig. 2.1) and an increase in efficiency (Fig. 2.4). It integrates the results of a generalized monitoring plan, prioritized monitoring plan, and predictive modeling (Fig. 2.2). After a generalized monitoring plan is used to identify areas most susceptible to invasion, prioritization is employed to narrow the search range within this area and to identify the species most important to monitor for. Predictive modeling is then applied to these high priority species to develop efficient monitoring plans for those species. In some instances, it may make sense to first predict which species are most likely to be introduced to a site or spread into areas of high conservation value. In either case, this would be the most efficient use of resources at both the planning and implementation stages of a monitoring program. These considerations are typically overlooked in most early detection programs (M. Brooks pers. obs.).

The payoff from investing in the optimized monitoring plan would be in implementation. Obviously, it would result in a minimum area being targeted for monitoring (Fig. 2.1). However, it will also increase the probability that the species most likely to be problematic and the sites where the species are most likely to occur and/or have the most negative effects have been identified. This would help identify the best type of monitoring and control efforts needed to reduce the likelihood of colonization, spread, and impacts of those high priority species.

2.6 An Example of How to Apply the Monitoring Framework

The framework described in this chapter can be used to increase the overall efficiency of early detection monitoring programs. With the addition of each successive monitoring approach, the extent of the area which is the focus of monitoring efforts can be reduced (Fig. 2.1). An example of how this process can work is presented below for a hypothetical management unit composed of typical landscape features (Fig. 2.5).

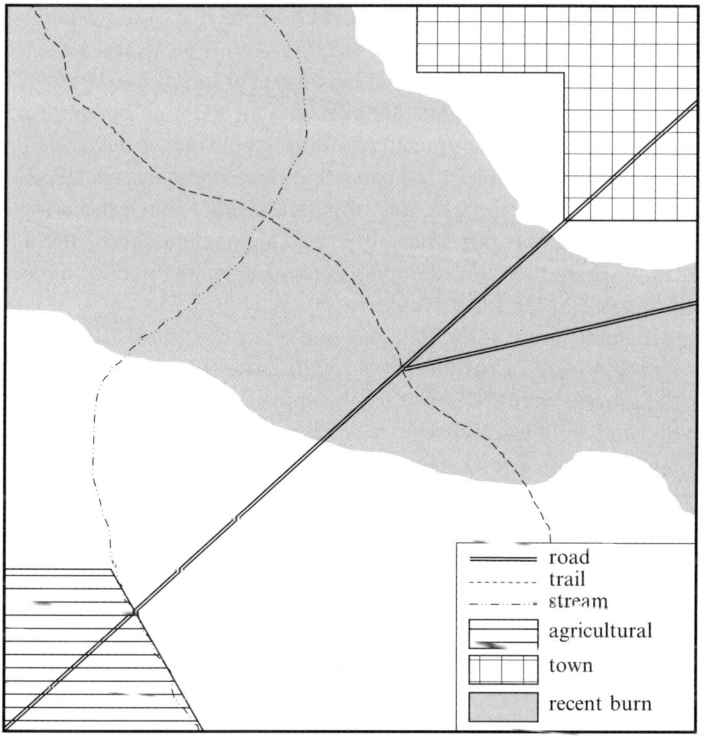

Fig. 2.5 A worked example of how the three types of monitoring plans can be applied to a management unit composed of common landscape features

If there are no species data available in or near the management unit, then a generalized approach is required (Fig. 2.2). The landscape features in this hypothetical management unit are associated with typical levels of propagule pressure and resource availability, which can be used to develop a generalized monitoring plan. Propagule pressure would be very high in the town and agricultural area, high along the roads, moderate along the trail and stream, and low elsewhere (Fig. 2.5). Resource availability would be very high in the town, high in the agricultural area, recent fire, and roadsides, moderate along the trails, and low elsewhere. The generalized monitoring priorities in this case would be as follows: *very high priority* in the town and agricultural area; *high priority* along the roads and in the burned area, especially where to two meet; *moderate priority* along the trail and stream; and *low priority* elsewhere. Thus, the areas of very high priority would comprise about 10% of the total area within the management unit, and if the high priority areas were added, the monitoring area would be about 50% of the total area.

If a species list is available, then a prioritized monitoring plan can be built upon the generalized plan (Fig. 2.2). The specifics of this plan will depend on the types of species that rank as highest priorities. For example, assuming that the species pool is dominated by highly invasive riparian plants, then monitoring efforts should be focused on the stream corridor, especially where it passes through the recent burn, agricultural field, and crosses the road (Fig. 2.5). This would reduce the monitoring area to less than 10% of the total management unit.

If georeferenced abundance data are available for the high priority riparian species in this example, then an optimized monitoring plan can be developed (Fig. 2.2). Assuming that habitat modeling indicates that these riparian species are typically associated with agricultural areas, then the monitoring effort can be focused even further on the stream corridor where it passes along the edge of the agricultural area, especially where it crosses the road. Accordingly, the monitoring areas would be reduced to < 1% of the total management unit (Fig. 2.5).

In most situations there will not be just one set of characteristics associated with potential invaders (e.g. riparian plants with affinities for agricultural areas). However, the process outlined above can be applied for each group of high priority species with similar characteristics to produce multiple components of an optimized monitoring plan. For example, assume that in addition to riparian plants of agricultural areas, the above example included high priority species that are often used as ornamentals in landscaping and others that are typical of roadsides in post-fire landscapes. In that more complicated example, the optimized monitoring plan would additionally include monitoring in the town (especially its interface with wildlands) and along the roadside within the burned area. This would increase the sampling area to about 5% of the management unit, but still well below the 10–50% associated with the other monitoring approaches.

The strength of the early detection monitoring framework presented in this chapter is in improving not only the efficiency of monitoring efforts, but also the efficiency of developing the monitoring plans themselves. In particular, by first developing a prioritized list of potential invaders, subsequent resources to develop predictive models can be most effectively allocated to those species that pose the greatest threat of invading and negatively affecting resource values. The framework also allows for realistic consideration of the extra effort needed to develop prioritized or optimized plans, so that more informed decisions can be made regarding the allocation of resources to develop early detection monitoring plans, implement them, and respond to new invaders with control treatments.

References

Anonymous (1993) Florida's most invasive species. Palmetto Fall 1993:6–7

Araujo MB, Luoto M (2007) The importance of biotic interactions for modelling species distributions under climate change. Glob Ecol Biogeogr 16:743–753

Austin MP (2002) Spatial prediction of species distribution: an interface between ecological theory and statistical modelling. Ecol Modell 157:101–118.

Austin MP, Meyers JA (1996) Current approaches to modelling the environmental niche of the eucalypts: implication for management of forest biodiversity. For Ecol Manage 85:95–106

AZ-WIPWG (2005) Invasive non-native plants that threaten wildlands in Arizona. Arizona Wildland Invasive Plant Working Group. http://www.swvma.org/invasivenonnativeplantsthatthreatenwildlandsinarizona.html. Accessed 6 June 2007

Bowen B, Shea A (1996) Exotic pest plants in Tennessee: Threats to our native ecosystems. Special Insert to Tennessee Exotic Pest Plant Council (TN-EPPC) News Fall 1996:1–4
Brooks ML (2007) Effects of land management practices on plant invasions in wildland areas. In: Nentwig W (ed) Biological Invasions: Ecological Studies 193. Springer, Heidelberg, Germany
Cal-IPC (2006) California Invasive Plant Inventory. Cal-IPC Publication 2006–02. California Invasive Plant Council. Berkeley, California
Champion PD, Clayton JS (2001) A weed risk assessment model for aquatic weeds in New Zealand. In: Groves RH, Panetta FD, Virtue JG (eds), Weed risk assessment. CSIRO, Collingwood, Victoria, Australia
Cousins R, Mortimer M (1995) Dynamics of weed populations. Cambridge University Press, Cambridge, United Kingdom
Daehler CC, Carino DA (2000) Predicting invasive plants: prospects for general screening system based on current regional models. Biol Invasions 2:93–102
Daehler CC, Denslow JS, Ansari S, Kuo HC (2004) A risk assessment system for screening out invasive pest plants from Hawaii and other Pacific Islands. Conserv Biol 18:360–368
Davis MA, Grime PJ, Thompson K (2000) Fluctuating resources in plant communities: a general theory of invasibility. J Ecol 88:528–534
Ejrnaes R, Aude E, Nygaard B, Munier B (2002) Prediction of habitat quality using ordination and neural networks. Ecol 12:1180–1187
Ferrier S, Guisan A (2006) Spatial modelling of biodiversity at the community level. J Appl Ecol 43:393–404
Florida Exotic Pest Plant Council Plant List Committee (2005) List of Florida's invasive species. Florida Exotic Pest Plant Council. http://www.fleppc.org/list/05List.htm. Accessed online 7 June 2007
Forman RTT, Sperling D, Bissonette JA, Clevenger AP, Cutshall CD, Dale VH, Fahrig L, France R, Goldman CR, Heanue K, Jones JA, Swanson FJ, Turrentine T, Winter TC (2003) Road ecology: science and solutions. Island Press, Washington, DC
Fox AM, Gordon DR, Dusky JA, Tyson L, Stocker RK (2000) IFAS Assessment of Non-Native Plants in Florida's Natural Areas. University of Florida Extension, Institute of Food and Agricultural Sciences, Gainesville, Florida
Fox AM, Gordon DR, Dusky JA, Tyson L, Stocker RK (2001) IFAS assessment of non-native plants in Florida's natural areas. SS-AGR-79. Agronomy Department, Florida Cooperative Extension Service, Institute of Food and Agricultural Sciences, University of Florida, Gainsville, Florida
Gould LL, Stuckey IH (1992) Plants invasive in Rhode Island. R I Wild Plant Soc Newsl 6:1–7
Grinnell JJ (1917) The niche-relationships of the California thrasher. Auk 34:427–433
Grotkopp E, Rejmanek M, Rost TL (2002) Toward a causal explanation of plant invasiveness: seedling growth and life-history strategies of 29 pine (*Pinus*) species. Am Nat 159:396–419
Groves RH (1986) Plant invasions of Australia: an overview. In: Groves RH, Burdon J (eds) Ecology of biological invasions: An Australian perspective. Australian Academy of Science, Canberra, Australia
Guisan A, Thuiller W (2005) Predicting species distribution: offering more than simple habitat models. Ecol Lett 8:993–1009
Guisan A, Weiss SB, Weiss AD (1999) GLM vs. CCA spatial modeling of plant species distribution. Plant Ecol 143:107–122
Heffernan KE, Coulling PP, Townsend JF, Hutto CJ (2001) Ranking invasive exotic plant species in Virginia. Natural Heritage Technical Report 01–13. Virginia Department of Conservation and Recreation, Division of Natural Heritage, Richmond, Virginia
Hiebert RD, Stubbendieck J (1993) Handbook for Ranking Exotic Plants for Management and Control. United States Department of the Interior, Natural Resources Report NPS/NRMWRO/NRR-93/08. National Park Service, Natural Resources Publication Office, Denver, Colorodo
Hiebert RD (1998) Alien Plant Species Ranking System. Unpublished document
Hobbs RJ, Huenneke LF (1992) Disturbance, diversity and invasion: implications for conservation. Conserv Biol 6:324–337
Holling CS (1978) Adaptive environmental management and assessment. Wiley, New York, New York

Hutchinson GE (1957) Concluding remarks. Cold Harbor symposium on quantitative biology 22:415–427
Klinger RC, Underwood EC, Moore PE (2006) The role of environmental gradients in non-native plant invasions into burnt areas of Yosemite National Park, California. Divers Distrib 12:139–156
Kolar CS, Lodge DM (2001) Progress in invasion biology: predicting invaders. Trends Ecol Evol 16:199–204
Krivanek M, Pysek P (2006) Predicting invasions by woody species in a temperate zone: a test of three risk assessment schemes in the Czech Republic (Central Europe). Divers Distrib 12:319–327
Lockwood JL, Cassay P, Blackburn T (2005) The role of propagule pressure in explaining species invasions. Trends Ecol Evol 20:223–228
Lonsdale WM (1999) Global patterns of plant invasions and the concept of invasibility. Ecol 80:1522–1536
MacArthur RH (1968) The theory of the niche. In: Lewontin RC (ed) Population biology and evolution. Syracuse University Press, Syracuse, New York
Mack MC, D'Antonio CM (1998) Impacts of biological invasions on disturbance regimes. Trends Ecol Evol 13:195–198
Mack RN, Simberloff D, Lonsdale WM, Evans H, Clout M, Bazzaz FA (2000) Biotic Invasions: Causes, epidemiology, global consequences, and control. Ecol Appl 10:689–710
Marty JT (2005) Effects of cattle grazing on diversity in ephemeral wetlands. Conserv Biol 19:1626–1632
Mehrhoff LJ (2000) Criteria for including a species as a non-native invasive species or a potentially invasive species in New England. George Safford Torrey Herbarium, University of Connecticut, Storrs
Mehrhoff LJ, Metzler KJ, Corrigan EE (2003) Non-native and potentially invasive vascular plants in Connecticut. Center for Conservation and Biodiversity, University of Connecticut, Storrs
Morse LE, Randall JR, Benton N, Hiebert R, Lu S (2004) An invasive species assessment protocol: Evaluating non-native plants for their impact on biodiversity. Version 1, NatureServe, Arlington Virginia. http://www.natureserve.org/getData/plantData.jsp. Accessed online 7 June 2007
Orr RL, Cohen SD, Griffin RL (1993) Generic nonindigenous pest risk assessment process. United States Department of Agriculture, Washington, DC
Pauchard A, Cavieres LA, Bustamante RO (2004) Comparing alien plant invasions among regions with similar climates: where to from here?. Divers Distrib 10:371–37
Peterson AT (2003) Predicting the geography of species invasions via ecological niche modeling. Q Rev Biol 78:419–433
Pheloung PC, Williams PA, Halloy SR (1999) A weed risk assessment model for use as a biosecurity tool evaluating plant introductions. J Environ Manage 57:239–251
Randall JM, Benton N, Morse LE (2001) Categorizing invasive weeds: the challenge of rating the weeds already in California. In: Groves RH, Panetta FD, Virtue JG (eds) Weed risk assessment. CSIRO Publishing, Collingwood, Australia
Randall JM, Morse LE, Benton N, Hiebert R, Lu S, Killeffer T (2008) The invasive species assessment protocol: A new tool for creating regional and national lists of invasive non-native plants that negatively impact biodiversity. Invasive Plant Sci Manage 1:36–49
Reichard S, Schuller R, Isaacson D, Whiteaker L, Kruckeburg A, Kagan J, Chappell C, Old R (1997) Non-native pest plants of greatest concern in Oregon and Washington as of August 1997. Pacific Northwest Exotic Pest Plant Council, Redmond, Oregon
Reichard SH, Hamilton CW (1997) Predicting invasions of woody plants introduced into North America. Conserv Biol 11:193–203
Rejmanek M (1989) Invasibility of plant communities. In: Drake JA, Mooney HA, DiCastri F, Groves RH, Kruger FJ, Rejmanek M, Williamson M (eds) Biological invasions: A global perspective. Wiley, New York, USA
Rejmanek M (2000) Invasive plants: approaches and predictions. Austral Ecol 25:497–506
Rejmanek M, Pitcairn MJ (2002) When is eradication of exotic pest plants a realistic goal. In: Veitch CR, Clout MN (eds) Turning the tide: The eradication of Invasive Species. Invasive Species Specialist Group of The World Conservation Union (IUCN), Auckland, New Zealand

Rejmanek M, Richardson DM (1996) What attributes make some plant species more invasive? Ecol 77:1655–1660
Rejmanek M, Richardson DM, Higgins SI, Pitcairn MJ, Grotkopp E (2005) Ecology of invasive plants: state of the art. In: Mooney HA, Mack RN, McNeely JA, Neville LE, Schei PJ, Waage JK (eds) Invasive alien species: A new synthesis, SCOPE 63. Island Press, Washington, DC
Richardson DM, Pysek P, Rejmanek M, Barbour MG, Pannetta FD, West CJ (2000) Naturalization and invasion of alien plants: concepts and definitions. Divers Distrib 6:93–107
Rouget M, Richardson DM (2003) Inferring process from pattern in alien plant invasions: a semi-mechanistic model incorporating propagule pressure and environmental factors. Am Nat 164:713–724
Rouget M, Richardson DM, Nel JL, Le Maitre DC, Egoh B, Mgidi T (2004) Mapping the potential ranges of major plant invaders in South Africa, Lesotho and Swaziland using climatic suitability. Divers Distrib 10:475–484
Schwegman J (1994) Exotics of Illinois forests. Erigenia 13:65–67
Scott JM, Heglund PJ, Morrison ML, Haufler JB, Raphael MG, Wall WA, Samson FB (2002) Predicting species occurrences: Issues of accuracy and scale. Island Press, Washington
Thorp JR, Lynch R (2000) The determination of weeds of national significance. National Weeds Strategy Executive Committee, Launceston, Australia. http://www.weeds.org.au/docs/WONS. Accessed online 7 June 2007
Thuiller WD, Richardson DM, Pysek P, Midgley GF, Hughes GO, Rouget M (2005) Niche-based modelling as a tool for predicting the risk of alien plant invasions at a global scale. Glob Chang Biol 11:2234–2250
Thuiller W, Albert C, Araujo MB, Berry PM, Cabeza M, Guisan A, Hickler T, Midgely GF, Paterson J, Schurr FM, Sykes MT, Zimmermann NE (2008) Predicting global change impacts on plant species' distributions: Future challenges. Perspect Plant Ecol Evol Syst 9:137–152
Timmins SM, Williams PA (1987) Characteristics of problem weeds in New Zealand's protected natural areas. In: Saunders DA, Arnold GW, Burridge AA, Hopkins AJM (eds) Nature conservation and the role of native vegetation. Surrey Beatty and Sons, Chipping Norton, Australia
Timmins SM, Owens SJ (2001) Scary species, superlative sites: assessing weed risk in New Zealand's protected natural areas. In: Groves RH, Panetta FD, Virtue JG (eds) Weed risk assessment. CSIRO Publishing, Collingwood, Australia
Underwood EC, Klinger R, Moore PE (2004) Predicting patterns of non-native plant invasion in Yosemite National Park, California, USA. Divers Distrib 10:447–459
Virginia Department of Conservation and Recreation and Virginia Native Plant Society (2003) Invasive alien plant species of Virginia. Virginia Department of Conservation and Recreation. Richmond, Virginia
Virtue JG, Groves RH, Panetta FD (2001) Towards a system to determine the national significance of weeds in Australia. In: Grove RH, Panetta FD, Virtue JG (eds) Weed risk assessment. CSIRO, Collingwood, Victoria, Australia
Wainger LA, King DM (2001) Priorities for weed risk assessment: Using a landscape context to assess indicators of functions, services, and values. In: Groves RH, Panetta FD, Virtue JG (eds) Weed risk assessment. CSIRO, Collingwood, Victoria, Australia
Warner PJ, Bossard CC, Brooks ML, DiTomaso JM, Hall JA, Howald A, Johnson DW, Randall JM, Roye CL, Ryan MM, Stanton AE (2003) Criteria for Categorizing Invasive Non-Native Plants that Threaten Wildlands. California Exotic Pest Plant Council and Southwest Vegetation Management Association. http://ucce.ucdavis.edu/files/filelibrary/5319/6657.doc. Accessed online 7 June 2007
Weiss J, McLaren D (1999) Invasive assessment of Victoria's State Prohibited, Priority and Regional Priority weeds. Keith Turnbull Research Institute, Agriculture Victoria, Frankston, Victoria, Australia
Williamson M (1999) Invasions. Ecography 22:5–12
Zimmermann NE, Kienast F (1999) Predictive mapping of alpine grasslands in Switzerland: species versus community approach. J Veg Sci 10:469–482

Chapter 3
Eradicating Plant Invaders: Combining Ecologically-Based Tactics and Broad-Sense Strategy

Richard N. Mack and Sara K. Foster

Abstract Eradication, i.e., the complete destruction, of all individuals of an Invasive Alien Species (IAS) in a new range is universally viewed as the permanent solution to damaging these plants, assuming any reentry of the target species is reliably prevented. Yet eradication is often deemed impractical if not effectively impossible, except when the alien species occurs in very low numbers in a single circumscribed beachhead. This contention may have arisen in part because many immigrant species upon first detection already consist of numerous, spreading populations. Consequently, control (i.e., containment) of the alien species is viewed as the only feasible alternative. We contend this view of eradication dwells disproportionately on the failures to eradicate nonnative species without comprehensive examination of total (or near total) eradications. A more balanced view of the feasibility of eradications may emerge if we trace the events, features, and circumstances that successful eradications hold in common. Most useful will be careful application of strategies that have led to eradications, of which winning and keeping public support is likely the single most important contributor to success.

Keywords *Berberis vulgaris* • Eradication • *Miconia calvescens* • Plant invasions • *Striga asiatica*

3.1 Introduction

Combating potentially Invasive Alien Species (IAS) (sensu Mooney et al. 2005) through postentry measures addresses the reality of our inability to devise fail-proof quarantine at political or natural geographic boundaries. Even though much effort is directed at thwarting unwanted alien plant entry through cargo inspection (McCullough et al. 2006) and regulatory action to prohibit the importation of

R.N. Mack (✉) and S.K. Foster
School of Biological Sciences, Washington State University, Pullman, WA 99164 USA
rmack@wsu.edu

Inderjit (ed.), *Management of Invasive Weeds*,

known or suspected damaging species (FAO 2000; Shine et al. 2005), some unwanted nonnative species always gain entry. Postentry detection and some prescription of response then become the inevitable second line of defense against these immigrant species (Wittenberg and Cock 2005).

Postentry response falls into two categories (aside from an ignominious third option, "do nothing"): control (i.e., curtail, contain, minimize, or otherwise limit) the alien species or eradicate it. Eradication refers to the complete destruction of all individuals (sensu Simberloff 2003a) and properly expresses a permanency of action that control does not achieve. For IAS or potentially invasive species, eradication involves the complete elimination of the species in a new range. This action is valued second only to the prevention of the alien species' entry through quarantine as an effective deterrent (Mack et al. 2000). Eradication does not of course preclude future reentry of the unwanted species (Simberloff 2003a and references therein), but it does mean that any new arrivals will not supplement (numerically, spatially, or genetically) an existing population.

Here we review and evaluate the record for eradication, as opposed to control or no-control, of alien plant species' postentry into a new range, regardless of whether this effort is initiated at the point of entry (the proverbial "beachhead") or much later in the plant invasion. The frequency of survey and the recent justified emphasis on early detection of the potential invader (Harris et al. 2001; Westbrooks 2004) are not per se objects of our review, although the speed and thoroughness with which any immigrant species is detected in a new range directly affects the likelihood of success of any eradication effort (Mack and Lonsdale 2002). Rather, we investigate here evidence for the feasibility of eradication. Much is at stake in this debate: the tide of immigrant species is increasing worldwide (U.S. Congress, Office of Technology Assessment 1993; Levine and D'Antonio 2003), adding to the worldwide pool of invasive and incipient invasive species. Making correct decisions on when eradication is feasible, as opposed to control, will largely affect the diversity, scope, and impact of plant invasions throughout the twenty-first century and beyond.

Informed opinion on plant eradication stretches along a continuum from infeasible to feasible with each opinion couched in provisos based on the species' traits and environmental circumstances in the new range (Groves and Panetta 2002). For example, Rejmánek and Pitcairn (2002) examined the outcome of eradication projects for 18 species in California. They found that eradication was often successful where the new range was <1 ha; a sharply lower success rate was recorded as the size of the treated area and number of foci increased. Although they apparently did not examine data for any campaigns that involved new ranges much larger than 1,000 ha, they concluded that eradication for plant infestations that occupy >1,000 ha "is very unlikely." Panetta and Timmins (2004) basically agree with this conclusion and propose "the 1,000 ha rule" based on this California study, even though Rejmánek and Pitcairin (2002) did not assign the term "rule" to their conclusion. Given that the California eradications occurred mainly in agricultural settings, Panetta and Timmins are even less optimistic for successful eradications in natural ecosystems. In these cases, they reason that the effective areas for eradication may be an order of magnitude less, although they offer no data or case histories to support their assertion.

Simberloff (2002, 2003a, b) as well as Mack and Lonsdale (2002) and Mack and Foster (2004) have offered counterpoint to these somewhat pessimistic (or at least conservative) views on the feasibility of plant eradications. They cite a growing array of studies that demonstrate eradication, or at least containment to the cusp of eradication, for a taxonomically diverse array of plants in terrestrial, freshwater, and even brackish environments and in new ranges > 1,000 ha. We dissect here case histories for which adequate information is available on the tactics and strategies employed (sensu Moody and Mack 1988) as we seek generalities that consistently distinguish successful from failed eradication efforts.

3.2 Early Attempts at Eradication

The attraction of being permanently rid of a problem is basic to human values; it is not surprising then that attempts at eradicating an invasive plant, as opposed to controlling its damage, or even enduring its on-going damage, have a long history. Deliberate, if poorly documented, attempts to clean grain of extraneous and potentially weedy species stretch back into antiquity (Leviticus 19:19). We have chosen to explore the history of alien plant eradication, in terms of known deliberate action, only from the late nineteenth century onward, although there are likely older, scattered attempts at eradication.

As agriculture in the US grew both in geographic extent and the collective value of agricultural products, the US Department of Agriculture (U.S.D.A.) by the late nineteenth century began to systematically report the spread of destructive alien plants (Dewey 1897). The U.S.D.A. was alarmed to see that in addition to a well known list of alien plants combated by farmers for > 200 years (Mack 2003), new invaders were appearing. One of the earliest to attract federal attention was *Salsola iberica* (Russian thistle), a native of western and central Asia, which reputedly arrived in the US in the mid-nineteenth century as a seed contaminant. *Salsola iberica* is a weed in cereal crops, and it rapidly began appearing in cereal fields in the Northern Great Plains. Within two decades after its putative entry near Scotland, South Dakota, it had become established at numerous locales in Iowa, Minnesota, South Dakota, Nebraska, and North Dakota (Dewey 1894).

Although Russian thistle already occurred in five-states by 1892, the initial federal goal was its eradication because it provides fuel for wildfires in addition to competition for cereal crops. Russian thistle's still local distribution (e.g. cereal fields, along railroad-right-of ways) probably contributed to the guarded optimism that eradication, rather than control, was possible, and various "remedies" were proposed for curbing its spread and persistence (Dewey 1894). Furthermore, action was taken to eliminate new foci of the invasion. For example, Russian thistle rapidly rode the rails and soon appeared in the western US. Piper (1894a, b) waged a determined but short-lived campaign in all towns and their connecting rail-lines in eastern Washington (>1,500 km from the locales where it was first detected, Dewey 1894), so as to detect and destroy all Russian thistle.

These late nineteenth century attempts at eradication, even locally, failed, consequences of the feeble methods (usually excavation and burning of the plants) and total lack of a discernible strategy for destroying many plants across a growing new range. In fairness to these early would-be plant eradicators, the ease with which Russian thistle was carried throughout the US along the rail lines plus its broad ecological amplitude and abundant seed production would have presented difficult challenges for eradication even today. *Salsola iberica* remains widespread and locally abundant in the western half of the US (Whitson et al. 1996), despite the short-lived campaign against it in the 1890s.

Eichhornia crassipes (water hyacinth) arrived in the US almost contemporaneously with Russian thistle and also attracted early federal interest in eradication. The circumstances surrounding the entry of this subtropical/tropical South America macrophyte have been repeatedly reported (Klorer 1909; Barrett 1989): it likely first arrived as a deliberate introduction at the World's Industrial and Cotton Centennial Exposition in New Orleans in 1884, and many attendees reputedly returned home with cuttings of this showy aquatic plant (Klorer 1909). Within a decade the plant had escaped repeatedly into rivers in the southern US, most notably in Louisiana and Florida (Webber 1897; Klorer 1909). Water hyacinth produces dense, floating vegetative mats that render boat traffic difficult to impossible. Its invasion triggers a cascade of environmental changes that affect the native biota, and it has also been linked to serving as a habitat for the water-residing vectors of human parasites (Barrett 1989).

Here again, tools available more than 100 years ago for combating this invader were meager, although there was clear public support for eradicating, not simply reducing water hyacinth (Klorer 1909). Control consisted of mechanical dredging of the vegetative mats. But these mats soon regrew, thanks to the plant's high productivity. Even then it was known that natural enemies could have a devastating effect on alien plants, Webber (1897) suggested that practical hope of destroying water hyacinth could lie with detecting such a parasite in water hyacinth's native range. Despite these setbacks, these early eradication advocates had recognized a key ingredient to any eradication effort – public cooperation (Dewey 1894; Klorer 1909). Unless the public supported the endeavor, eradication could not be achieved.

3.3 The Campaign Against *Berberis vulgaris*

3.3.1 Barberry Eradication on a Continental Scale

Given the failure of even local extermination, much less eradication of Russian thistle and water hyacinth, the likelihood that the US would embark on a continent-wide eradication campaign early in the twentieth century would seem nil. Yet just such an undertaking was initiated in 1918 across a 13-state region that eventually encompassed six more states. The target was *Berberis vulgaris*, the European barberry (Kempton 1921). Although widely known as a common escapee from cultivation – it had been used since colonial times as a source of fruit for jams, wood for axe handles, and an ornamental hedge (Mack 2003) – these uses for the shrub paled by comparison to the damage it wrought as the alternate host for the stem rust (*Puccinia graminis*) of cereals, such as wheat and oats. The scientific link between European barberry, cereals, and *P. graminis* had been conclusively demonstrated by the end of the nineteenth century (Hutton 1927). Unfortunately by then, 300 years of the shrub's dissemination and naturalization meant that *B. vulgaris* had become a common resident in towns and farms throughout the northern half of the country (Mack 2003 and references therein). During its history in the US (as well as its much longer association with cereal crops in its native western European range), barberry had contributed to chronic devastation of wheat. These losses reached a crisis for the US by 1916, when it was estimated that > 180 × 10^6 bushels of wheat had been lost to stem rust (Kempton 1921). Given that these losses would likely grow in the future and coupled with a severe worldwide demand for wheat, the US acted on an unprecedented scale. Since no means was known to directly attack the stem rust fungus itself, the bold decision was made to break its life cycle by totally eradicating its alternate host, *B. vulgaris* (Kempton 1921).

The US track record of combating invasive plants as well as the scale of the proposed undertaking likely gave architects of this eradication campaign some basis for hesitation: not only was the shrub abundantly widespread, but also was still sold widely by nurseries, and its fruits were dispersed locally by birds (Meier 1933). These would-be eradicators did however have reason for optimism: much of western Europe had already been rendered free of barberry in the early twentieth century through eradication campaigns in England, Scotland, The Netherlands, Germany, and especially Denmark. Within 10 years of the passage of a national law prohibiting barberry, Denmark had almost totally eradicated the shrub. Even though the scale of the project contemplated in the US dwarfed the eradication projects in Europe, the US planners drew considerable confidence from the European outcome (Stakman 1923).

Two early steps in the US eradication strategy proved essential to success. In 1919, Federal Quarantine Regulation No. 38 was enacted to prohibit the interstate transport of barberry (Meier 1933), thus eliminating an important source of reinfestation. And from the outset, the federal government effectively proclaimed the goals and public benefits of Regulation 38 through posters, fliers, and local talks. Even the new medium of motion picture films was employed to reach the citizenry (Kempton 1921).

The US Barberry Eradication Program operated from 1918 to 1978 and eventually involved the thorough survey of *all* land holdings in or near agriculture in the initial 13-state area as well as much of the terrain in those states that supported forests (Michigan, Minnesota, Montana, Wisconsin) (Fig. 3.1). Lack of modern tools of surveillance and land inspection (remote sensing, aerial photography, GIS-based systems) was probably an advantage: the investigators instead relied totally on ground surveys (termed in the parlance of the day, "foot-scouting') to detect barberry

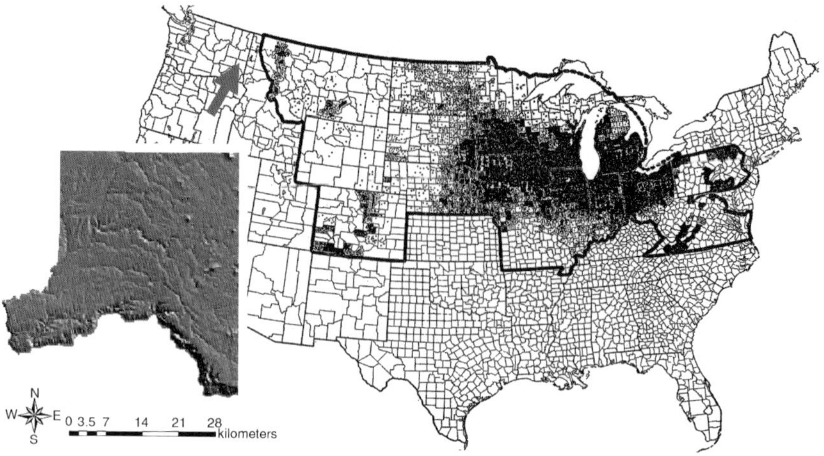

Fig. 3.1 *Berberis vulgaris* (European barberry) was nearly eradicated between 1918 and 1978 across a huge swath of the wheat-growing regions of the US (area enclosed in *heavy black outline*). A smaller, but equally effective campaign was conducted between 1944 and 1978 in eastern Washington, including Whitman County (inset). A Whitman County re-survey in 2002–2003 detected only nine plants in 100 sites where barberry had last been seen pre-1978 (Foster 2003)

(Hutton 1927). With a thoroughness that seems astounding today, teams from the Barberry Eradication Program searched every town, village, hamlet, and the area around all farm dwellings as well as walked along the boundaries of all agricultural fields, searching for barberry. More than $2.0 \times 10^6\,km^2$ were eventually searched. Once a plant was found, it was promptly removed. In the Program's beginning, plants were destroyed by excavation, even by dynamite (Kempton 1921)! Kerosene and rock salt – crude tools by today's standards – were also employed (Morris and Popham 1925). Furthermore, the investigators knew that a single reconnaissance through an area that had supported barberry was unlikely to detect all plants; barberry adults are 1–2 m tall, but seedlings can easily escape detection. Consequently, each land parcel from which barberry had been removed was thoroughly examined on a routine basis that stretched over years – long enough that initially missed seedlings grew and became conspicuous (Kempton 1921) (Fig. 3.2).

Excellent recordkeeping was essential to safely release parcels from further inspection, thereby marshalling the hunt for remaining plants. Working without aerial photographs, the teams prepared detailed hand-drawn maps of farms; each building and the location of each barberry removed was indicated as an aid to resurveys. The recordkeeping also included tallies of the plants destroyed (> 99,000,000 by 1967) (Roelfs 1982); as these numbers rose, the incidence of stem rust fell. By 1950, the incidence of stem rust on cereals across this middle third of the US had plummeted (Campbell and Long 2001).

Despite the extraordinary diligence in the Barberry Eradication Program, it fell short of its ultimate goal; *B. vulgaris* still occurs in the US, even though sale of stem rust-susceptible varieties of European barberry have been prohibited for 80 years.

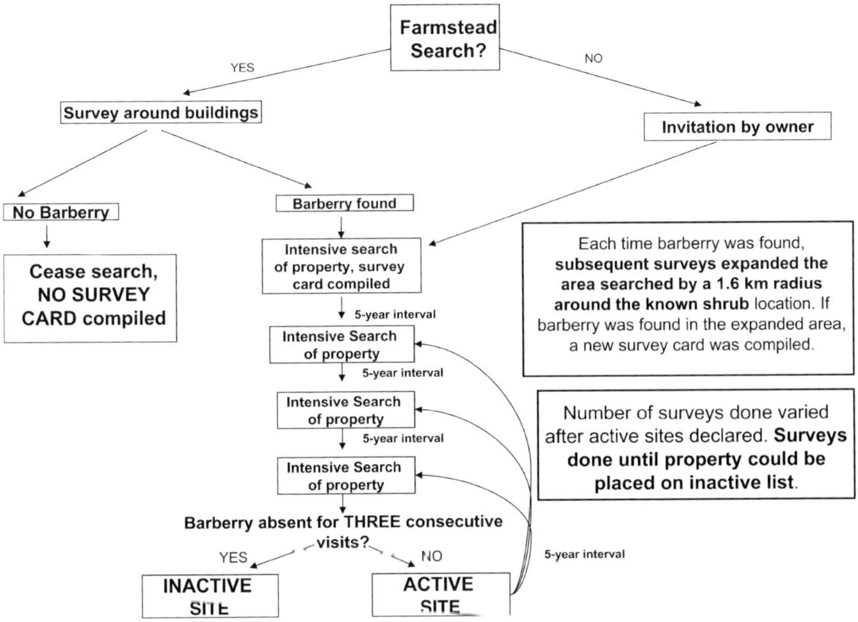

Fig. 3.2 The Barberry Eradication Campaign in the US involved careful, documented annual resurveys of farmsteads from which barberry had been removed. This so-called L survey card from the eradication campaign in Whitman County, Washington illustrates the methodical year-by-year assessment of barberry occurrences on each inspected property. Any property on which barberry was detected within five years of all shrubs' removal was surveyed for an additional five years (Whitman County Barberry Eradication Program, 1943–1978)

Plants that escaped detection have now had about 50 years in which to grow and reproduce. These plants are still however sufficiently rare and that stem rust is uncommon, admittedly in part through the subsequent development of rust-resistant varieties of wheat (Leonard 2001). Nevertheless, European barberry has been reduced to a rarity, such that the most comprehensive modern treatment of the Great Plains flora does not even list *B. vulgaris* (Barkley et al. 1986).

3.3.2 Eradication Campaign Against Berberis vulgaris in the Pacific Northwest: Its Long-Term Effect

The US Barberry Eradication Program launched in 1918 did not initially encompass all cereal growing regions; much wheat has long been grown in the Pacific Northwest, east of the Cascade Range (Meinig 1968). Wheat in this region did not escape stem rust attack, and by the mid-1940s, the U.S.D.A. extended barberry eradication to Washington state (Busdicker 1946). Similar to the central US, the topographic relief varies: rolling hills that once supported native steppe had been rapidly converted to wheat farms by 1900 (Meinig 1968) and were surrounded by forests (Daubenmire 1970) in which barberry could also reside.

Execution of this later campaign profited from lessons learned in the 1920 and 1930s. Admirable emphasis was placed on inspecting every land parcel, whether in private or public ownership (Busdicker 1945). As a result, adequate attention was placed on detecting isolated plants as well as obvious populations near buildings. Only after a once infested site was found to harbor no barberry plants for three consecutive years was it declared barberry-free (D.S. Jackson, pers. comm.). Here as in the Great Plains, the public was engaged in the effort, and all this work was underpinned by exceptionally detailed recordkeeping and hand-drawn maps of the surveyed areas (Whitman County Barberry Eradication Program, 1943–1978) (Fig. 3.2).

Herbicides that had been developed since 1940 were widely used, so plant excavation or application of rock salt or kerosene to destroy barberry was minimal (Whitman County Barberry Eradication Program, 1943–1978, D.S. Jackson, pers. comm.). Inspection across the uneven terrain was simply accomplished through the determination of the field workers. The Ponderosa pine-dominated forests that border much of the region's agricultural fields often support almost impenetrable understories of tall shrubs (e.g. *Physocarpus malvaceus*) on steep slopes (Daubenmire and Daubenmire 1968). Under these circumstances, barberry could be easily overlooked. Consequently, team members walked parallel transects in sight of each other, sweeping back and forth across these sites (D.S. Jackson and J.W. Burns, pers. comm.). Only the smallest barberry seedlings would have escaped such close order drill!

Whitman County (5,592 km^2) is typical of this western US wheat-growing area in which the Eradication Program operated from ca. 1944 until 1978. Approximately 36% of the county's area supports cereal agriculture (wheat, oats, barley) (National Agricultural Statistics Service 2002). Using the current occurrence of barberry as a proxy for the region, we investigated the effectiveness of the eradication campaign in the mid-twentieth century. With the original survey records and the enthusiastic cooperation of long-retired personnel from the Eradication Program (D.S. Jackson and J.W. Burns), a resurvey was conducted in 2002–2003. The resurvey concentrated on the approximately 100 sites that had not been officially released from further inspection when the Program ended in 1978, i.e., sites most likely to still harbor barberry. Each site was surveyed by walking along transects that were 10 m apart; any barberry plants were recorded by GPS (Foster 2003).

The results obtained in 2002–2003 are striking. A grand total of just nine barberry plants on six properties were found in Whitman County, compared with at least 49,313 plants that were removed during the 33 years of the Program's operation. All these remaining plants were adults (1–3 m tall), and on the basis of ring counts, all were at least 25-years old. Some were probably seedlings that were still too small for detection by the Program's end in 1978; others likely germinated post-1978 (Foster 2003). Given that all the detected plants are adults, we of course searched thoroughly in their vicinities for seedlings; none were found. We have no explanation for this quite fortuitous lack of barberry recruitment.

Neither in the Great Plains nor in Whitman County, WA was eradication sensu stricto achieved; some plants escaped detection. But the low to nil incidence of stem rust across these areas today, almost 30 years after suspension of the Program, is graphic testimony to its effectiveness (Leonard 2001). However, these results do illustrate an often unappreciated benefit of an eradication campaign: even if eradication is not achieved, the population(s) of the target species are reduced to a numerical range in which environmental/demographic stochasticity becomes an on-going threat to persistence (Mack 2000). Lessons learned during this long campaign are relevant today: a strictly-enforced prohibition of reentry by the target species into the treatment area, thorough ground searching for all individuals over multiple years, accurate recordkeeping, and public acceptance of the project's necessity and benefits at the outset (Kempton 1921) pay enormous dividends.

3.4 Imminent Eradication: The 50-Year Campaign Against *Striga asiatica* (Witchweed)

Large-scale eradication efforts are not confined to the past: perhaps the largest on-going eradication program is being waged against witchweed (*Striga asiatica*) in the southeastern US Reasons for continuing a program that has been in continuous operation since the late 1950s are readily apparent: similar to the impetus for the European barberry campaign, witchweed is a huge threat to agriculture. Unlike barberry, witchweed is a direct agent of damage, as all members of *Striga* are hemi-parasitic plants that attack and usurp resources from their host plants, including corn and sorghum (Eplee 1981).

The native range of the genus *Striga* is Eurasia and Africa. Consequently, there was no question that the species had been introduced when it was discovered in cornfields in North Carolina in 1956. Left untreated, this invasion would have almost certainly spread far across the US and eventually parasitized corn and sorghum in the Great Plains. Fortunately (and all too rarely) in two years, a federal/state program dedicated to eradicating, not simply controlling, witchweed was launched in North Carolina and neighboring South Carolina (Westbrooks 1993). Its initiation was none too soon: the minute seeds of *S. asiatica* were being unwittingly spread in contaminated farm implements from field to field. Removing the invader mechanically (cf. eradication of barberry) would have been infeasible. Instead an effective multipronged tactical approach has been taken. In addition to direct herbicide treatment, witchweed's premature germination is induced through soil injections of ethylene or sowing "false hosts" (e.g., cotton and soybeans), i.e., plants that induce germination but are not parasitized by witchweed. Either technique induces witchweed to germinate, and the plants are then destroyed with herbicide (Eplee 1981; Westbrooks 1993).

Equally important with tactics has been the strategy (sensu Moody and Mack 1988) employed to eradicate *S. asiatica*. First, a strict embargo was placed on the movement of farm equipment from counties infested with witchweed, thereby minimizing the plant's further spread. Second, the Program wisely concentrated its first eradication effort on the small, outlier populations (Eplee 1979). Consequently, new populations could not readily arise while the infested area was being steadily

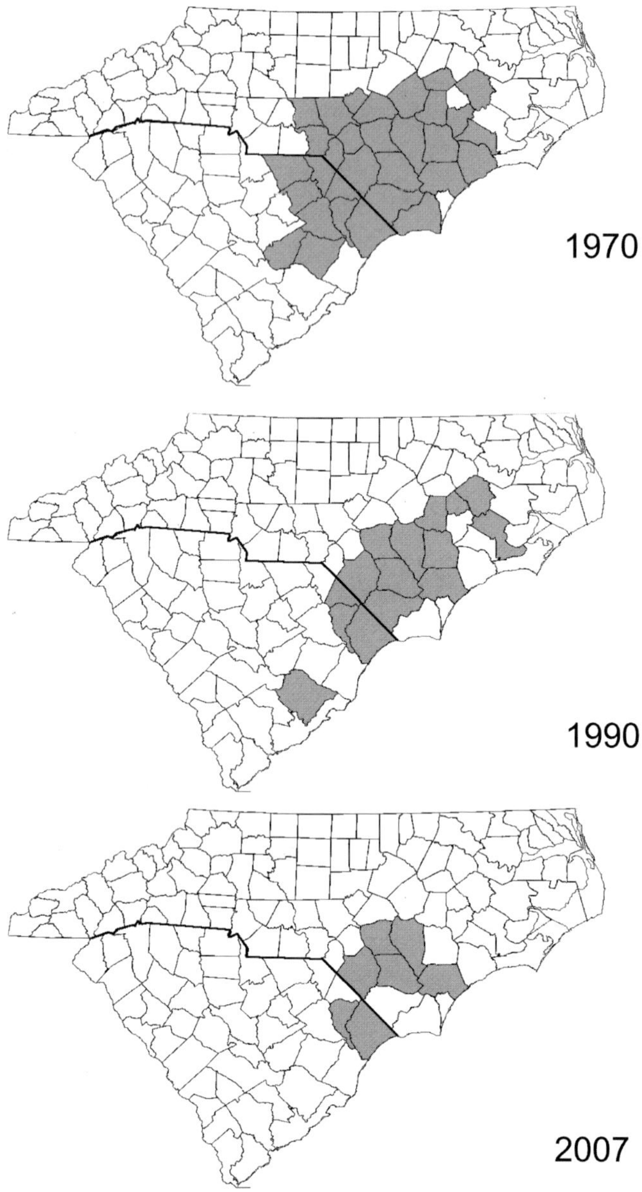

Fig. 3.3 The hemiparasite, *Striga asiatica* (witchweed), occupied > 160,000 ha in North and South Carolina (USA) by 1970 (*shaded counties* in each state). Stringent application of inter-county quarantine of contaminated farm implements coupled with an effective strategy to eradicate witchweed on infested sites had progressively reduced the invasion to < 1,000 ha by early 2007 (map sources: http://www.lib.ncsu.edu/exhibits/sodfather/images/timeline58.jpg; http://ceris.purdue.edu/napis/pests/ww/imap/ww03nc.html)

shrunk inward through treatment. The strict protocol of detailed survey and equally detailed recordkeeping of sites that had proven so effective with barberry eradication was implemented with *S. asiatica*. Furthermore, sites are currently released from further surveys, i.e., deemed now free of witchweed, after only multiple consecutive years without detectable plants. If witchweed reappears on a land parcel even five years after all plants had been destroyed, the parcel is reinserted into the yearly survey scheme (APHIS/PPQ 1992, R.C. Horne, pers. comm.).

Strict application of these principles since the late 1950s has borne stellar results: total area still occupied by *S. asiatica* as of mid-2007 had shrunk from a maximum range of 162,000 ha (Westbrooks 1993) to a total of < 1,000 ha in the Carolinas (Witchweed Eradication Project Status Report 2007) (Fig. 3.3). At this rate of range reduction, witchweed may be eradicated in the next 5–10 years – a seemingly long time on what has already been a long (49 years) and expensive (estimated at $250 million, Eplee 2001) program. But this total cost is obviously a small fraction of even the annual value of the US corn crop, which was valued at $52.1 billion in 2007 (http://www.nass.usda.gov/Statistics_by_State/Michigan/Publications/Current_News_Release/nr0817.txt). However, it is fair to point out that eradication, while in sight, has not yet been achieved, and there is the residual concern that witchweed has colonized as yet undetected locales outside the shrinking ring of active eradication (R. Westbrooks and R.C. Horne, pers. comm.). Eradication is an admittedly elusive goal, but the benefits can, nonetheless, be seen even before the last invasive plant has been destroyed.

In addition to having incorporated salient components of the earlier barberry program, the Witchweed Eradication Program is praiseworthy for its rapid decisions on appropriate tactics and strategy – only two years after first detection of witchweed, funds were appropriated for its eradication, not simply its control. Effective plant destruction techniques (analogous to tactics, sensu Moody and Mack 1988) were developed from the Program's outset, even as the plant's biology was investigated, rather than emerging through sequential investigations. Continued, methodical persistence, especially after decades of action, will be essential to the Program's eventual success.

3.5 Eradication of Kochia (*Bassia scoparia*) in Western Australia: A Rare Product Recall

Most plant naturalizations stem from deliberate introductions (Mack and Erneberg 2002), and the simple majority of plant invasions have likely arisen from this pool of species chosen for importation. Introduced plants have been envisioned as sources of food, fiber, medicine, seasonings, and most commonly as indoor or landscape ornamentals (Mack 1991). Forage species, turned naturalized or even invasive, form another class of these immigrants (e.g., *Agropyron cristatum, Bothriochloa ischaemum, Cynodon dactylon, Trifolium repens*) (Heady 1975; Gabbard and Fowler 2007 and references therein). Preintroduction assessment of these species, if indeed any assessment was performed at all, has consistently dwelled on the species' palatability, ease of cultivation, and any other purportedly beneficial aspects. Detrimental features, such as the potential for the species to become invasive or a source of fuel for wild fires, have often been ignored (Lonsdale 1994).

Western Australia contains immense treeless areas that have been long viewed as potential grazing land for livestock. The region's aridity not only sets a limitation on forage production but also exacerbates soil salinity (Burvill 1979). The impetus has been clear for graziers to foster species (native or introduced) that are palatable and grow vigorously on these salt-laden soils. This need was thought to have been met in part by Kochia (*Bassia scoparia*), a Eurasian chenopodiaceous shrub. Seeds of the shrub were introduced in 1990 onto 52 properties in Western Australia. Signs appeared within a year that the introduction was going awry when a landowner noticed *B. scoparia* seedlings becoming established well beyond the point of Kochia's introduction. By 1993 the shrub was behaving clearly as an incipient invader: dead Kochia formed seed-laden tumbleweeds, and these blew across the treeless landscape, spreading numerous seeds (Randall 2001).

An eradication program rapidly took form after the landowner's report. Further introductions were halted within 12 months, thanks to the state and national funding that was sustained throughout the eradication campaign. Even with such swift action, *B. scoparia* eventually occupied 2,281 ha; most of this area occurred as two parcels (140 and 1,000 ha) (Dodd 2004). Eradication site-by-site with herbicides, mechanical removal, and intense sheep grazing proved effective, and sites were resurveyed yearly. A site was considered Kochia-free if no plants were detected for three consecutive years after the last plant had been eliminated. Despite the shrub's conspicuous features and a short-lived seed bank (coupled with sustained, diligent plant removal), some plants were detected as much as seven years after initiation of the eradication campaign. By 2004, Western Australia was, however, considered free of Kochia (Dodd 2004).

This eradication project was successful through the remarkable confluence of events, species' traits, and circumstances. First and foremost, the problem was rapidly reported to government authorities by a conscientious landowner. Consequently, the problem was detected before Kochia had dispersed to thousands of widely separated sites, and luckily fences minimized the dead, seed-laden tumbleweeds from spreading without constraint (Dodd and Randall 2002). Furthermore, adult Kochia are large and readily detected, and the shrub's seed bank is long-lived (Dodd 2004). Moreover, the authorities were able to take prompt action without the need to search for public funds for eradication. And finally, the public was alerted by the press to Kochia's threat (Randall 2001).

3.6 A Warning Unheeded: The Failure to Eradicate *Miconia calvescens* in Hawai'i

The Hawaiian Islands are simultaneously among the most remote and most invaded landmasses. The Islands' long involvement as a way station for oceangoing ships and its eager colonization by the US (Daws 1968) has produced a naturalized flora that now rivals its unique native flora in species richness (Wagner et al. 1990). Many alien species (e.g., *Clidemia hirta, Hedychium gardnerianum, Pennisetum setaceum, Psidium cattleianum*) have become invasive over the last 150 years (Stone et al. 1992).

Among these invaders is the small tree *Miconia calvescens* (Melastomataceae), a native of Central and South America (Meyer and Florence 1996). Miconia's history of introduction into Hawaii has been thoroughly reviewed by Medeiros et al. (1997), and we only summarize those events here. Most relevant to the failure to eradicate this tree once it was detected in Hawaii is its known record of invasion in French Polynesia, where it was introduced into a botanical garden in Tahiti in 1937. It escaped, dispersal probably facilitated by birds that could carry its seeds to inaccessible sites on the craggy volcanoes on Tahiti and Moorea. The tree's tolerance of light regimes inside the islands' native forests meant that it could not only persist inside these forests (Fig. 3.4), but also that its rapid spread remained virtually undetected, including in aerial photographs (J-Y Meyer, pers. comm.). Prospects for Miconia's control, much less eradication, now seems daunting. Much of the forests on Tahiti and Moorea is now dominated by *M. calvescens,* and the consequences of this aggressive invader will almost certainly be played out in the future in a general

Fig. 3.4 (**a**) *Miconia calvescens* has been dispersed by birds and storms across much of Tahiti, French Polynesia, including steep mountaintops; these sites are inaccessible by humans, thereby virtually precluding Miconia removal. (**b**) Once Miconia invades native forest it soon totally dominates the understory and eventually the overstory as well (Note the person in background is largely obscured by Miconia juvenile plants) (photos by R.N. Mack)

degradation of French Polynesia's native communities and biodiversity (Meyer and Florence 1996).

M. calvescens was introduced as an ornamental into Hawai'i (on Oahu) in 1961; similar to its history on Tahiti, a botanical garden was at least one site of its early introduction (Medeiros et al. 1997). Wagner et al. (1990) noted Miconia as becoming naturalized locally and cautioned about its future potential for spread. Much stronger signals of the invasive potential of Miconia had already been received, informally by the early 1970s (Meyer and Florence 1996) with repeated warnings through the 1980s, based on the environmental disaster unfolding in French Polynesia. The local press did sound an alarm as early as 1991 about the potential damage that could be caused by Miconia (Altonn 1991 as cited in Medeiros et al. 1997), but these public alerts unfortunately did not spark a concerted search-and-destroy campaign. Instead plants were removed on Maui, Oahu, and Hawaii beginning in the mid-1990s, even as new foci of the invasion were being detected (Medeiros et al. 1997).

Action on Maui exemplifies the consequences of not moving rapidly into an eradication mode. For example, by May 1991 on East Maui, any response was "still in the information-gathering stage" (Loope et al. 1992), while a cooperative effort to eradicate Miconia was being developed (Gagne et al. 1991 as cited in Loope et al. 1992). However, in all likelihood the invasion on Maui by that time was far greater than then perceived – a consequence of a limited survey for the invader on the island. Effective control, much less eradication, is hamstrung without accurate information on the geographic boundaries of the invasion (Panetta and Lawes 2005) (cf. successful campaign against *B. scoparia* in Australia). As much more extensive reconnaissance took place, beginning in approximately 1995, the investigators found many more foci of *Miconia* spread across a much wider area than detected earlier (Medeiros et al. 1997, J. Gooding, pers. comm.). A concerted plant removal effort in the last 10 years has probably curbed further wholesale range expansion by destroying fruiting trees in outlier populations. But large infestations, such as two populations (400 ha and 800 ha each) on the north side of Maui near Hana remain, in addition to innumerable outliers. Although the opportunity for eradication on Maui may not be irretrievably lost, the chances have been lowered appreciably through a combination of inadequate early detection/rapid response, erratic financial support and the difficult search terrain (numerous steep, isolated valleys with dense vegetation). Public cooperation and understanding of the threat held by an unchecked *Miconia* invasion is, however, encouragingly high (J. Gooding, pers. comm.).

Whether eradication is still feasible for all islands in Hawai'i is likely subject to an island-by-island assessment and, of course, the intensity of the eradication effort. Clearly, the cost/benefit ratio, using the fate of French Polynesia as a worse case scenario, strongly favors a much more intensive eradication campaign than has been employed for the past 10–15 years. Ironically, the current expenditures to combat Miconia may not even be containing the invader statewide, much less achieving eradication. On the basis of an economic model,

Burnett et al. (2006) conclude that too little effort, i.e., funding, is being spent. Yet the cost of increasing funding to an "optimal program", i.e., a permanent severe reduction in Miconia, although not eradication, would require only a onetime sixfold increase over current expenditures. Such an optimal program Kaiser (2006) would reap enormous gains in obviating plant removal costs in the future, by several orders of magnitude, plus preventing further environmental damage by Miconia.

The optimal control envisioned by Kaiser (2006) and Burnett et al. (2006) takes no account, however, of the spatial distribution of Miconia, so that the cost/benefit of removing a tree is the same as any other tree. This simplification of the contribution a single tree may make in establishing a new focus – far removed from the main foci of the invasion – does not account for the ability or likelihood of the descendants of an isolated plant to become established in heretofore unoccupied territory. One product of this simplification is that where an invader is widespread but in low numbers – exactly the circumstances under which new foci arise (Moody and Mack 1988) – the cost of detecting and destroying these individuals is not cost effective. The optimal goal is then interpreted as maintaining the invader's populations in low numbers but not seeking total eradication (Burnett et al. 2006; Kaiser 2006). Although permanently holding an invader at low numbers may be functionally equivalent to its eradication (e.g. the current status of *B. vulgaris* in the northern US), the risks are not identical. First, maintaining these low levels requires a consistent control effort. And policy-makers anywhere are inclined to withdraw funds once the perceived hazard has been reduced to a low level, e.g. *Hydrilla verticillata* in Florida (Mack and Foster 2004). Total eradication makes the consequences of such withdrawal of funds less hazardous. Second, continuing to destroy all known foci of an invasion and any new populations as they are detected may not achieve eradication, but it may come very close.

3.7 "Weeds Won't Wait" to be Eradicated: The Aborted Eradication of *Crupina vulgaris*

As the spread of *Miconia* in French Polynesia and Hawai'i illustrates, delay in initiating an eradication program yields ominous consequences. As Westbrooks has long warned "Weeds won't wait!" (R.G. Westbrooks, pers. comm.), their invasions inexorably unfold, unless we apply determined intervention. Westbrooks' epigram is especially relevant to the timeliness of an eradication campaign; yet irreparable delay has occurred far too often, even if the underlying reasons differ. The invasion of *Crupina vulgaris* (Common crupina) in the western US has become a case study in which well intended but prolonged assessment of an unfolding invasion, coupled with last minute and equivocal environmental concern, eventually blocked an ambitious eradication project at probably the last feasible opportunity for success. Simberloff (2003b) has recently outlined the sequence of events in the spread of Common crupina, so only a brief synopsis is necessary here.

C. vulgaris, a European annual, was first identified in north-central Idaho in 1968 – its first detection in North America (Stickney 1972). A cursory survey in 1970 found the plant confined to approximately 16 ha – a small infestation that could have been swiftly eradicated. But no action was taken. Instead nine years elapsed before even an eradication feasibility study was initiated, and six more years slipped by before it was determined that Common crupina could indeed be eliminated. Nevertheless, a detailed strategy of eradication was developed (Zamora et al. 1989a), and an eradication manual was distributed (Zamora et al. 1989b), all in preparation for the pending action. In the meantime, the plant had been added to the Federal Noxious Weed List (Westbrooks 1993), which made it eligible for federal funds for eradication. Finally in 1991, a federal/state task force proposed spending five million dollars for Common crupina eradication, but by this time the invasion had ballooned to cover > 25,000 ha. With a determined effort and appropriate strategy, eradication was perhaps still possible, but this effort was never mounted as isolated public objection arose to the proposed use of the herbicide picloram (Tordon®) for eradication. (The unsubstantiated claim was made that the herbicide would leach into adjacent rivers and detrimentally affect native salmon.) (R. Westbrooks, pers. comm.). At that point any attempt to control, much less eradicate *C. vulgaris*, except locally, collapsed (D. Thill, pers. comm.). Consequently, the spread of this invasive species in Idaho continues (T. Prather, unpublished data).

The unfortunate tale played out with the failure of Common crupina eradication holds several lessons. First, the time between initial detection and the marshalling of resources for eradication was far too long, partially a consequence of a cumbersome governmental response to a readily perceived environmental hazard. Once assembled, the strategy and tactics for eradication might still have been successful, except for the crucial failure to win strong public support.

3.8 Two High Stakes Current Eradication Campaigns

3.8.1 Chromolaena odorata in Queensland: High Stakes in an On-Going Eradication Campaign

Chromolaena odorata (Siam weed, Triffid), a native of South America, has been a scourge for decades in much of the African and Asian subtropics and tropics (Holm et al. 1977). The shrub is an aggressive competitor, and its seeds can be dispersed by wind, water, and attached to cargo as well as a seed contaminant (McFadyen and Skarratt 1996). Once established in a new range, it can rapidly spread under forest canopies as well as along roadsides and stream courses. Siam weed's vegetative growth is so prolific that not only does it provide competition for light, but also its high inflammability exacerbates its ability to alter communities. Removing *C. odorata* from arable fields is so difficult that it has rendered agricultural fields functionally worthless (R.N. Mack, pers. observ.). Consequently, its discovery in northern Queensland in 1994 (Waterhouse 1994) sparked a justified outcry for its eradication: its potential new range in Australia includes much of the eastern country's coast (McFadyen and Skarratt 1996) and possibly much more (Queensland Department of Primary Industries and Fisheries 2007a).

The National (Australian) Siam Weed Eradication Program is attempting to eliminate *C. odorata* from Queensland. Elimination of Siam weed in several hundred small infestations was once thought to be a difficult but attainable goal; it has unfortunately proved much more complicated. As is so often the case with a rapidly spreading invader, the shrub has appeared in multiple and widely separated locales in Queensland – aided in part by its dispersal via severe tropical storms, such as Cyclone Larry in 2006. Invasion so far consists mainly of populations occupying < 50 ha, although the largest infestation covers 1,400 ha in or near Townsville and Thuringowa. Siam weed is being destroyed at many of the known sites, even as detection continues for any newly established or previously undetected populations (Queensland Department of Primary Industries and Fisheries 2007a).

The conditions and circumstances encountered in eradicating *C. odorata* seem extraordinarily challenging. Much of the survey must be conducted on foot, as effective aerial reconnaissance is limited to a two-week flowering period when the shrub's white inflorescences can be detected. Eradication is conducted commonly in remote tropical forests on steep terrain that have few roads or trails. Furthermore, the use of vehicles can be counter-productive to eradication, as *C. odorata* seeds readily attach to vehicles. Consequently, herbicide sprayers must be hand-carried to infestations or the plants must be hand-pulled. Finally, crocodiles, cassowaries, feral pigs, and lethally poisonous snakes can be encountered during fieldwork, and there is the added risk to workers of contracting Leptospirosis or Scrub Typhus (Queensland Department of Primary Industries and Fisheries 2007a)!

As indicated above, this eradication program may hold the dubious distinction of being conducted under the worse possible conditions. Yet the stakes are enormously high for Australia. The potential damage for much of coastal Australia is readily apparent in the horrific damage this South American shrub has already inflicted in India and West Africa (Day and McFadyen 2004). If not eradicated, Siam weed has the potential to alter much coastal forest in Australia, and the nation would be hard pressed to wage a long-term campaign of control, given so many other pressing biological invasions (Humphries et al. 1991).

To its credit, the Queensland Government is simultaneously conducting eradication campaigns against six other invasive species *Clidemia hirta*, *Limnocharis flava*, *Mikania micrantha*, *M. calvescens*, *Miconia racemosa*, and *Miconia nervosa*. None is as widespread as Siam weed, but each presents environmental problems that rival the reputation of *C. odorata*, especially *Clidemia hirta*, *Mikania micrantha*, and the aforementioned *M. calvescens*. Left unchecked each has the potential to invade otherwise undisturbed habitat in remote and largely uninhabited areas (Queensland Department of Primary Industries and Fisheries 2007b).

3.8.2 An Unequivocal Candidate for Eradication: *Heracleum mantegazzianum*

Reluctance to embark on a plant eradication program can draw from a pool of conventional wisdom: the target species is already too widespread across too many sites, it is inconspicuous, it occurs in inaccessible habitats, the public is unlikely to view the introduced plant as a widely shared hazard, and the eradication will be prohibitively costly. The case for eradicating *H. mantegazzianum* (Giant hogweed) in the US trumps all these reasons. This large perennial Umbellifer is native to the subalpine zone in the western Caucasus Mountains (Azerbaijan, Georgia, and southern Russia) (Mandenova 1951 and Otte and Franke 1998 as cited in Page et al. 2006). It was likely introduced deliberately in the US as an ornamental and as a spice in Iranian condiments (RNM, pers. obser.). Its introduction has been an egregious mistake as most aerial plant parts are festooned with stiff pustulate-based hollow trichomes that produce linear furanocoumarins. Humans and other mammals are highly sensitive to the phytophotodermatitis that these compounds produce, which are activated under UV. Contact with these trichomes produces painful blisters and even scarring (Page et al. 2006 and references therein). Any culinary/ornamental benefits of this plant are more than outweighed by its serious health risk.

Fortunately, Giant hogweed has been recognized as a Federal Noxious Weed in the US since the mid-1980s, and consequently its interstate transport is forbidden (Foy et al. 1983); 12 states further forbid its intrastate movement or even its occurrence (http://plants.usda.gov/java/noxiousDriver). Even though it now occurs in 16 US states, its total area of residence is probably < 100 ha (M.A. Bravo, pers. comm.) – seemingly well within the scope of even a modestly-funded eradication project. The ability to eradicate Giant hogweed is aided by the plant's ecological amplitude: it does not become established in communities with high plant cover. Instead it occupies disturbed sites (e.g., roadside ditches, old fields, gravel bars) as well as riparian areas (Page et al. 2006). And certainly destruction of Giant hogweed would seem well worth the cost – a case where a cost-benefit analysis seems needless, given the legitimate concern for public health.

In the US, Giant hogweed may be most widespread in Pacific Northwest, far removed from other populations in the Mid-West and along the East Coast. The western third of Washington, centering on metropolitan Seattle, has reportedly > 1,000 locations. Many of these locations likely support a single plant, but more extensive populations have become established along stream courses by water-borne seeds (King County DNRP/WLR 2006). These streamside populations warrant immediate attention because *H. mantegazzianum* spread initially in the Czech Republic along river courses (Pysek 1991). The legal requirement to destroy Giant hogweed coupled with its serious health risk to humans as well as its confinement to a small current new range in the US would seem to offer almost ideal circumstances for its eradication. Fortunately, eradication is seemingly the goal in Washington and elsewhere in the US.

3.9 Invasive Plant Eradication: Common Features of Success (and Failure)

Concerted attempts have been made to eradicate alien plants for more than one hundred years. All such attempts, whether successes or failures, form the basis by which we can detect features common to successful eradication campaigns. Some of the conclusions below admittedly appear as common sense; however, others appear initially counter-intuitive but nevertheless have a record of support. As a result, it appears possible to form a "to do" and the equally valuable "do not do" list of steps for eradication (Groves and Panetta 2002; Mack and Lonsdale 2002). Underlying this list is attention to both the tactics (i.e., the specific tools used to destroy the target species) as well as a comprehensive strategy (i.e., attention to the larger geographic scale issues of curbing and eventually totally destroying the target species) (sensu Moody and Mack 1988).

1. *Further entry into the new range must be completely curtailed.* It is pointless to initiate eradication as long as the opportunity persists for simultaneous reentry, whether accidental or deliberate. Early in the US eradication campaign against European barberry federal regulations were enacted to prohibit its sale (Kempton 1921). The *S. asiatica* eradication program in the US wisely imposed a quarantine of farm implement movement between counties as an effective first-step to minimize its inadvertent reintroduction into treated fields (Eplee 1981 and references therein). Deliberate dispersal of a target species, such as a species still valued in horticulture, must also be completely curtailed. Failure to stem such dispersal renders any eradication effort pointless, e.g., as long as *E. crassipes* can still be purchased anywhere in the US (Isaacson 1996), no comprehensive eradication of the plant in the US can be sensibly considered.
2. *The target species should be readily detectable.* As defined here, eradication unlike control ultimately involves the destruction of all plants. This specificity of action places a premium on detecting even isolated individuals. If target plants are small or inconspicuous, or both, detection is obviously much hampered. In contrast, successful eradications have commonly targeted large herbaceous perennials (*Hieracium pilosella, S. asiatica*) (Mack and Londale 2002) or shrubs (*B. scoparia)* (Dodd 2004). A plant's conspicuousness is not, however, a guarantee of its eradication (cf. early attempts to eradicate *E. crassipes*), only a highly useful feature.
3. *The search terrain should be readily accessible.* Here again, common sense is borne out in practice. Campaigns against *S. asiatica* and *B. scoparia* have been conducted in low relief terrain in southeastern US and Western Australia,

respectively (Sand and Manley 1990; Dodd 2004), where the target plants did not usually occupy inaccessible sites (cf. *M. calvescens* in French Polynesia, Meyer and Florence 1996).

4. *The target species' seed bank should be short-lived.* Duration of resurveys of sites after all visible plants have been removed relates directly to the longevity of the target species' seed bank (Groves and Panetta 2002). The proverbial ideal target species has a short seed bank life (e.g., < three years), and several successive annual searches of once infested sites could remove any remaining individuals. This species attribute relates directly to Item 7 below. As a corollary, eradication is strongly hampered, even rendered problematic, if the species routinely persists through vegetative propagation, i.e., adventitious roots and stems, rhizomes, or stolons. Consequently, eradication of *E. crassipes* or *Polygonum cuspidatum* (Japanese knotweed) would be exceedingly difficult because both species produce seeds and readily propagate vegetatively (Barrett 1980; Palmer 1994).
5. *Likelihood of eradication is inversely proportional to the size of the occupied new range and the number of locations* (Groves and Panetta 2002). The close, inverse relationship between eradication success and the extent of an invasive species is so strong as to persuade some that it is essentially an inviolate principal (Rejmánek and Pitcairn 2002; Groves and Panetta 2002; Panetta and Timmins 2004). And clearly, the bulk of plant eradications have occurred for species with very limited distribution in their new range (Groves and Panetta 2002; Mack and Lonsdale 2002). But the on-going *S. asiatica* eradication program as well as the earlier Barberry Eradication Campaign illustrate that while admittedly more difficult, eradication or near eradication can be achieved if key components of the eradication strategy are scrupulously followed.
6. *Emphasis should be directed first at the invasion's outlying nascent foci.* The initial remediation to environmental damage is often directed at the largest, most conspicuous source of the hazard, e.g. so-called Superfund sites of pollutants in the US (Cannon 2007). However, this approach fails in eliminating the hazard of a plant invasion because plants, unlike pollutants, perpetuate themselves. Furthermore, small, outlier populations are more likely to contribute recruits to previously unoccupied new range than plants within a large focus of the invasion (Moody and Mack 1988). As a result, destroying these outlying populations before they become new sites of plant dispersal is essential. Awareness of this eradication principle is apparent in the US Witchweed Eradication Campaign (Eplee 1979, 1981). In effect, the barberry campaign also emphasized the destruction of outlier foci: clear attention was paid to all plants, regardless of their local abundance (Kempton 1921).
7. *Resurvey of all searched sites for remaining individuals must be diligent and prolonged.* No matter how conspicuous the target species, all plants will not be initially detected. Furthermore, resurveys are mandatory to detect plants that emerged postsurvey (Mack and Lonsdale 2002 and references therein). Given the unlikelihood of detecting immigrants immediately upon their entry into the new range, a seed bank will have already been established by the time eradication is initiated.

Determining the length of the resurvey period can be problematic. Too short and undetected plants emerge after cessation of the surveys, thus largely off-setting the eradication effort. Too long, however, also holds risks, as eradication campaigns are expensive and potentially viewed as disruptive and intrusive by the public (Simberloff 2003a, b). Once the plant has not been detected for several years, public funds risk being withdrawn (Mack and Foster 2004).

As a consequence, devising effective minimal resurvey periods have consistently been an objective of eradication campaigns (Kempton 1921; APHIS/PPQ 1992; Panetta and Timmins 2004). The likelihood that the target species will reappear on each surveyed site declines steeply as years elapse with no detected plants. But the reappearance of even a single plant means that survey of the sites begins again on a year-by-year basis. Only after no target plants have been detected at a site for multiple years is the site released from further surveys (e.g., APHIS/PPQ 1992).

8. *Public cooperation and financial support must be sustained.* The importance of broad public understanding for the need and feasibility of eradication is probably the most important component of a successful eradication program. Without it, most eradication campaigns are doomed to failure, unless the invader is confined to a few small sites on public lands. The value of public support is well illustrated by comparing the outcomes of the campaigns against *S. asiatica*, *B. vulgaris*, and *C. vulgaris* in the US.

The largest campaign of which we are aware, the attempt to eradicate *B. vulgaris* in the US, contained a decidedly mixed bag of features that could have spelled success or failure. Although barberry is a conspicuous shrub, and the new range was largely in a readily traversed agricultural landscape, its seeds are persistent in the soil. Moreover, birds are effective seed dispersers, thereby opening the opportunity for reentry to cleared sites (Kempton 1921). The most serious aspect that could have stymied eradication was barberry's huge US range (Fig. 3.1). But public support overcame this seemingly insurmountable hurdle. Tens of thousands of people were eventually involved in the eradication effort, not just farmers but also seasonal eradication crews and others. Searching every parcel of land became a competition for school children that the government heartily encouraged by awarding medals and other recognition (Roelfs 1982)! Clear emphasis was placed on careful survey and resurvey of this immense area for barberry, so that newly emergent plants were eventually destroyed. Although the current witchweed campaign does not encompass nearly as large an area (and the public has not been asked to become directly involved), the program enjoys regional support in the Carolinas even 50 years after its initiation (R. Westbrooks, pers. comm.). The witchweed campaign would have failed utterly had not the public understood the need to ban intercounty movement of farm implements and the repeated search and herbicide treatment of private property – understanding reached through a deliberate outreach effort (Sand 1990).

In contrast, widespread public support has been neither speedy nor adequate for the eradication of *M. calvescens* in Hawaii. The public had been made repeatedly aware of the threat of Miconia in the early 1990s by the Hawaiian press, but the alarm was raised only after the plant had been in the state for about 20 years (Medeiros et al. 1997) – perhaps late in Miconia's lag phase in this new range but still early enough to achieve eradication had ample funds been promptly mustered, sustained, and administered. Instead, the campaign has needed to rely on a patch quilt of funding from county, state, and federal agencies. Funding in the last few years has been approximately one million dollars, which Kaiser (2006) concludes is too little as employed to halt the spread, much less reverse it. By 2007, even parts of this inadequate funding were in jeopardy (http://www.mauinews.com/page/content.detail/id/28162.html). Without the public's insistence to increase the resources devoted to this campaign, the spread of *Miconia* in Hawaii and its concomitant damage seem assured.

On the basis of the points outlined above, the feasibility of any plant eradication program could likely be distilled to a simple sum of "yes" and "no" answers to produce an accurate recommendation on whether to proceed with eradication. Although a decision of "proceed no further" could be reasonably drawn from a string of mostly negative responses (e.g., no, the target species is not readily detected in the landscape), the examples we have reviewed here show that aside from sustained public support, there is no apparent minimum number of features that spell success or failure. But without broad public support the rationale decision for a proposed eradication effort may well be "proceed no further."

3.10 Conclusions

We have reviewed the basis for whether invasive and potentially invasive plant species can be eradicated. This question is not new: the difficulty of totally destroying all members of an alien species has long been appreciated, especially when the species is widespread and in innumerable, inaccessible sites. We contend that this debate is best conducted by carefully examining cases histories, both successes and failures in eradications. In our view, measured optimism is warranted on prospects for successful eradications (Simberloff 2002), *provided* a series of steps are rigorously applied, e.g., rapid response when little of the potential new range has been occupied, first action directed at the outlier foci, diligent resurvey of treated sites to remove newly emergent or overlooked individuals. Perhaps most important here is a step that ecologists and land managers have not uniformly pursued – soliciting active public involvement in the eradication effort by (1) informing the public of the need for eradication in environmental and economic terms and (2) gaining adequate, sustained public funds and assistance for destroying all populations of the target species. Application of the ecological lessons of combating invasive species combined with a deliberate strategy for gaining public involvement are essential for success.

Acknowledgments We thank M.A. Bravo, J.W. Burns, S.J. Darbyshire, J. Dodd, J. Gooding, R. C. Horne, R. Iverson, D.S. Jackson, B. A. Kaiser, P. Krushelnycky, K. Galway, S. Lloyd, R. Old, T. Prather, J. Randall, D. Thill, B. Waterhouse, R. Westbrooks, and C. Wilson for valuable, often difficult to acquire, unpublished information and helpful discussions. Three anonymous reviewers also provided useful corrections and comments.

References

[APHIS/PPQ] Animal and Plant Health Inspection Service. Plant Protection Quarantine (1992) Domestic Program Manual Witchweed (*Striga asiatica* lour.) M301.80. United States Department of Agriculture

Barkley TM, Brooks RE, Schofield EK (eds) (1986) Flora of the Great Plains. University Press of Kansas, Lawrence, KS

Barrett SCH (1980) Sexual reproduction in *Eichhornia crassipes* (Water Hyacinth) II. Seed production in natural populations. J Appl Ecol 17:113–124

Barrett SCH (1989) Waterweed invasions. Sci Am 261:90–97

Burnett K, Kaiser B, Pitafi BA, Roumasset J (2006) Prevention, eradication, and containment of invasive species: illustrations from Hawaii. Agric Resource Econ Rev 35:63–77

Burvill GH (1979) Agriculture in Western Australia: 150 years of development and achievement, 1829–1979. University of Western Australia Press, Nedlands, Australia.

Busdicker HB (1945) Washington annual report, barberry eradication project. United States Department of Agriculture, Pullman, WA

Busdicker HB (1946) Washington annual report, barberry eradication project. United States Department of Agriculture, Pullman, WA

Campbell CL, Long DL (2001) The campaign to eradicate the Common Barberry in the United States. Pages 16–43 In: Peterson PD (ed) Stem rust of wheat: from ancient enemy to modern foe. APS Press, St. Paul, MN

Cannon JZ (2007) Overview of the Superfund Program. Pages 25–48 In: Macey GP, Cannon JZ (eds) Reclaiming the land: rethinking Superfund institutions, methods, and practices. Springer, New York, NY

Daubenmire R (1970) Steppe vegetation of Washington. Wash Agric Exp Stn Tech Bull 62

Daubenmire R, Daubenmire JB (1968) Forest vegetation of eastern Washington and Northern Idaho. Wash Agric Expt Stn Tech Bull 60

Daws G (1968) Shoal of time; a history of the Hawaiian Islands. Macmillan, New York, NY

Day MD, McFadyen RE (eds) (2004) Chromolaena in the Asia-Pacific region: proceedings of the 6th International Workshop on Biological Control and Management of Chromolaena. Australian Centre for International Agricultural Research. Canberra, A.C.T., Australia

Dewey LH (1894) The Russian thistle. Its history in the United States, with an account of the means available for its eradication. U S Dep Agric, Division of Botany Bull 15

Dewey LH (1897) Migration of weeds. Pages 263–286. In: 1896. U S Dep Agric Yearb Agric

Dodd J (2004) Kochia (*Bassia scoparia* (L.) A.J. Scott) eradication in western Australia: a review. Pages 496–499. In: Sindel BM, Johnson SB (eds) Proceedings of the Fourteenth Australian Weeds Conference. Weed Science Society of New South Wales, Sydney

Dodd J, Randall RP (2002) Eradication of kochia (*Bassia scoparia* (L.) A.J. Scott, Chenopodiaceae) in Western Australia. Pages 300–303. In: Spaford Jacob H, Dodd J, Moore JH (eds) Thirteenth Australian Weeds Conference. Plant protection society of western Australia. Victoria Park, W.A., Australia

Eplee RE (1979) The *Striga* eradication program in the United States of America. Pages 269–272 In: Musselman LJ, Worsham AD, Eplee RE (eds) Proceedings of the second symposium on parasitic weeds. North Carolina State University, Raleigh, NC

Eplee RE (1981) Striga's status as a plant parasite in the United States. Plant Dis 65:951–954

Eplee RE (2001) Co-ordination of witchweed eradication in the USA. Page 36 In: Wittenberg R, Cook MJW (eds) Invasive alien species: a toolkit of best practices and management practices. CAB International, Wallingford, UK

FAO (2000) Multilateral trade negotiations on agriculture: a resource manual. vol. 3. SPS and TBT agreements. Food and Agriculture Organization of the United Nations, Rome

Foster SK (2003) The barberry eradication program in Whitman County, Washington: a reassessment. MS thesis, Washington State University, Pullman

Foy CL, Forney DR, Cooley WE (1983) History of weed introductions. Pages 65–92. In: Wilson CL, Graham CL (eds) Exotic plant pests and North American agriculture. Academic Press, New York, NY

Gabbard BL, Fowler NL (2007) Ecological amplitude of a diversity-reducing invasive grass. Biol Invasions 9:149–160

Groves RH, Panetta FD (2002) Some principles for weed eradication programs. Pages 307–310. In: Spaford Jacob H, Dodd J, Moore JH (eds) Thirteenth Australian Weeds Conference. Plant protection society of western Australia. Victoria Park, W.A., Australia

Harris S, Brown J, Timmins S (2001) Weed surveillance – how often to search? Science for Conservation 175. Department of Conservation, Wellington, New Zealand

Heady HF (1975) Rangeland management. McGraw-Hill, New York, NY

Holm LG, Plucknett DL, Pancho JV, Herberger JP (1977) The world's worst weeds: distribution and biology. East-West Center/University Press of Hawaii, Honolulu

Humphries SE, Groves RH, Mitchell DS (1991) Plant invasions: the incidence of environmental weeds in Australia. Kowari 2. Australian National Parks and Wildlife Service, Canberra

Hutton L (1927) Barberry eradication reducing stem rust losses in wide area. Pages 114–118 In: U S Dep Agric Yearb Agric

Isaacson RT (1996) Andersen horticultural library's source list of plants and seeds. 4th edn. Andersen Horticultural Library, University of Minnesota, Chanhassen, MN

Kaiser BA (2006) Economic impacts of non-indigenous species: Miconia and the Hawaiian economy. Euphytica 148:135–150

Kempton FE (1921) Progress of barberry eradication. U S Dep Agric Circ 188

King County DNRP/WLR (2006) King County Department of Natural Resources and Parks Visual Communication and Web Unit. 0612no WEEDmap06.ai wgab. Seattle, WA, USA. http://dnr.metrokc.gov/wlr/LANDS/Weeds/maps.htm. Accessed 04 June 2008

Klorer J (1909) The water hyacinth problem. J Assoc Eng Soc 42:42–48

Leonard KJ (2001) Stem rust – future enemy? Pages119–146 In: Peterson PD (ed) Stem Rust of wheat from ancient enemy to modern foe. APS Press, St. Paul, MN

Levine JM, D'Antonio CM (2003) Forecasting biological invasions with increasing international trade. Conserv Biol 17:322–326

Lonsdale WM (1994) Inviting trouble: introduced pasture species in northern Australia. Aust J Ecol 19:345–354

Loope LL, Nagata RJ, Medeiros AC (1992) Alien plants in Haleakala National Park. Pages 551–576 In: Stone CP, Smith CW, Tunison JT (eds) Alien plant invasions in native ecosystems of Hawaii management and research. University of Hawaii, Manoa

Mack RN (1991) The commercial seed trade: an early disperser of weeds. Econ Bot 45:257–273

Mack RN (2000) Cultivation fosters plant naturalization by reducing environmental stochasticity. Biol Invasions 2(2):111–122

Mack RN (2003) Plant naturalizations and invasions in the eastern United States: 1634–1860. Ann Mo Bot Gard 90:77–90

Mack RN, Simberloff D, Lonsdale WM, Evans H, Clout M, Bazzaz FA (2000) Biotic invasions: causes, epidemiology, global consequences and control. Ecol Appl 10:689–710

Mack RN, Erneberg M (2002) The United States naturalized flora: largely the product of deliberate introductions. Ann Mo Bot Gard 89:176–189

Mack RN, Lonsdale WM (2002) Eradicating invasive plants: hard-won lessons for islands. Pages 164–172 In: Veitch D, Clout M (eds) Turning the tide: the eradication of invasive species. Invasive Species Specialty Group of the World Conservation Union (IUCN): Auckland, New Zealand

Mack RN, Foster SK (2004) Eradication or control? combating plants through a lump sum payment or on the installment plan. Pages 56–61. In: Sindel BM, Johnson SB (eds) Proceedings of the Fourteenth Australian Weeds Conference. (Weed Science Society of New South Wales, Sydney)

McCullough DC, Work TT, Cavey JF, Liebhold AM, Marshall D (2006) Interceptions of nonindigenous plant pests at US ports of entry and border crossings over a 17-year period. Biol Invasions 8(4):611–630

McFadyen RC, Skarratt B (1996) Potential distribution of *Chromolaena odorata* (Siam Weed) in Australia, Africa and Oceania. Agric Ecosyst Environ 59:89–96

Medeiros AC, Loope LL, Conant P, McElvaney S (1997) Status, ecology and management of the invasive plant, *Miconia calvescens* DC (Melastomataceae) in the Hawaiian Islands. Bishop Mus Occas Pap 48:23–36

Meier FC (1933) The stem rust control program. J Econ Entomol 26:653–659

Meinig DW (1968) The great Columbia Plain: a historical geography. University of Washington Press, Seattle

Meyer J-Y, Florence J (1996) Tahiti's native flora endangered by the invasion of *Miconia calvescens* DC. (Melastomataceae). J Biogeogr 23:775–781

Moody ME, Mack RN (1988) Controlling the spread of plant invasions: the importance of nascent foci. J Appl Ecol 25:1009–1021

Mooney HA, Mack RN, McNeely JA, Neville LE, Schei PJ, Waage JK (eds) (2005) Invasive alien species: a new synthesis. Island Press, Washington, DC

Morris HE, Popham WL (1925) The barberry eradication campaign in Montana. Univ Montana Agric Exp Stn Bull 180

National Agricultural Statistics Service (2002) Agricultural Statistics Database http://www.nass.usda.gov:81/ipedb. Accessed 04 June 2008

Page NA, Wall RE, Darbyshire SJ, Mulligan GA (2006) The biology of invasive alien plants in Canada. 4. *Heracleum mantegazzianum* Sommier & Levier. Can J Plant Sci 86:569–589

Palmer JP (1994) *Fallopia japonica* (Japanese knotweed) in Wales. Pages 159–171. In: De Waal LC, Child LE, Wade PM, Brock JH (eds) Ecology and management of invasive riverside plants. Wiley, New York, NY.

Panetta FD, Timmins SM (2004) Evaluating the feasibility of eradication for terrestrial weed incursions. Plant Prot Q 19:5–11

Panetta FD, Lawes R (2005) Evaluation of weed eradication programs: the delimitation of extent. Diversity Distrib 11:435–442

Piper CV (1898a) The Russian thistle in Washington. Wa Agric Exp Stn Bull 34

Piper CV (1898b) The present status of the Russian thistle in Washington. Wa Agric Exp Stn Bull 37

Pysek P (1991) *Heracleum mantegazzianum* in the Czech republic: dynamics of spreading from the historical perspective. Folia Geobot Phytotaxon 26(4):439–454

Queensland Department of Primary Industries and Fisheries (2007a) National Siam Weed eradication program: annual report July 2006-June 2007. South Johnstone, Australia

Queensland Department of Primary Industries and Fisheries (2007b) National four tropical weeds eradication program: annual report July 2006-June 2007. South Johnstone, Australia

Randall R (2001) Eradication of a deliberately introduced plant found to be invasive. Pages 174. In: Wittenberg R, Cook MJW (eds) Invasive alien species: a toolkit of best practices and management practices. CAB International, Wallingford, UK

Rejmánek M, Pitcairn MJ (2002) When is eradication of exotic pest plants a realistic goal? Pages 249–253. In: Veitch D, Clout M (eds) Turning the tide: the eradication of invasive species. Invasive Species Specialty Group of the World Conservation Union (IUCN): Auckland, New Zealand

Roelfs AP (1982) Effects of barberry eradication on stem rust in the United States. Plant Dis 66:177–181

Sand PF (1990) Discovery of witchweed in the United States. Pages 1–7. In: Sand PF, Eplee RE, Westbrooks RG (eds) Witchweed research and control in the United States. Weed Science Society of America, Champaign, IL

Sand PF, Manley JD (1990) The witchweed eradication program survey, regulatory and control. Pages 141–150. In: Sand PF, Eplee RE, Westbrooks RG (eds) Witchweed research and control in the United States. Weed Science Society of America, Champaign, IL

Shine C, Williams N, Burhenne-Guilmin F (2005) Legal and institutional frameworks for invasive alien species. Pages 233–284. In: Mooney HA, Mack RN, McNeely JA, Neville L, Schei PJ, Waage JK (eds) Invasive alien species: a new synthesis. Island Press, Washington DC

Simberloff D (2002) Today Tiritiri Matangi, tomorrow the world! Are we aiming too low in invasives control. Pages 4–12. In: Veitch D, Clout M (eds) Turning the tide: the eradication of invasive species. Invasive Species Specialty Group of the World Conservation Union (IUCN): Auckland, New Zealand

Simberloff D (2003a) Why not eradication? Pages 541–548. In: Rapport DJ, Lasley BL, Rolston DE et al. (eds) Managing for healthy ecosystems. Lewis Publishers, Boca Raton, LA

Simberloff D (2003b) Eradication—preventing invasions at the outset. Weed Sci 51: 247–253

Stakman EC (1923) Barberry eradication prevents black rust in Western Europe. U S Dep Agric Circ 269

Stickney PF (1972) *Crupina vulgaris* (Compositae: Cynareae), new to Idaho and North America. Madroño 21:402

Stone CP, Smith CW, Tunison JT (eds) (1992) Alien plant invasions in native ecosystems of Hawaii: management and research. Cooperative National Parks Resources and Studies Unit and University of Hawaii, Manoa

U S Congress, Office of Technology Assessment (1993) Harmful non-indigenous species in the United States. OTA-F-565. U.S. Government Printing Office, Washington, DC

Wagner WL, Herbst DR, Sohmer SH (1990) Manual of the flowering plants of Hawai'i. University of Hawaii Press: Bishop Museum Press, Honolulu

Waterhouse BM (1994) Discovery of *Chromolaena odorata* in northern Queensland, Australia. Chromolaena Newsletter No. 9. Univ Guam Agric Exp Stn

Webber HJ (1897) The water hyacinth. U S Dep Agric Division of Botany Bull 18

Westbrooks RG (1993) Exclusion and eradication of foreign weeds from the United States by USDA APHIS. Pages 225–241. In: McKnight BN (ed) Biological pollution. The control and impact of invasive exotic species. Indiana Academy of Science, Indianapolis, IN

Westbrooks RG (2004) New approaches for early detection and rapid response to invasive plants in the United States. Weed Technol 18: 1468–1471

Whitman County Barberry Eradication Program (1943–1978) Form L survey cards, Whitman County Extensions Office, Colfax, WA

Whitson TD, Burrill LC, Dewey SA et al. (eds) (1996) Weeds of the West. Western Society of Weed Science in cooperation with the western United States land grant universities cooperative extension services, Laramie, WY

Witchweed Eradication Project Status Report (2007) North Carolina Department of Agriculture and Consumer Services. Plant Industry Division-Plant Protection Section. October 1, 2006 – March 31, 2007. Raleigh, NC

Wittenberg R, Cock MJW (2005) Best practices for the prevention and management of invasive alien species Pages 209–232. In: Mooney HA, Mack RN, McNeely JA, Neville L, Schei PJ, Waage JK (eds) Invasive alien species: a new synthesis. Island Press, Washington DC

Zamora DL, Thill DC, Eplee RE (1989a) An eradication plan for plant invasions. Weed Technol 3:2–12

Zamora DL, Thill DC, Eplee RE (1989b) Eradication manual for Common Crupina (*Crupina vulgaris* Cass.) Univ Idaho Agric Exp Stn Bull 701

Chapter 4
Management of Plant Invaders Within a Marsh: An Organizing Principle for Ecological Restoration?

James O. Luken and Keith Walters

Abstract Controlling plant invaders is often one aspect of ecological restoration. However, the planning and application of control measures can lead to difficult questions regarding project goals and measures of success. We present a case study of a coastal wetland system in South Carolina, USA, where two plant invaders, *Phragmites australis* and *Typha domingensis*, were targeted for control. As project participants gradually accepted the concept that success must be measured in terms of long-term system parameters rather than short-term invader control, the methods and approaches changed. As an alternative to applying herbicides, a method of reconnecting the system to the ocean was pursued. Instead of simply measuring plant control, a before-after-control-impact monitoring design was implemented that allowed comparison among restored and multiple reference systems in the immediate area. Attempts to reestablish tidal flow and modify environmental conditions to alter system attributes were variable with both unplanned positive and negative effects. Most of these impacts were associated with the fact that the wetland existed in a state park used by large numbers of people for passive recreation. The case study demonstrates that plant invasion and the willingness of people to control plant invaders can provide a useful starting point for eventual development and implementation of scientifically meaningful attempts at ecological restoration.

Keywords BACI • Ecological restoration • *Phragmites australis* • Reference site • *Tidal reconnection* • *Typha domingensis*.

4.1 Introduction

Salt marshes along the eastern USA coast are susceptible to dramatic changes in community composition and structure when hydrological connections to the ocean are restricted (Warren et al. 2002). Restrictions can result from natural,

J.O. Luken (✉)[1] and K. Walters[2]
[1]Department of Biology,
[2]Department of Marine Science,
Coastal Carolina University, Conway, SC 29528, USA,
joluken@coastal.edu

nearshore geological processes or from human activities such as marsh filling and diking. Generally, salt marshes cut-off from the ocean are converted to fresh or brackish marshes that can be invaded by perennial monocots including the common reed grass (*Phragmites austalis*) and cattails (*Typha* spp.). Ecological effects of *Phragmites* invasions and attempts at removal have been studied extensively in the Northeast USA from Massachusetts to Maryland (Boumans et al. 2003; Gratton and Denno 2006; Kimball and Able 2007; Teal and Weishar 2005). *Phragmites* invasions generally are less common and less well studied in the southeastern US.

The economic value of various ecological functions associated with tidal salt marshes (Costanza et al. 1997) has increased the interest of coastal managers to restore modified and/or invaded salt marsh sites. Restoration goals for impounded marshes typically, when stated, include reestablishment of tidal exchange, elimination of salt-intolerant and/or invasive grasses, and eventual development of ecological attributes similar to natural or reference tidal marshes (Roman et al. 2002; Warren et al. 2002). However, many problems and questions associated with marsh restoration remain because of difficulties in defining which ecological characteristics and/or which reference marshes should be considered. Evidence to date indicates certain marsh characteristics remain dissimilar to natural marshes decades after restoration (Craft et al. 2003; Zedler and West 2008). Coastal systems also can be unstable and characterized by a long history of switching from one ecological state to another (Booth et al. 1999; Zedler and West 2008), making determination of an appropriate restoration target difficult. Furthermore, coastal marshes increasingly are fringed by residential and commercial development making marsh conditions inextricably connected to and dependent on the social, political, and economic environment of coastal communities. The connection between marshes and local human communities means that any restoration activities will be influenced by a diverse array of stakeholders.

Plants frequently form the foundation for categorizing systems (e.g., *Spartina*-marsh), and plant management emerges as a primary activity in most restoration projects (Young 2000; Young et al. 2005). When plant invasions are present, plant eradication or control generally are required. Although the successful management and restoration of plant invasions must be guided by ecological theory and accepted research approaches (Neckles et al. 2002), managers universally recognize that a strong theoretical background is only one part of the restoration process. Along with ecological theory, a range of economic, political, and social factors can influence efforts to manage plant invasions as was shown in the case study involving knapweed, an introduced species invading pastures and rangeland in Colorado (Luken and Seastedt 2004)

This chapter focuses on what ostensibly is a salt marsh restoration project located in coastal South Carolina, USA. We use the restoration project to illustrate processes common to many efforts involving habitat restoration and invasive species, namely how an initially simple instance of controlling plant invaders eventually developed into a complex case of ecosystem management. We attempt to clarify the frequently conflicted nexus among researchers focused on testing for scientific generalizations, resource managers mandated to change or manipulate a system for the benefit of others, and the general public or stakeholders who see the environment as a resource that supports various recreational activities.

4.2 Methods and Approaches

4.2.1 Study Site

Sandpiper Pond is a 15 ha brackish to freshwater marsh located within the barrier beach system at Huntington Beach State Park, South Carolina, USA (Fig. 4.1). Prior to the 1980s, Sandpiper Pond was connected to the ocean by a narrow channel. The pond proper was composed of open water, tidal mud flats, and salt marsh vegetation (e.g., *Spartina alterniflora*, *Juncus roemerianus*). After the 1980s the channel closed. Sandpiper Pond became a mostly freshwater system and was invaded by two perennial monocot species, the common reed (*Phragmites australis*) and southern cattail (*Typha domingensis*) (Fig. 4.2).

4.2.2 Project Inception

Located within a state park, Sandpiper Pond (33:30:53 N, 79:03:06 W) is managed for multiple uses that include recreation and tourism. The name Sandpiper Pond was derived from the diversity of wading birds that historically used the site, and the state park is considered one of the premier birding sites on the southeast

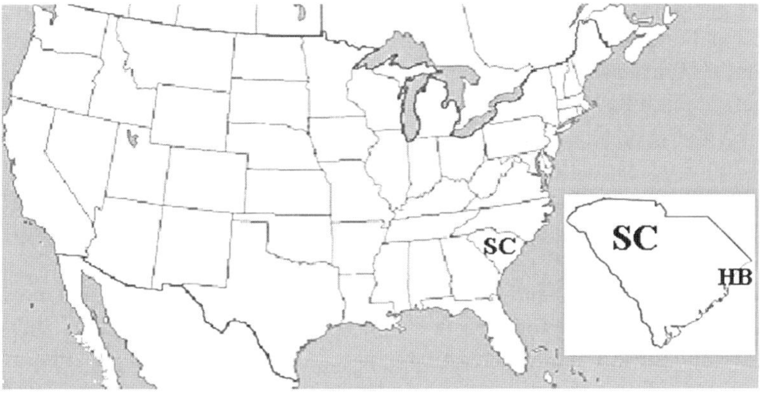

Fig. 4.1 A map of South Carolina, USA (SC) and the location of Huntington Beach State Park (HB) within South Carolina

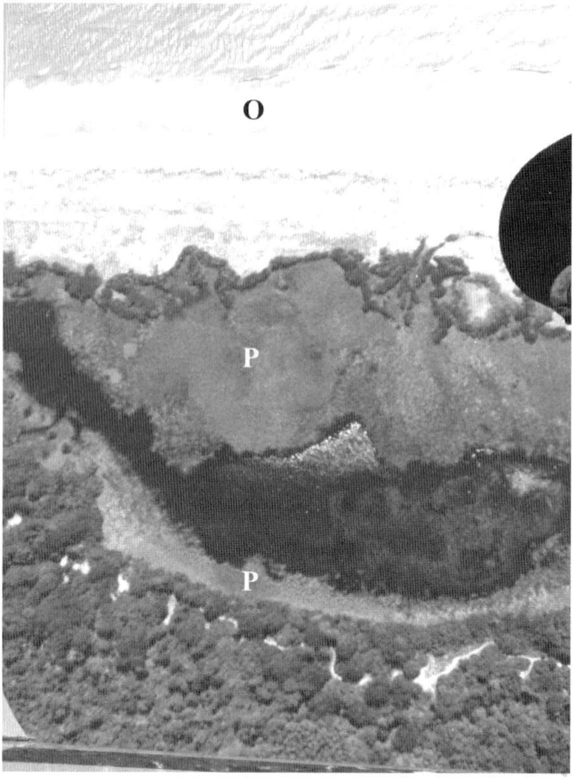

Fig. 4.2 The central portion of Sandpiper Pond long after closure of the channel to the ocean. Fringing areas (P) surrounding the central water are stands of *P. australis* and *Typha* sp. The open ocean (O) is at the *top*

coast (Luken and Moore 2005). Initial concerns about the ecological condition of Sandpiper Pond were expressed to park personnel by a group of volunteers, the Friends of Huntington Beach State Park. The Friends, composed mostly of bird watchers, concluded that avian use of Sandpiper Pond had declined in recent years. Evidence for the decline in bird sightings mainly was anecdotal, although observations occasionally were recorded in a communal log located near the site. The Friends obtained a small grant from the US Environmental Protection Agency to restore the site focusing primarily on control of the two plant invaders.

Once funding was secured, park personnel arranged a meeting among park managers, Friends of Huntington Beach, and faculty from Coastal Carolina University to discuss the project and to determine how to proceed. There was a wide range of opinions regarding how to proceed and how to define success. Two proposed approaches for plant control, spraying of herbicide and excavation of a channel reconnecting Sandpiper to the ocean, were debated. Participants eventually came to the conclusion that simply controlling plants would not adequately define restoration success and that long-term monitoring to characterize trends in various ecological parameters was essential.

4.2.3 Project Design

Previous research suggested that salt stress may be sufficient to control certain plant invaders (Burdick et al. 2001; Bart and Hartman 2002; Farnsworth and Meyerson 2003), and thus it was decided that a new channel would be excavated to bring salt water into the system. A modified before-after-control-impact (BACI) monitoring design was chosen to assess the effects of reconnection on various ecosystem characteristics. BACI designs typically are applied in situations where the before condition represents a predisturbance state (Green 1979; Stewart-Oaten et al. 1992; Wiens and Parker 1995). The question in most BACI studies is whether changes in an unimpacted environment can be attributed to an identified disturbance yet to occur. At Sandpiper Pond we were interested in whether the current reconnection would lead to system changes.

Selection of a BACI experimental design required both a delay in channel excavation to allow collection of "before" data and the selection of reference or control sites. A relatively short 3–4 months delay in channel excavation was agreed to because of funding and scheduling concerns. For example, channel construction had to be completed before the beginning of spring loggerhead sea turtle nesting. The limited delay only allowed for collection of seasonally restricted, winter to early spring, before data. The selection of control sites and ability to optimize the detection of significant impact effects in typical BACI studies is a subject of much concern (Underwood 1994; Benedetti-Cecchi 2001). For the Sandpiper study, we selected two control sites: a saltwater pond (Jetty Pond) created when the Murrells Inlet jetties were constructed and at approximately the same time as the impoundment of Sandpiper Pond, and a salt marsh (Huntington Marsh) located on the backside of Huntington Beach State Park just west of Sandpiper Pond. Jetty Pond (33:31:30 N, 79:02:12 W) was approximately the same size as Sandpiper but remained connected to the ocean and was surrounded by developing salt marshes predominated by *Spartina alterniflora*, *Spartina patens*, *Salicornia virginica*, and *Limonium carolinianum*. Huntington Marsh (33:30:48 N, 79:03:22 W) specifically was chosen to represent a high-marsh environment and was predominated by *Juncus roemerianus*. Rationale for selecting a high-marsh site was driven by the observation that remnant populations of *J. roemerianus* still existed around Sandpiper Pond, and *J. roemerianus* represented a likely early "colonist" if opening the channel had the desired effect. Multiple control or reference locations also were chosen to reduce limitations of the BACI design (Underwood 1994) and because a priori information did not suggest selection of a "correct" restoration target in view of the potentially strong system modification by the plant invaders.

4.2.4 Inlet Construction and Maintenance

Excavation of the new inlet to the ocean began in April 2005 (Fig. 4.3). Equipment operators began excavating near the pond with the goal of developing a channel ca. 30m wide and 400m long. Large quantities of sand were pushed laterally up or down the beach during excavation because the channel cut through a 3+ m tall barrier dune system. Provisions for removing sand from the site were not economic. The inlet was opened successfully on 17 April 2005 during a low tide. Water immediately began flowing from Sandpiper Pond and into the ocean. However, subsequent wave action deposited sand at the mouth of the inlet, and tidal exchange was stopped except for times of very high tides. The inlet was reopened again on 14 September 2005 after turtle nesting season was over, but the opening was again closed shortly thereafter.

During the brief April period when the inlet was open, a coastal storm in combination with a high tide flooded Sandpiper Pond with sea water. On the subsequent low tide, large quantities of detritus were mobilized and washed out of Sandpiper Pond into the ocean. The wrack spread along the coast and eventually was deposited on the beach in long windrows (Fig. 4.4). The unanticipated wrack deposition was a problem as large

Fig. 4.3 The initial reexcavation of a channel from Sandpiper Pond through the barrier beach and extending into the ocean in April 2005. Stands of *P. australis* surround either side of the inlet and the excavator is located just in front of the ocean

Fig. 4.4 Wrack deposited along Huntington Beach as a result of inlet excavation and mobilization of Sandpiper Pond detritus

numbers of people utilize the beach for passive recreational activities. Fortunately, subsequent high tides eventually broke up and transported the wrack off site.

In 2006, sea-beach amaranth (*Amaranthus pumilus*), a federally threatened and endangered plant species, was found growing in the excavated inlet. Presumably excavation activities stimulated germination of dormant seeds. The emergence of *A. pumilus* quickly ended any further attempts to excavate the inlet and plants were flagged while human traffic was restricted in the areas around the plants.

During 2006/2007 park personnel redirected their activities to the opposite side of Sandpiper Pond to increase tidal exchange between the pond and the salt marsh. Efforts focused on a small culvert and tidal creek connecting Sandpiper Pond to the salt marsh on the backside of the barrier beach. The small culvert eventually was replaced with a larger culvert and at present allows tidal exchange during high tides.

4.2.5 Monitoring Efforts

Additional funding through South Carolina's Sea Grant Program was obtained to monitor the effects of the restoration and control efforts. Permanent vegetation and soil monitoring stations were established at Sandpiper Pond (n = 12), Jetty Pond (n = 6), and Huntington Marsh (n = 4). At each monitoring station, samples were collected from 3 to 4 elevations above, at, and below the terrestrial-marsh boundary. Stations were monitored twice yearly, once during the winter or dormant growing season (November to January) and once during the summer or active growing season (May to July). Plant species richness and cover were determined from within 1 m^2 quadrats. Plant biomass was collected from either 0.25 (in *J. roemerianus* stands only) or 0.5 m^2 quadrats in which all above-ground stems were clipped at the sediment surface and later dried at 60°C for 2 + d before determining dry mass. Sediment cores (2.1 cm dia., 5 cm depth) were collected from each station and elevation and processed to determine soil pore water salinity and organic content. Pore water salinity was measured by adding a know volume of deionized water (DW) to the sediment sample, agitating multiple times, determining salt content of the supernatant using a refractometer, and calculating total salinity by standardizing to the total water (DW + soil pore water) in the sample. Organic content was determined by placing a known amount of sediment in a furnace at 500°C for 4 + h and calculating the ash-free dry mass (AFDM) by subtraction.

Along with plant and soil monitoring, CCU faculty and students and Huntington Beach volunteers monitored a variety of other system characteristics such as fish communities, bird use, and basic water chemistry. Some sampling efforts are ongoing, but others have ceased as grants expired and students graduated.

4.2.6 Analyses

Data were analyzed using a variety of parametric, nonparametric, and multivariate approaches. The lack of extensive before and, to a lesser extent, after sampling times by necessity limits the actual application of BACI analyses (Underwood 1994). Instead, we applied one- and two-way ANOVA models and nonparametric Kruskal-Wallis tests to the data where appropriate (e.g., normality and homogeneity assumptions satisfied). To compare among treatment levels (e.g., Sandpiper, Jetty, Huntington locations) appropriate pairwise comparison tests for parametric, Ryan-Einot-Gabriel-Welsch F or REGW F (Day and Quinn 1989), and nonparametric, Dunn's multiple comparison test (Hollander and Wolfe 1973), were applied. All tests, where possible, were run using SPSS v14. A nonmetric multidimensional scaling (NMDS) ordination approach was used to analyze species composition data (Cox and Cox 2001). Sorenson's distance measure weighed by relative coverage was calculated for species compositions from the before or winter season samples, and PC-ORD v5 used for NMDS.

4.2.7 Educational Efforts

In concert with the restoration and monitoring, there also were educational and outreach activities. A new observation deck and interpretive display were constructed during 2005. The goal of these structures was to inform park visitors about the ongoing restoration efforts (Fig. 4.5). In addition, Sandpiper Pond regularly is used as an educational resource both in park programs and courses taught at CCU.

Fig. 4.5 Placard constructed near an observation deck explaining the goals and procedures of the Sandpiper Pond restoration project

4.3 Preliminary Results

Species richness and predominant species composition for the plant communities within each marsh prior to reopening the inlet are presented in Table 4.1. Although predominated by *P. australis* and *T. domingensis*, Sandpiper Pond supported greater numbers of plant species than either reference location (Table 4.1). Jetty Pond and Huntington Marsh were relatively species poor and supported species characteristic of high marsh communities (Table 4.1). Ordination of the initial species compositions for each location using NMDS also suggested that the plant community at Sandpiper Pond was different from the reference locations (Fig. 4.6). Jetty Pond and Huntington Marsh were more similar to each other (clusters C and D), except for a few sites where *Spartina patens* predominated, than to Sandpiper Pond (clusters A and B, Fig. 4.6).

Initial sediment pore water salinity and organic content for samples collected from within each marsh also are presented in Table 4.1. Pore water salinity was significantly different among marshes (Kruskal-Wallis $\chi^2 = 18.96$, df = 2) with the freshwater condition at Sandpiper Pond significantly different from either reference marsh (Dunn's multiple comparisons test, $p < 0.05$). Sediment AFDM also was significantly different among marshes ($F_{2,19} = 4.89$, $p < 0.02$). Organic content was significantly greater within Sandpiper compared with Jetty Pond sediments (REGW F, $p < 0.05$).

Table 4.1 Plant species richness and predominant species composition and mean (±SE) soil salinity and organic content (ash-free dry mass) from below shoreline samples at the three marsh locations within Huntington Beach State Park, SC, USA

	Location		
Characteristics	Huntington Marsh	Jetty Pond	Sandpiper Pond
Plant species richness	8	5	17
Predominant species	*Borrichia frutescens*	*Salicornia virginica*	*Phragmites australis*
	Juncus roemerianus	*Spartina patens*	*Typha domingensis*
Soil salinity (ppt)	22.4 ± 3.6^{a}	27.5 ± 3.2^{a}	0.0 ± 0.0^{b}
Soil AFDM (mg/g)	10.5 ± 5.0^{a}	18.0 ± 4.5^{ab}	43.2 ± 8.1^{b}

Superscripts (e.g., a, b) indicate significant subsets determined either by REGW-F (soil AFDM) or Dunn's multiple comparison procedure (soil salinity)

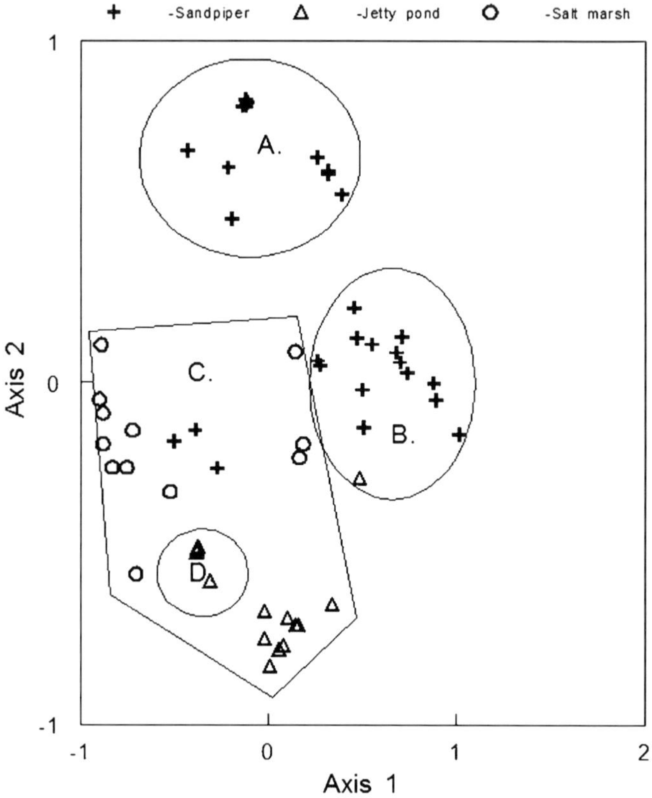

Fig. 4.6 Nonmetric multidimensional scaling ordination of community samples from three locations (indicated by *symbols*) at Huntington Beach State Park, South Carolina, USA. Groups identified by Sorenson's distance and NMDS at a final stress of 10.3 are clustered and identified by *letters*

Changes in pore water salinity and above-ground biomass in the winter after opening the channel were modest or not consistent with significant "after impact" effects. Pore water salinities rose slightly at Sandpiper Pond (5.9 ± 0.9 ppt) remaining

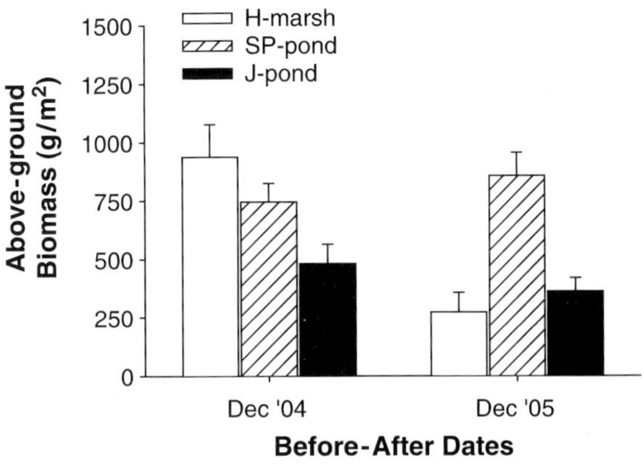

Fig. 4.7 Mean (±SE) live and dead above-ground biomass from Huntington marsh and Jetty pond control locations and Sandpiper pond impacted location before, December 2004, and after, December 2005, restoration of channel flow

fresh to brackish and not approaching salinities in the reference marshes, which remained similar to the before levels (Table 4.1). Above-ground biomass exhibited a significant interaction ($F_{2, 38} = 6.16$, $p < 0.006$) between date (before, after) and location (Sandpiper, Jetty, Huntington), but results did not suggest an effect of increasing salinity on the vegetation in Sandpiper Pond (Fig. 4.7). Instead, Fig. 4.7 results suggest that the modified system, Sandpiper Pond, was more stable than either reference systems. Total above-ground biomass declined from 24.3% to 74.9% in Jetty Pond and Huntington Marsh compared with the 14.7% increase at Sandpiper Pond between winter 2004 and 2005.

Although enough quantitative data do not yet exist to conduct a full BACI analysis, qualitatively Sandpiper Pond has changed appreciably between spring 2004 and the most recent visit in 2007 (Fig. 4.8). A comparison of the images in Fig. 4.8 suggest that open water area has increased and previous sections of live *T. domingensis* and/or *P. australis* contain noticeable increases in standing-dead stems. However, the invasive species and extensive detrital mat are still present 3 year after initial efforts at control.

4.4 Sandpiper Pond in Retrospect

Original motivations for the Sandpiper Pond project were based on perceived, negative differences between the current invaded marsh and the historical uninvaded marsh. Notably, the project was initiated on the basis of anecdotal (e.g., greater wading bird use prior to invasion) and observational information (e.g., tidal flats now covered with vegetation). However, critical attributes of the invaded system were never characterized fully before plans to manage the system were conceived.

Fig. 4.8 The north end of Sandpiper Pond in spring 2005 after channel opening (**a**) and spring 2007 after additional restoration efforts involving installation of a culvert connecting Murrells Inlet Marsh to Sandpiper pond (**b**). *Arrows* are pointing to the same relative point in each picture

Obviously, wading birds were precluded from shallow open water and mud flats that were no longer part of the system in 2004, but a complete census of bird communities was never conducted. The perceived decline in overall bird observations also could be the result of invasion, a global decline in bird populations or even a decline in the ability of birders to make observations because of visual impairment from the invasive plants. The initial lower pore water salinity, higher soil organic content, higher plant species richness, and biomass measured at Sandpiper compared with control marshes also could have been confounded. Reliance on a single sampling time or even short-term (e.g., months) monitoring periods to characterize system attributes can be misleading (Underwood 1994; Chapman 1999; O'Connor and Crowe 2005). Regardless of the lack of convincing data on system degradation, funds were provided for plant control under a grant program that stressed broad involvement of the public in on-the-ground restoration activities.

Sandpiper Pond represents an unambiguous demonstration of the inherent difficulties associated with integrating research/monitoring objectives into a management/restoration project. For example, park and volunteer personnel could only delay construction of the channel until April/May 2004, long enough to collect one set of "before" data but not long enough to provide even one complete set of seasonal data for a rigorous BACI study design (Underwood 1994). Difficulties in melding research and management efforts also become more acute when funds are limited (typically a universal factor), treatments or management activities are not controlled fully, and/or many people are involved in a volunteer capacity. Clearly, attempts to manage plant invasions without commitments (i.e., funding) for long-term monitoring may not lead to anticipated restoration goals (Luken 1997). The initial inlet construction at Sandpiper did not lead to a measurable control of invasives during the 1 year of research funding available. Only after park personnel followed an adaptive management approach and restored tidal flow through the culvert did changes to the system become visible. Unfortunately, no funding currently is available to document if the changes to Sandpiper four years after initial efforts are measurable or not. However, Sandpiper Pond also demonstrates the potential positive outcomes associated with involving diverse groups of people in large-scale ecosystem manipulations. Numerous CCU faculty and students, park personnel, community stakeholders, and park visitors have participated either actively or passively in the Sandpiper restoration activities. The educational experiences associated with community-based restoration projects alone can represent a measurable outcome and result in a more scientifically and environmentally aware public (Brumbaugh et al. 2000a, b).

A fundamental problem associated with marsh restoration is that pre-restoration conditions may set limits (i.e., restoration thresholds) on postrestoration development (Hobbs and Harris 2001). For example, Lindig-Cisneros et al. (2003) found that sediment sterility interfered with attempts to influence the height of a restored California *Spartina* marsh through repeated nitrogen fertilization. Impounded marsh dominance by *P. australis* and *Typha* spp. also affected the pace at which salt marsh species colonized after reconnection with the ocean (Warren et al. 2002). The widespread existence of restoration thresholds reinforces the need for rigorous assessment protocols (Hobbs and Harris 2001) that include appropriate experimental designs and accurate indicators of success. In the absence of such protocols, resource managers may assume that management goals are being achieved when in fact the restored systems are simply making the transition from one degraded state to another degraded state. Even the existence of appropriate indicators may not be sufficient to assess the progress of a project as it often is necessary to recalibrate protocols when applied outside the "home" region (Pennings et al. 2003).

Rigorous assessment protocols for complex ecological systems require true replication of treatments and controls. When such replication is not possible, as was the case with Sandpiper Pond, alternative experimental designs are required (e.g., BACI). Difficulties and assumptions associated with applying alternative designs and statistical approaches continue to be identified (Underwood 1994; Walters and Coen 2006). The use of a BACI approach to design Sandpiper sampling efforts overcame the lack of restored treatment replication but suffered from a single before sampling and a limited number of control sites. Even if sampling was more extensive, a BACI design assumes that the impact is identifiable in terms of time, intensity, and spatial distribution. Unfortunately, the "impact" at Sandpiper Pond turned out to be variable in all aspects. Impact variation often was unpredictable and uncontrollable and included the recent prevalence of extremely high tides and coastal storms that have accelerated beach erosion leading to occasional overwash events at sites other than the excavated inlet. The very method of applying the impact (i.e., inlet vs. culvert) changed during the course of the study and is likely to change again if park personnel determine original restoration goals have not been met.

The initial monitoring of Sandpiper and control marshes at best may provide background for developing a set of new hypotheses that can be tested with more precise approaches. Salt tolerance is an important factor limiting the invasion of *Phragmites australis* (Burdick et al. 2001; Bart and Hartman 2002). Greenhouse experiments examining plant growth responses to various salt concentrations and species combinations would increase understanding about the responses of marsh vegetation to the reestablishment of tidal influence. Marshes in transition from brackish conditions to salt water conditions also may provide excellent opportunities for field experiments aimed at understanding the structure and composition of salt marsh communities (Pennings et al. 2005). Previous research suggests that salt marsh recovery from *Phragmites* invasion can be slow (Warren et al. 2002), and one possible explanation could be the significantly greater amounts of invader biomass and detritus that persist in invaded marshes (e.g., Sandpiper Pond). Experiments designed to tease out the role of detritus accumulation in inhibiting rapid community change and the connections to fundamental differences in patterns of senescence between invaders (e.g., *P. australis*) and native marsh plants (e.g., *S. alterniflora*) easily could be conducted. The unexpected establishment of sea-beach amaranth, a threatened and endangered species, at the site of inlet excavation also solicits further studies on the interaction between marsh and dune seed banks and soil disturbance. Finally, the Sandpiper project provides an excellent arena for addressing the issue of appropriate restoration targets and the value of adaptive management to achieving stated targets.

Acknowledgments We thank the folks at Huntington Beach State Park including Mike Wolf and Steve Roff who, without which, the Sandpiper Pond Project would not have happened, the Friends of Huntington Beach for organizing and seeking funding for portions of the project, Danielle Zoellner and a raft of undergraduate and graduate students at CCU for help in the field and lab, and two anonymous reviewers for comments on earlier drafts. Financial support was provided by an EPA 5-Star matching grant (C# 2004–0017–018), NOAA's South Carolina Sea Grant Consortium (Seed P/M-2D-V310) and the Department of Parks, Recreation and Tourism.

References

Bart D, Hartman JM (2002) Environmental constraints on early establishment of *Phragmites australis* in salt marshes. Wetlands 22:201–213

Benedetti-Cecchi L (2001) Beyond BACI: Optimization of environmental sampling designs through monitoring and simulation. Ecol Appl 11:783–799

Booth RK, Rich FJ, Bishop GA et al (1999) Evolution of a freshwater barrier-island marsh in coastal Georgia, USA. Wetlands 19:570–577

Boumans RMJ, Burdick DM, Dionne M (2002) Modeling habitat change in salt marshes after tidal restoration. Rest Ecol 10:543–555

Brumbaugh RD, Sorabella LA, Johnson C et al. (2000a) Small scale aquaculture as a tool for oyster restoration in Chesapeake Bay. Mar Tech Soc J 34:79–86

Brumbaugh RD, Sorabella LA, Garcia CO et al. (2000b) Making a case for community-based oyster restoration: an example from Hampton Roads, Virginia, U.S.A. J Shellfish Res 19:467–472

Burdick DM, Buchsbaum R, Holt E (2001) Variation in soil salinity associated with expansion of *Phragmites australis* in salt marshes. Environ Exp Botany 46:247–261

Chapman MG (1999) Improving sampling designs for measuring restoration in aquatic habitats. J Aq Ecosys Stress Recov 6:235–251

Craft CB, Megonigal JP, Broome SW et al. (2003) The pace of ecosystem development of constructed *Spartina alterniflora* marshes. Ecol Appl 13:1317–1432

Costanza R, d'Arge R, de Groot R et al. (1997) The value of the world's ecosystem services and natural capital. Nature 387:253–260

Cox TF, Cox MAA (2001) Multidimensional scaling. Chapman & Hall, London

Day RW, Quinn GP (1989) Comparisons of treatments after an analysis of variance in ecology. Ecol Mono 59:433–463

Farnsworth EJ, Meyerson LA (2003) Comparative ecophysiology of four wetland plant species along a continuum of invasiveness. Wetlands 23:750–762

Gratton C, Denno RF (2006) Arthropod food web restoration following removal of an invasive wetland plant. Ecol Appl 16:622–631

Green RH (1979) Sampling design and statistical methods for environmental biologists. Wiley Interscience, Chichester

Hobbs RJ, Harris JA (2001) Restoration ecology: repairing the earth's ecosystems in the new millennium. Restor Ecol 9:239–246

Hollander M, Wolfe DA (1973) Nonparametric statistical methods. Wiley, New York

Kimball ME, Able KW (2007) Nekton utilization of intertidal salt marsh creeks: Tidal influences in natural *Spartina*, invasive *Phragmites*, and marshes treated for *Phragmites* removal. J Exp Mar Biol Ecol 346:87–101

Lindig-Cisneros R, Desmond J, Boyer KE et al. (2003) Wetland restoration thresholds: can a degradation transition be reversed with increased efforts. Ecol Appl 13:193–205

Luken JO (1997) Management of plant invasions: implicating ecological succession. In: Luken JO, Thieret JW (eds) Assessment and Management of Plant Invasions. Springer-Verlag, New York

Luken JO, Seastedt TR (2004) Management of plant invasions: the conflict of perspective. Weed Technology 18:1514–1517
Luken JO, Moore R (2005) 101 wild things along the Grand Strand. PFC Press, Spartenburg
Neckles HA, Dionne M, Burdick DM et al. (2002) A monitoring protocol to assess tidal restoration of salt marshes on local and regional scales. Restor Ecol 10:556–563
O'Connor NE, Crowe TP (2005) Biodiversity loss and ecosystem functioning: distinguishing between number and identity of species. Ecology 86:1783–1796
Pennings SC, Selig ER, Houser LT et al. (2003) Geographic variation in positive and negative interactions. Ecology 84:1527–1538
Pennings SC, Grant MB, Bertness MD (2005) Plant zonation in low-latitude salt marshes: disentangling the roles of flooding, salinity and competition. J Ecol 93:159–167
Roman CT, Raposa KB, Adamowicz SC et al (2002) Quantifying vegetation and nekton responses to tidal restoration of a New England salt marsh. Restor Ecol 10:450–460
Stewart-Oaten A, Bence JR, Osenberg CW (1992) Assessing effects of unreplicated perturbations – no simple solutions. Ecology 73:1396–1404
Teal JM, Weishar L (2005) Ecological engineering, adaptive management, and restoration management in Delaware Bay salt marsh restoration. Ecol Eng 25:304–314
Underwood AJ (1994) On beyond BACI: sampling designs that might reliably detect environmental disturbances. Ecol Appl 4:3–15
Walters K, Coen LD (2006) Approaches to analyzing community compositional change over ecological time: examples from natural and constructed intertidal oyster reefs. J Exp Mar Biol Ecol 330:81–95
Warren RS, Fell PE, Rozsa R et al. (2002) Salt marsh restoration in Connecticut: 20 years of science and management. Rest Ecol 10:497–513
Wiens JA, Parker KP (1995) Analyzing the effects of accidental environmental impacts: approaches and assumptions. Ecol Appl 5:1069–1083
Young TP (2000) Restoration ecology and conservation biology. Biol Conserv 92:73–83
Young TP, Petersen DA, Clary JJ (2005) The ecology of restoration: historical links, emerging issues and unexplored realms. Ecol Lett 8:662–673
Zedler JB, West JM (2008) Declining diversity in natural and restored salt marshes: a 30-year study of a Tijuana Estuary. Rest Ecol 16:249–262

Chapter 5
A Habitat-Classification Framework and Typology for Understanding, Valuing, and Managing Invasive Species Impacts

Christoph Kueffer and Curtis C. Daehler

Abstract It is frequently lamented that invasion biology has not been very successful in developing reliable generalizations for management. In particular, there is an urgent need to improve the understanding and assessment of impacts of invasive species. We argue that a refined conceptualization of biotic invasion derived from a management perspective, rather than purely from ecological theory, can help to better understand, value and manage impacts of invasive species. We propose a habitat-classification framework on the basis of four habitat types that are defined by their differences in type and degree of human modification, and differences in human valuation. The first type, *anthropogenic habitat*, encompasses highly disturbed and anthropogenic areas such as agriculture, plantation forestry, or urban areas. The second type, *reference habitat*, represents relatively undisturbed habitat dominated by native species. The third type, *abandoned habitat*, involves habitats that currently experience relatively little human interference but that have been highly disturbed or managed in the past, e.g., old fields or abandoned plantation forests. The fourth habitat type, *designed habitat*, involves situations where humans deliberately and strongly manipulate a habitat to create a new habitat that primarily suites conservation objectives (e.g., restoration of a former native habitat). These four habitat types differ in invader characteristics, invader impacts, management strategies, and research needs. Our typology may help stimulate more interdisciplinary research yielding improved conceptual and practical understanding of impacts of invasive species.

Keywords Applied research • Land use • Natural • Novel ecosystems • Oceanic island • Post-normal science • Restoration ecology • Secondary succession • Socioecological research • Valuation

---

C. Kueffer(✉) and C.C. Daehler
Department of Botany, University of Hawaii at Manoa, USA
kuffer@hawaii.edu

5.1 Introduction

There is broad agreement among scientists and conservationists that alien invasive species pose one of the foremost threats to global biodiversity today (Millennium Ecosystem Assessment 2005). However, the scientific uncertainties regarding biotic invasions remain high (National Academies of Sciences 2002), which hinders effective management of invasive species. In particular, there is an urgent need to improve the understanding and assessment of invasive species impacts (Ewel et al. 1999; Parker et al. 1999; Strayer et al. 2006; Kueffer and Hirsch Hadorn, 2008). For instance, the role of biotic invasions in species extinctions is neither well-understood nor well-documented (Gurevitch and Padilla 2004). At the same time, potential positive functions of alien species in habitat restoration are increasingly discussed (D'Antonio and Meyerson 2002; Ewel and Putz 2004; Kueffer et al. 2007b). Removal of invasive species may lead to negative as well as positive effects (Zavaleta et al. 2001). Philosophers argue that invasion biologists have no well-defined concept of "harm by invasive species" and often only weak empirical data are available to support the need for control of invasive species (Sagoff 2005). Consequently, nature conservation managers state that an insufficient understanding of negative impacts of various invasive species is a major obstacle to priority setting of management actions (Kueffer et al. 2007a).

The disconnect between research and management may be related to the way invasion biology has been historically framed. During the past decades, invasion biology research has been strongly influenced by a conceptual framework derived from only a few key publications (Davis 2006), notably Elton's classical book "The Ecology of Invasions by Animals and Plants" (1958) and the first SCOPE Program on invasive species between 1982 and 1988 (Drake et al. 1989). The basic assumption was that biological research on the causal processes underlying biotic invasions would produce the necessary knowledge to tackle the environmental problem of biotic invasions. However, recent research on the effectiveness of science for environmental problem solving (Funtowicz and Ravetz 1993; Pohl and Hirsch Hadorn 2007; Kueffer and Hirsch Hadorn, 2008) indicates that this assumption may be flawed in two important ways. First, impacts were addressed without explicitly clarifying human valuation. Basically, it was assumed that any detectable effect on an ecosystem property by an alien species is problematic (Kueffer and Hirsch Hadorn, 2008). Second, research questions were derived from the intrascientific perspective of population and community ecology instead of being framed from a management perspective. In contrast, an alternative research approach would formulate research questions based on the goals, options, and restrictions of the problem-solvers (Pohl and Hirsch Hadorn 2007). Relevant starting questions may then be: What are the management goals in a particular management context? What are appropriate management options? What are the benefits and costs of different management options? A framing of research questions that accounts for the management context from the beginning may facilitate ecological generalizations in ways more useful for management (Orians et al. 1986; Hirsch 1995; Kueffer 2006b).

An instructive example is recent emphasis on the study of transportation vectors or pathways in alien invasions (vector science) (Ruiz 2003; Mooney et al. 2005; Kowarik and von der Lippe 2007) to compliment earlier biological research that emphasized postarrival invasion processes (Richardson and Pysek 2006). Understanding biotic invasions through the lens of different transport pathways allows framing of research in a way that is tailored to the institutional context of particular pathways. Recent research has, for instance, focused on the transport of alien species through water ballast in marine shipping (Minton et al. 2005) or the horticultural trade (Reichard and White 2001). The framing of research based on pathways has improved risk assessments for managers by suggesting new forms of context dependence. For instance, risk management of Mediterranean fruit fly introductions to the US, through trade of pink tomatoes from Northern Africa, combines regulations based on the origin of the product, trade restrictions based on the season, and preventative measures during production and transport (Hallman 2007). Although vector science has helped to better understand and manage the spread of alien species, it does not solve the problem of managing impacts in invaded habitats.

To reduce disconnect between managers and researchers interested in invasive species impacts, we suggest that greater emphasize should be placed on context-dependent management and research based on careful consideration of human values. We identify a typology of four generalizable habitat types that, according to our experience, represent typical contexts of invasions. We discuss how the traits of successful invaders and their impacts may differ between these four habitat types, and how a consideration of these types may therefore help guide invasive species research and facilitate transfer of management experiences between individual cases. We illustrate the typology mainly using examples from oceanic islands, which have long been recognized as model systems for invasion biology (Elton 1958). Although we mostly discuss plant invasions, the proposed typology can also be applied to other taxonomic groups. We define invasive species as species that have been introduced and spread outside of their native range, presenting some problem according to the relevant experts and stakeholders.

5.2 A Framework for Invasive Species Research

To better understand, value and manage impacts of invasive species, we propose a habitat-classification framework based on four habitat types that are defined by their differences in type and degree of human habitat modification, and differences in human valuation (Fig. 5.1, Table 5.1). The first type, *anthropogenic habitat*, encompasses highly disturbed and anthropogenic areas such as agriculture, plantation forestry, or urban areas. In anthropogenic habitats, cultivated plants have replaced most natural vegetation, and management emphasizes socioeconomic gains from production and/or ecosystem services. The second type, *reference habitat*, represents relatively undisturbed habitats dominated by native species. Reference habitat is typically part of a protected area or a wilderness area, and the

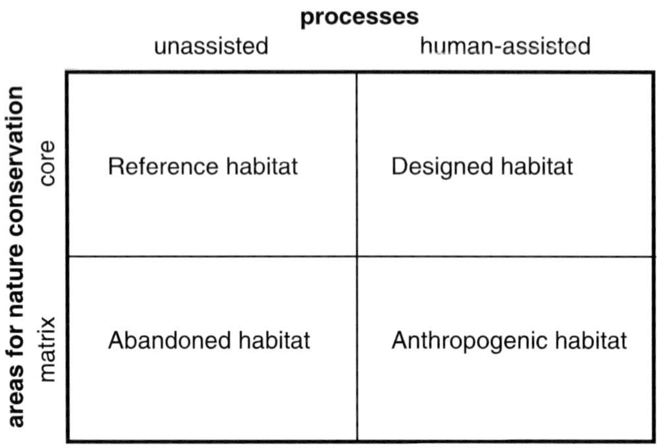

Fig. 5.1 Four habitat types (anthropogenic, reference, abandoned, designed) that represent different environmental and management contexts of relevance for biotic invasions. The habitats are arranged according to degree of human influence (unasssisted vs. human-assisted processes) and relevance for nature conservation (core vs. matrix nature conservation areas). See text for further explanation

primary management objective is to conserve it in its current state. We use the term "reference" to indicate that the area has a special value for nature conservation because of some habitat quality that is considered a reference state for "high quality nature." We do not mean that reference habitat is untouched by man. Therefore we do not use terms such as "pristine" or "undisturbed" commonly found in the literature. The third type, *abandoned habitat*, involves habitats that currently experience relatively little human interference but that have been highly disturbed or managed in the past, e.g., old fields or abandoned plantation forests. They may be of high conservation value but they are not reference habitat because their habitat state is not considered a reference for "high quality nature." The fourth habitat type, *designed habitat*, involves situations where humans deliberately and strongly manipulate a habitat to create a new habitat that primarily suites conservation objectives (e.g., restoration of a former native habitat). Designed habitats are characterized by their ongoing dependence on management (conservation-reliant *sensu* Scott et al. 2005). Ecological restoration areas that are not conservation-reliant are considered either reference or abandoned habitat, depending on the extent to which the restoration has been successful. Differences among the four habitat types, their human valuation, and their invaders allow us to define a typology (Table 5.1), which can be instructive in informing management and research.

The four habitat types reflect the four main strategies used in nature conservation today: maintaining biodiversity in cultural and urban landscapes (mainstreaming biodiversity, Petersen and Huntley 2005), protection of natural areas, nature conservation on abandoned land, and habitat restoration. Conservation on abandoned land may not yet be widely considered a distinctive type of conservation

Table 5.1 Four habitat types of relevance to biotic invasions

Habitat type	Habitat characteristics	Traits of invader	Impacts of invader	Management action	Research focus
Anthropogenic habitat	Disturbed or anthropogenic	"Weedy" invaders such as ruderal, or early successional species	Reservoir for invasions into biodiversity areas	Reduction of propagule pressure	Upscaling of control methods to large areas
			Competition with native ruderal species	Mainstreaming invasive species control in different production sectors	Socio-ecological research on plant invasions, e.g. urban and agricultural ecology
			Increase of biodiversity of anthropogenic habitat, and substitute for native species		
Reference habitat	Undisturbed, "high quality nature"	Traits similar to native species, but possibly with some novel traits	May be low, if no novel traits are involved	Early detection and eradication	Long-term and indirect impacts
			May be indirect or cryptic	Monitoring of native biota and ecosystem functioning	Indicators of ecosystem health
			Long-term effects uncertain		Ecology of rare native species and communities
Abandoned habitat	Abandoned land after prior exposure to strong anthropogenic disturbance	Depends on initial state and successional stage of the habitat (time since disturbance)	Positive, negative, or both	Directing secondary succession according to management goals	Stability and functioning of novel assemblies of native and alien species

(continued)

See text for further explanation

Table 5.1 (continued)

Habitat type	Habitat characteristics	Traits of invader	Impacts of invader	Management action	Research focus
				Sustaining ecosystem functioning rather than restoring native dominated habitat	Provision of ecosystem services and products including native biodiversity
Designed habitat	Strongly managed for nature conservation and constantly depending on management	Management and habitat conditions may select for specific types of invaders	Depends on management goals and stage in the restoration process	Ecosystem design: Manipulation of native and alien species to attain specific conservation-oriented objectives	Restoration techniques
		Deliberately introduced alien species with traits that facilitate restoration			Ecology of community assembly
					Risks versus benefits of alien species use
					Socio-ecological research on natural/artificial dichotomy

action, but abandoned land is increasingly discussed as a unique and relevant habitat type in restoration ecology (novel ecosystems *sensu* Hobbs et al. 2006).

The four habitat types are prototypical in the sense that they can be classified according to two basic dimensions of relevance to nature conservation (Fig. 5.1). The first dimension, *deliberate human interference*, differentiates between habitats where ecological processes are frequently and deliberately manipulated by people (anthropogenic habitat, designed habitat), and others where this does not happen on a substantial scale (reference habitat, abandoned habitat). This dichotomy reflects the difference between wild and domesticated, where wild means "untamed" or "not cultivated." The second dimension, *nature conservation focus*, distinguishes between core nature conservation area (reference habitat, designed habitat) and matrix areas (anthropogenic habitat, abandoned habitat) that surround the core areas. In core conservation areas, the availability of nature conservation resources and management capacity tend to be higher, and conflicts with alternative management objectives, such as production, tend to be lower than in matrix habitat. The prototypical nature of the four habitat types may be an indication that they are a solid basis for theoretical thinking about biotic invasions from a habitat modification perspective.

5.3 Anthropogenic Habitat

5.3.1 Habitat Characteristics

Anthropogenic habitats include agricultural and urban land, plantation forestry, ruderal and waste sites, or roadsides. These sites are characterized by high levels of unused resources (especially light and nutrients), frequent or large disturbances, and high inputs of alien species propagules (Fig. 5.2). Species diversity is often low, and empty niche opportunities for invasive species are common (cf. Dietz and Edwards 2006). Dietz and Edwards (2006) suggest the term *primary invasions* to characterize invasions in anthropogenic habitat in contrast to *secondary invasions* of less disturbed habitat.

Fig. 5.2 Invasions in anthropogenic habitat. Weedy alien species typically invade persistently disturbed, anthropogenic habitats. In Hawaii, weed communities along roadsides at high elevations above c. 1,500 m asl. are dominated by European weeds that were historically introduced through agriculture (e.g. *Anthoxanthum odoratum, Holcus lanatus,* or *Verbascum thapsus*). Lowland and high elevation individuals of these species apparently differ ecologically (Gabi Jakobs, unpublished data), and it may be that they have adapted to local conditions in roadside vegetation (Photo Eva Schumacher)

5.3.2 Functional Type of Invasive Alien Species

Invasive species associated with anthropogenic habitat are characterized by traits typical of agricultural weeds (Baker 1974). They are resource-demanding, fast-growing, and fast-reproducing ruderal or early successional species. They usually have high seed output, and an efficient seed dispersal mechanism (particularly wind or birds). Invasive plants with high qualitative defense (toxins such as alkaloids) against generalist herbivores but low quantitative defense (e.g. lignins, tannins) may particularly profit from the high nutrient levels and presence of mainly generalist herbivores in anthropogenic habitat (compare Müller-Schärer et al. 2004; Blumenthal 2006). Invasive species characteristic for anthropogenic habitat represent a paradigmatic type of invader that may be called a weedy invader. Species with weedy traits that profit from resource-rich and highly disturbed anthropogenic habitat have been important in the development of invasion biology (Baker 1974; Rejmanek 1996; Davis et al. 2000). Within anthropogenic habitat, traits of invasive species may differ between agricultural and urban land. For instance, there are often more annuals in weed communities on arable land and more biennial and perennials in ruderal vegetation around settlements (Lososova et al. 2006). Further, weeds on arable land may be associated with particular crops and their management systems (Pyšek et al. 2005).

5.3.3 Impacts

From a nature conservation perspective, anthropogenic habitat is of major concern as potential pathways into natural areas, and as reservoirs, where alien species may build up propagule pressure and adapt to local conditions (Pysek and Jarosík 2005; Dietz and Edwards 2006; Didham et al. 2007). After evolutionary change (cf. Dietz and Edwards 2006) or due to source-sink dynamics (cf. Didham et al. 2007), some species associated with anthropogenic habitat may be able to enter relatively undisturbed, reference habitats (see below). With continuing expansion of human land use, this scenario will require increasing attention.

The impacts of alien species on biodiversity within anthropogenic habitat may also be relevant. Farmers may in the future be more commonly forced to actively coproduce biodiversity values on their land, and similarly biodiversity is considered a product of sustainable forestry. In areas with a long human land-use history, many native ruderal species depend on anthropogenic habitat, and such areas may be threatened by the invasion of alien species. However, it seems that in these regions the land-use history has also selected for native species that are able to coexist with alien ruderals, and invasive alien species seem to be only infrequently able to threaten native ruderal biodiversity (Klingenstein and Diwani 2005; Maskell et al. 2006). In fact, alien species generally increase the species diversity in such habitats (Maskell et al. 2006). Alien weed diversity may not be considered a biodiversity value, but, for instance, in Europe many of the early introduced alien species (archaeophytes) are now listed as threatened species on national red lists (Klingenstein and Diwani 2005). Alien species may also provide important resources for native species in urban areas – e.g. fruits of alien trees as food source for native birds (Corlett 2005). In heavily disturbed sites, ecosystem impacts by invasive species are difficult to disentangle from impacts by other disturbance factors (Didham et al. 2005), and often invasive species are probably not the driver of ecosystem change (Maskell et al. 2006).

5.3.4 Management Action

In anthropogenic areas, weed management has been long practiced, with a focus on reducing economic and nuisance weeds, and these experiences from weed science are also relevant for invasive species research and management (Smith et al. 2006). The integration of control activities in everyday professional management of agriculture, forestry, tourism, landscaping, or maintenance of public infrastructure should be a priority, as this will reduce overall propagule pressure and invasion threat to surrounding habitats in a way that is economically efficient and is tailored to the production systems (mainstreaming invasive species control *sensu* Petersen and Huntley 2005).

5.3.5 Research Focus

Ecosystem integrity and the direct interactions of invasive species with the native biota are of lesser research interest in the case of anthropogenic habitat. Rather, the frameworks of agricultural, urban, and landscape ecology that integrate human action into ecological theories should inspire research on plant invasions in anthropogenic systems (Kueffer and Hirsch Hadorn, 2008). The study of the urban ecology of alien species has a long tradition in Europe (cf. Davis 2006). Especially for a better understanding of the dispersal of weeds in anthropogenic landscapes, an appreciation of the crucial role of human activity is essential (Benvenuti 2007). The rise of extensive agriculture and agri-environmental schemes in many European countries may have important consequences for spread of invasive species (Donald and Evans 2006). Research is needed to understand how similar changes in other parts of the world are influencing invasive species patterns on agricultural land, including areas that practice traditional agriculture (Schneider and Geoghegan 2006). Another main research focus should be the development of effective, efficient, and environmentally friendly control measures that allow upscaling to large areas. Weed scientists have much experience in this area of research, and invasion biology would benefit from better integration of concepts and principles from weed science (Smith et al. 2006).

5.4 Reference Habitats

5.4.1 Habitat Characteristics

Reference habitats have a special value for nature conservation because of some habitat quality that is considered a reference state for "high quality nature" (Fig. 5.3). They typically are relatively undisturbed habitats dominated by native species. Reference habitats are often characterized by high functional and species diversity and low levels of unused resources. Empty niche opportunities are relatively low and pest loads and diversity are high. Nowadays, many reference habitats are on marginal land that is characterized by harsh environmental conditions (e.g. high altitude or dry habitats). All these factors in combination may make reference habitats more resistant to invasions than disturbed habitats (Elton 1958; Richardson and Pysek 2006). However, today alien species can be found in even the most remote and pristine habitats such as boreal forests (Rose and Hermanutz 2004) or Antarctica (Frenot et al. 2005). Furthermore, the fragmentary nature of many undisturbed habitats increases their vulnerability to invasion (Pysek et al. 2002). Over time, reference habitat can become degraded due to invasion or anthropogenic influences, leading to a shifting baseline in our notions of "high quality nature" (Knowlton and Jackson 2008). If degradation is perceived to be extreme, reference habitat may be abandoned as a reference, and it will have the same features as abandoned habitat (see later).

Fig. 5.3 Invasions in reference habitat. Relatively undisturbed habitat such as native palm forests on an oceanic island of the Seychelles group (*picture*) are often relatively resistant to invasions. Palm forests in Seychelles have recently also been invaded by alien vines, such as *Merremia peltata*, and it is not known if an increase in air CO_2 levels may have contributed to this. Invaders with novel traits, such as rats, may have strong impacts on palm forest, for instance by predating on seeds (Photo Eva Schumacher)

5.4.2 Traits of Invasive Alien Species

In reference habitat, invasive species may be functionally similar to the local native species (cf. Dietz and Edwards 2006; Kueffer 2006a). In the case of plant invaders, shade-tolerance is an example of a trait that facilitates invasion into undisturbed forests (Daehler et al. 2004); this trait is also common in native forest species. Some invasive plants are also well adapted to low levels of unused soil resources prevailing in undisturbed habitats (compare Kueffer 2006a; Funk and Vitousek 2007; Schumacher 2007). Invasive woody plants invading dry inselberg habitats have been shown to be as well adapted as native species to drought stress (Schumacher et al. 2008). Such traits are expected to be uncommon among invasive plants associated with anthropogenic habitat.

However, alien plants invading reference habitat may also profit from having one or more novel traits not (substantially) present in the native community, for instance the ability to fix atmospheric nitrogen (Vitousek et al. 1987) or release allelopathic root exudates (Callaway and Maron 2006). New functional groups such as large predators (Fritts and Rodda 1998; Courchamp et al. 2003), large herbivores (Courchamp et al. 2003; Culliney 2006), new groups of invertebrates, e.g. earthworms (Hendrix and Bohlen 2002), ants (Holway et al. 2002), or new diseases (Benning et al. 2002) can rapidly spread into reference habitat. The nature of the trait novelty may be subtle. For instance, allelopathy may be a common mechanism among native species but the native species may not be adapted to the particular allelopathic substances released by an alien species (Callaway and Maron 2006). Or, native species may be adapted to respond to predators but not to the specific predatory strategies of an alien invader (Schlaepfer et al. 2005). The exploitation of the lack of certain traits or functional groups in an invaded region because of biogeographic barriers is probably relevant in all four habitat types, but in reference habitat these species are perhaps of most concern because of their permeating impacts on native ecosystems.

5.4.3 Impacts of Invasive Alien Species

As discussed in the previous paragraph, invaders of reference habitat may have a combination of traits similar to the native biota and novel traits. It can be expected that species that lack the latter may have minor ecological impacts. For instance, shade-tolerant tropical invaders in the Seychelles have leaf litter properties and decomposition rates similar to native species and consequently have only minor impacts on soil fertility (Kueffer et al. 2008). Species with novel traits, in contrast, can have substantial impacts (Vitousek 1990) even on undisturbed habitat. Introduced diseases threaten rare bird species (Benning et al. 2002), and predators with novel behaviors substantially reduce population sizes of native prey species (Fritts and Rodda 1998). Invasive ants and earthworms can markedly affect soil biota and processes (Hendrix and Bohlen 2002; Holway et al. 2002).

A major open question is how these impacts may unfold in the long-term (Strayer et al. 2006). Impacts can lead to positive feedbacks among invasive species that lead to a deterioration of reference habitat values (invasional meltdown, Simberloff and Von Holle 1999). Synergistic interactions with other global change drivers such as climate change or habitat fragmentation (Didham et al. 2007) may increase the impact of species with novel traits. But it may also be that the impact of the alien species decreases with time (Morrison 2002). In particular, the native biota may be able to adapt relatively quickly to novel traits (cf. Dietz and Edwards 2006), for instance thanks to phenotypic plasticity (Peacor et al. 2006). Or some advantage of the invader may diminish, for instance because it accumulates pests.

An important aspect for evaluating impacts of alien species in reference habitat is that the management objective is to conserve specific natural processes, communities, and rare species that characterize the reference habitat. Such rare, endemic species or biotic interactions may be most sensitive to impacts of invasion, and therefore even invasive species that generally only weakly impact native biota may be of major concern. For instance, Krushelnycky (2007) concluded that "[arthropod] communities that had already lost many endemic species [...] were relatively resistant to further species loss upon [invasive] ant arrival, whereas more intact communities were vulnerable to substantial declines in richness when ants invaded."

Initially small impacts may be more difficult to detect and understand, especially when considering the feasibility of conducting regular surveys in reference habitat. Hidden but potentially very important impacts in reference habitats could include hybridization of alien plants with rare native species (Mooney and Cleland 2001), an increase of disease incidence in native plants promoted by the presence of an alien host plant (Malmstrom et al. 2005), or disruption of belowground mutualisms (Stinson et al. 2006).

5.4.4 Management Action

Where feasible, eradication of small infestations of new alien species should be a priority in consideration of the precautionary principle. However, for most natural areas, it will not be possible to regularly search for new introductions of alien species. Besides, financial constraints, knowledge about the taxonomy, distribution and impacts of all but the most disruptive invasive species is usually limited. Rather than comprehensive invasive species monitoring, rare species and ecosystem health monitoring programs could be used to indicate invasive species problems that require management. It may also be promising to develop composite indicators that track trends in a suite of alien species with similar life histories (Hulme 2006), and then relate this to indicators of ecosystem health, identify interactions with other global change drivers, and determine possible management actions. There is a high potential that control action in reference habitat will have negative impacts on other ecosystem properties; therefore, invasive species control has to be carefully considered in a whole-ecosystem context (Zavaleta et al. 2001).

5.4.5 *Research Focus*

Research is foremost needed that addresses the functioning of natural ecosystems and the ecology and functional roles of common and rare native species in the light of a multitude of global change agents and their synergetic interactions (Didham et al. 2007). Studies are needed to carefully assess the relative importance of invasive species vs. other environmental stressors on management objectives in reference habitat so as to identify the most efficient use of management resources (Hulme 2006). Indirect human impacts on reference habitat, such as climate change or increase in CO_2 partial pressure in the air, may increase the vulnerability of reference habitat to biotic invasions, and these effects need to be better understood. For instance, increased atmospheric CO_2 may accelerate the spread of native and alien invasive vines in deep shade (Granados and Koerner 2002), or mountains ecosystems that are currently relatively resistant to biotic invasions may become vulnerable to invasions with global change (Dietz et al. 2006; Pauchard et al., in press and see http://www.miren.ethz.ch). Further, a better understanding of evolutionary responses of invasive and native species is needed to determine whether such responses are more likely to increase or decrease impacts of invasive species in undisturbed habitat (cf. Dietz and Edwards 2006).

5.5 Abandoned Habitat

5.5.1 *Habitat Characteristics*

Abandoned habitats are areas that have been heavily disturbed or intensively managed in the past, e.g. old fields (Cramer et al. 2008) or abandoned plantation forests (Lugo 2004), or they are former reference habitats that have been highly degraded due to anthropogenic influence or invasion. Abandoned habitat often contains new combinations of native and alien species, and such ecosystems have been termed *novel ecosystems* or *emerging ecosystems* (Hobbs et al. 2006). The terms *abandoned habitat* and *novel ecosystem* are similar, but *abandoned habitat* does not imply that ecosystem properties have to be novel. The species of novel ecosystem have often not coevolved (Wilkinson 2004; Hobbs et al. 2006), and abiotic conditions may also be considerably different than in the typical native vegetation at a site (Lugo 2004). Ecosystem functioning and species assembly and interactions are difficult to predict because they depend on the disturbance history of the site, assembly history, and the local pool of alien and native species (Chazdon 2003;

Suding et al. 2004; Hobbs et al. 2006; Cramer et al. 2008). In Seychelles, for instance, initial colonization of deforested land by the alien tree *Cinnamomum verum* has probably strongly shaped most secondary successions on these tropical oceanic islands for the past 200 years (Kueffer et al. 2007b) (Fig. 5.4).

5.5.2 Functional Type of Invasive Alien Species

Most commonly, abandoned habitat will initially have conditions similar to anthropogenic habitat, and alien species invading anthropogenic habitat may also most commonly invade abandoned habitat in an early phase of secondary succession. Initially, the established vegetation is often partly or mostly composed of alien species that have been planted for particular traits (e.g., nitrogen fixation, fast-growth, adaptation to grazing). Such selected traits may have a strong influence on community functioning and ecosystem processes. The (contingent) initial species composition may therefore have an important influence on community development and lead to alternative succession scenarios (compare e.g. Suding et al. 2004). Similar to native ecosystems, in abandoned habitat traits of dominant species are expected to change over time due to succession (Prach et al. 1999), but patterns may be less predictable than in native ecosystems (Hobbs et al. 2006). This is because a reduced but

Fig. 5.4 Secondary successions in abandoned habitat. Large areas on the oceanic island of the Seychelles are covered by a secondary forest dominated by the invasive tree *Cinnamomum verum*. This novel ecosystem formed after major deforestation in the early nineteenth century, and provides important ecosystem services such as erosion control and protection of water catchments. The fruits of *C. verum* are a major food source for several endemic frugivorous animals, and the strong belowground root competition of the species apparently facilitates the regeneration of native species compared to invasive species (Kueffer et al. 2007b) (Photo Eva Schumacher)

variable complement of mutualists in abandoned habitat may lead to greater variability in success among both introduced and native plant species.

5.5.3 Impact of Invasive Alien Species

Invasive species may trigger deteriorating ecosystem processes such as increased fire frequency (D'Antonio and Vitousek 1992). However, invasive species can also play positive roles in habitat recovery from abandoned land. Facilitation may happen if alien species replace the function of a native species that is no longer present, or if they assist in restoring degraded abiotic habitat conditions (Rodriguez 2006). Alien species may, for instance, enhance seed dispersal services to native species (Buckley et al. 2006). Certain alien species may also help to mitigate negative impacts of other alien species. In Seychelles, native species apparently regenerate relatively well under the canopy of the alien tree *Cinnamomum verum*, while the regeneration of other alien, and potentially thicket-forming, species such *Psidium cattleianum* is hindered (Kueffer et al. 2007b). In Mauritius, the nesting success of a rare endemic passerine bird is greatest in alien *Cryptomeria* tree stands, probably because of avoidance of the habitat by alien mammalian predators (Safford and Jones 1998). Often invasive species will have positive effects on some habitat values and negative effects on others (Sax 2005).

5.5.4 Management Action

In abandoned habitat, clear and realistic objectives should be defined prior to making costly management decisions. Abandoned habitat is often large in area but marginal in its productivity or restoration potential. Alien species control can have unintended consequences. In Hawaii, feral sheep began to thrive after feral dogs were substantially reduced by hunting (Culliney 2006). The removal of sheep and cattle led to an explosive expansion of several invasive plants on Santa Cruz Island (Zavaleta et al. 2001). In novel ecosystems, alien species may be regarded as an integral part of a habitat management strategy. The facilitation of secondary succession and the restoration of self-sustained ecosystem functioning may be more important than the restoration of native-dominated vegetation (Zavaleta et al. 2001; D'Antonio and Meyerson 2002; Cramer et al. 2008).

5.5.5 Research Focus

The functioning of novel ecosystems as well as their importance in providing ecosystem goods and services is not well studied, and a number of research priorities have been identified (Hobbs et al. 2006). Novel ecosystems may provide unique research opportunities for understanding ecological interactions and processes (Wilkinson 2004; Young et al. 2005). They may allow for real-time observation of the effects of immigration history or rapid evolutionary processes for community assembly and the occurrence of alternative ecosystem states (Suding et al. 2004; Young et al. 2005; Strayer et al. 2006), or help to clarify the roles of biota-environment feedbacks, such as plant-soil feedbacks, in shaping ecosystem functioning (Ehrenfeld et al. 2005). Abandoned habitat may be particularly valuable to test hypothesis in invasion biology. For example, it may be expected that an alien species profits less from the release from specialist enemies (enemy release, Keane and Crawley 2002) in abandoned habitat compared with reference habitat because competitors are more likely to be alien and likewise benefit from enemy release, or native competitors may also benefit from enemy release in recently assembled communities.

5.6 Designed Habitat

5.6.1 Habitat Characteristics

Designed habitat is deliberately and strongly manipulated by humans to create a new habitat that suites conservation objectives (e.g., Conservation Management Areas in Mauritius and Rodrigues, Mauremootoo and Payendee 2002; Kaiser et al. 2008) (Fig. 5.5). In our classification, designed habitat is characterized by its constant dependence on management (conservation-reliant *sensu* Scott et al. 2005). In an early management phase, designed habitat will typically be ecologically similar to anthropogenic habitat insofar as ecosystem patterns and processes are often simplified, and light availability will typically be high because of the removal of former vegetation. In contrast to anthropogenic habitat, however, soils will often be degraded and fertility low. With time designed habitat may, depending on habitat quality and intensity of management, become similar to either abandoned or reference habitat; and if management stops it will become, by definition, either abandoned or reference habitat.

5.6.2 Functional Type of Invasive Alien Species

Alien species present in designed habitat may either invade the habitat without assistance or be deliberately introduced. Deliberate introductions to designed habitat are not considered invasive because they are part of the design. However, deliberate introductions of alien species may later become invasive. In the case of unassisted invasions, the specific land management regime (e.g., low intensity but

Fig. 5.5 The role of alien species in designed habitat. In Mauritius, since the early 1980s c. 40 ha of forest have been weeded and partly fenced. These plots of designed habitat, known in Mauritius as Conservation Management Areas (CMA), typically cover only a few hectares and are embedded in a matrix landscape dominated by alien species. These CMAs are intended to mimic the prehuman habitat states, are almost completely composed of native vegetation, but are dependent on constant human intervention such as weeding or fencing against invasive animals such as rats or monkeys (Photo Christoph Kueffer)

high frequency management) and habitat conditions (e.g., high light availability combined with low soil fertility) may provide invasion opportunities for specific types of invaders.

5.6.3 Impact of Alien Species and Management Action

In the case of designed habitat, alien species may be deliberately introduced to restore degraded soils, remove surplus nutrients from abandoned agricultural land, provide perches for birds or shading and nurse trees for regenerating plants, or enhance seed dispersal, among many others (D'Antonio and Meyerson 2002; Ewel and Putz 2004). A major concern is the possibility that deliberately introduced species of an ecological design project may spread to other conservation areas. This risk may be particularly high because alien species best suited for ecological design projects will be well adapted to local conditions, will interact with native mutualists, and may be selected to have a strong influence on ecosystem properties.

The choice between native and alien species for designed habitat has a pragmatic and a more fundamental dimension. From a pragmatic perspective, we may ask if and when alien species are more effective as tools for a particular purpose in an ecological design project than native species. Arguments for alien species may be that there is a larger pool of species available (especially species with novel traits such as nitrogen-fixation that may not be present in the native flora), their ecology is better known, and alien species may perform better because they are less attacked by pests or herbivores (i.e. enemy release). Arguments against the use of alien species may be that the chances are higher that they have negative side-effects (e.g., form mono-dominant stands and are difficult to control) and lower that they have nontarget positive side-effects (e.g., provide food and habitat to a wide variety of native species). On a more fundamental level, it is uncertain if and when alien species may be valued positively as long-term constituents of the designed habitats (Donlan et al. 2005; Nicholls 2006; McLachlan et al. 2007). The idea to introduce alien species as ecological analogs of extinct species has gained momentum (Donlan et al. 2005; Nicholls 2006). In Mauritius for instance, the translocation of Aldabran giant tortoises as ecological analogs for the extinct Mauritian giant tortoises is currently being investigated (cf. Zavaleta et al. 2001; Nicholls 2006). Another consideration in the context of designed habitat may be promotion of species that are threatened in their native range because of global change dynamics such as climate change. Such species may be translocated into designed habitat (assisted migration, McLachlan et al. 2007). More generally, in an era of global change, introduction of alien species for ecological design may become increasingly relevant. It may, for instance, be necessary to introduce alien species to a habitat to increase its resilience to climate change, or assist the migration of whole habitats.

5.6.4 Research Focus

Invasions into designed habitat have received little scientific study. Many basic ecological questions related to diversity-ecosystem functioning, seed limitation, or the importance of soil microbe communities are of high relevance to ecosystem design (Young et al. 2005). In fact, their intermediate complexity between artificial model systems, such as microcosms, and natural ecosystems provides a convenient compromise between tractability and real-world applicability for ecological research. If a designed habitat is isolated, e.g., on a small island, deliberate introduction of alien species may be a convenient system for real-world risk screening prior to the species' broader introduction in a region. Because alien species that have a high potential to assist rehabilitation of degraded sites also have a high potential to become disruptive (Richardson et al. 2004), research is needed to identify effective strategies for reducing these risks. The safest option may be to breed or genetically engineer sterile varieties of these species (Ewel et al. 1999). Designed habitat may be foremost an opportunity for experimentally addressing the natural/artificial dichotomy from both a natural and social sciences perspective. How do alien and native species differ in their effects on biotic communities and ecosystems? Will these differences between native and alien species change with global change and native species loss, and how does this affect how we define and value "high quality nature"?

5.7 Discussion

Our typology for understanding, valuing, and managing invasive species impacts is based on different habitat types (anthropogenic, reference, abandoned, and designed). The four habitat types represent prototypes, but in reality the differences between them are sometimes blurred and may become more so in the future. For instance, land may be managed simultaneously for biodiversity and other products or services, and these habitats are hybrids of anthropogenic and designed habitat. In some places, abandoned land may transform into reference habitat over time, but the exact transition point is not clearly defined. In some cases, additional refinement of categorizations could help to clarify biotic invasion issues. For instance, in anthropogenic habitat, urban areas, intensive agriculture and extensive or traditional agriculture differ as contexts for biotic invasions. Nevertheless, we think that our typology represents a valuable first sketch, demonstrating how a habitat-based framework could advance invasive species research and management.

5.7.1 Toward a Habitat-Based Framework for Invasive Species Research

Initially, invasive species research has attempted to separately generalize traits of problematic invaders (invasiveness) and characteristics of vulnerable habitats (invasibility) (Drake et al. 1989; Williamson 1996; Lonsdale 1999; Kolar and Lodge 2001). However, although this approach helped to identify heuristics to predict the invasiveness of species and the invasibility of habitats, it did not provide an integrative framework to understand the interactions of species, environment, and human action or the dynamics of invasions in space and time (Kueffer and Hirsch Hadorn, 2008). Phase-transition models were a first successful attempt to gain a more integrative understanding of biotic invasions (Richardson et al. 2000; Kolar and Lodge 2001; Dietz and Edwards 2006; Facon et al. 2006; Richardson and Pysek 2006; Theoharides and Dukes 2007). In these models, an invasion is characterized as a sequence of distinct phases. These phases typically include introduction, establishment, and spread in a new area. Phase-transition models have allowed specifying relevant ecological processes for different phases (Dietz and Edwards 2006;

Theoharides and Dukes 2007). Further, human action was included as a major driver of the transportation of alien species (vector science) (Ruiz 2003; Mooney et al. 2005; Kowarik and von der Lippe 2007). This enabled the integration of different disciplines such as social sciences, economics, law, and natural sciences for the study of biotic invasions. However, although vector science has helped us understand and manage the spread of alien species, it did not solve the problem of understanding and managing impacts in invaded habitats.

Our typology (Table 5.1) attempts to frame invasion issues based on habitat states and their human valuation. There are several lines of previous research that point in a similar direction. Research has attempted to generalize knowledge about the vulnerability of habitats to invasions, based on general ecological processes such as disturbance regimes, resource availabilities, or species diversity (Lonsdale 1999; Levine et al. 2004; Fridley et al. 2007), but interactions between species traits and habitat characterization have not been at the central focus. It has long been argued that the spread and impact of invasive species in a habitat has to be understood and generalized as a species-by-environment interaction (Richardson 1990; Morse 2004; Callaway 2006). A multitude of studies have shown the relevance of anthropogenic habitat modification and land use for explaining biotic invasions (Hobbs 2000; Maskell et al. 2006). However, such research on the interactions between species traits and habitat characteristics and/or land use has so far mainly compiled case examples. Interestingly, social geographers have also worked on a theoretical understanding of land use changes and biotic invasions (Robbins 2001; Schneider and Geoghegan 2006). Robbins (2001) for instance discussed abandoned land and proposed the term "quasiforests" to account for such hybrid landscapes that fall in-between the natural and the social. Our framework is derived from the realities that managers are confronted with and emphasizes the context-dependence of invader impacts. We suggest our typology may be a good starting point to bring together theoretical thinking from the natural and social sciences and improve our conceptual and practical understanding of impacts of invasive species.

Acknowledgments We thank Jake Alexander, Paul Krushelnycky, and two anonymous reviewers for very helpful comments on earlier versions of the article, and Eva Schumacher for assistance with the preparation of the figures. CK was supported by USDA NRI Cooperative Research, Education, and Extension Service Grant # 2006–35320–17360.

References

Baker HG (1974) The evolution of weeds. Ann Rev Ecol Syst 5:1–24

Benning TL, LaPointe D, Atkinson CT, Vitousek PM (2002) Interactions of climate change with biological invasions and land use in the Hawaiian Islands: modeling the fate of endemic birds using a geographical information system. PNAS 99:14246–14249

Benvenuti S (2007) Weed seed movement and dispersal strategies in the agricultural environment. Weed Biol Manag 7:141–157

Blumenthal DA (2006) Interactions between resource availability and enemy release in plant invasion. Ecol Lett 9:887–895

Buckley YM et al. (2006) Management of plant invasions mediated by frugivore interactions. J Appl Ecol 43:848–857
Callaway RM, Maron JL (2006) What have exotic plant invasions taught us over the past 20 years? Trends Ecol Evol 21:369–374
Chazdon RL (2003) Tropical forest recovery: legacies of human impact and natural disturbances. Perspect Plant Ecol Evol Syst 6:51–71
Corlett RT (2005) Interactions between birds, fruit bats and exotic plants in urban Hong Kong, South China. Urban Ecosyst 8:275–283
Courchamp F, Chapuis J-L, Pascal M (2003) Mammal invaders on islands: impact, control and control impact. Biol Rev 78:347–383
Cramer VA, Hobbs RJ, Standish RJ (2008) What's new about old fields? Land abandonment and ecosystem assembly. Trends Ecol Evol 23:104–112
Culliney JL (2006) Islands in a Far Sea. The fate of nature in Hawaii. Revis Ed. University of Hawai'i Press, Honolulu, USA
D'Antonio CM, Meyerson LA (2002) Exotic plant species as problems and solutions in ecological restoration: a synthesis. Restor Ecol 10:703–713
D'Antonio CM, Vitousek PM (1992) Biological invasions by exotic grasses, the grass fire cycle, and global change. Ann Rev Ecol Syst 23:63–87
Daehler CC, Denslow JS, Ansari S, Kuo H-C (2004) A risk-assessment system for screening out invasive pest plants from Hawaii and other Pacific Islands. Conserv Biol 18:360–368
Davis MA (2006) Invasion biology 1958–2005: the pursuit of science and conservation. In: Cadotte MW, McMahon SM, Fukami T (eds) Conceptual ecology and invasion biology. Springer, Berlin
Davis MA, Grime JP, Thompson K (2000) Fluctuating resources in plant communities: a general theory of invasibility. J Ecol 88:528–534
Didham RK, Tylianakis JM, Gemmell NJ, Rand TA, Ewers RM (2007) Interactive effects of habitat modification and species invasion on native species decline. Trends Ecol Evol 22:489–496
Didham RK, Tylianakis JM, Hutchison MA, Ewers RM, Gemmell NJ (2005) Are invasive species the drivers of ecological change? Trends Ecol Evol 20:470–474
Dietz H, Edwards PJ (2006) Recognition that causal processes change during plant invasion helps explain conflicts in evidence. Ecol 87:1359–1367
Dietz H, Kueffer C, Parks CG (2006) MIREN: a new research network concerned with plant invasion into mountain areas. Mountain Res Dev 26:80–81
Donald PF, Evans AD (2006) Habitat connectivity and matrix restoration: the wider implications of agri-environment schemes. J Appl Ecol 43:209–218
Donlan J et al. (2005) Re-wilding North America. Nature 436:913–914
Drake JA et al. (1989) Biological invasions: a global perspective. Wiley, Chichester
Ehrenfeld JG, Ravit B, Elgersma K (2005) Feedback in the plant-soil system. Ann Rev Environ Resour 30:7.1–7.41
Elton CS (1958) The ecology of invasions by animals and plants. Methuen, London
Ewel J et al. (1999) Deliberate introductions of species: research needs. BioScience 49:619–630
Ewel JJ, Putz FE (2004) A place for alien species in ecosystem restoration. Front Ecol Environ 2:354–360
Facon B, Genton BJ, Shykoff J, Jarne P, Estoup A, David P (2006) A general eco-evolutionary framework for understanding bioinvasions. Trends Ecol Evol 21:130–135
Frenot Y et al. (2005) Biological invasions in the Antarctic: extent, impacts and implications. Biol Rev 80:45–72
Fridley JD et al. (2007) The invasion paradox: reconciling pattern and process in species invasions. Ecol 88:3–17
Fritts TH, Rodda GH (1998) The role of introduced species in the degradation of island ecosystems: a case history of Guam. Ann Rev Ecol Syst 29:113–140
Funk JL, Vitousek PM (2007) Resource-use efficiency and plant invasion in low-resource systems. Nature 446:1079–1081
Funtowicz SO, Ravetz JR (1993) Science for the post-normal age. Futures 25:739–755

Granados J, Koerner C (2002) In deep shade, elevated CO_2 increases the vigor of tropical climbing plants. Global Change Biol 8:1109–1117
Gurevitch J, Padilla DK (2004) Are invasive species a major cause of extinctions? Trends Ecol Evol 19:470–474
Hallman G (2007) Phytosanitary measures to prevent the introduction of invasive species. In: Nentwig W (ed) Biological invasions. Springer, Berlin, pp. 367–384
Hendrix PF, Bohlen PJ (2002) Exotic earthworm invasions in North America: ecological and policy implications. BioScience 52:801–811
Hirsch G (1995) Beziehungen zwischen Umweltforschung und disziplinärer Forschung. Gaia 4:302–314
Hobbs RJ (2000) Land-use changes and invasions. In: Mooney HA, Hobbs RJ (eds) Invasive species in a changing world. Island Press, Washington, DC, pp. 55–64
Hobbs RJ et al. (2006) Novel ecosystems: theoretical and management aspects of the new ecological world order. Global Ecol Biogeogr 15:1–7
Holway DA, Lach L, Suarez AV, Tsutsui ND, Case TJ (2002) The causes and consequences of ant invasions. Ann Rev Ecol Syst 33:181–233
Hughes RF, Denslow JS (2005) Invasion by a N_2-fixing tree alters function and structure in wet lowland forests of Hawai'i. Ecol Appl 15:1615–1628
Hulme PE (2006) Beyond control: wider implications for the management of biological invasions. J Appl Ecol 43:835–847
Kaiser CN, Hansen DM, Müller CB (2008) Habitat structure affects reproductive success of the rare endemic tree *Syzygium mamillatum* (Myrtaceae) in restored and unrestored sites in Mauritius. Biotropica 40:86–94
Keane RM, Crawley MJ (2002) Exotic plant invasions and the enemy release hypothesis. Trends Ecol Evol 17:164–170
Klingenstein F, Diwani T (2005) Invasive alien species from a nature conservation point of view in Germany. In: Secretariat IPPC (ed) Identification of risks and management of invasive alien species using the IPPC framework. Proceedings of the workshop on invasive alien species and the International Plant Protection Convention, Braunschweig, Germany, 22–26 September 2003. FAO, Rome, Italy, pp 133–145
Knowlton N, Jackson JBC (2008) Shifting baselines, local impacts, and global change on coral reefs. PLOS Biol 6:e54.
Kolar CS, Lodge TS (2001) Progress in invasion biology: predicting invaders. Trends Ecol Evol 16:199–204
Kowarik I, von der Lippe M (2007) Pathways in plant invasions. In: Nentwig W (ed) Biological invasions. Springer, Berlin, pp. 29–47
Krushelnycky P (2007) The effects of invasive ants on Hawaiian arthropod communities. In: Hawai'i conservation conference. Conference program with abstracts. Hawai'i Conservation Alliance Foundation, Honolulu, USA. http://hawaiiconservation.org/2007hcc.asp.
Kueffer C (2006a) Impacts of woody invasive species on tropical forests of the Seychelles. Diss. ETH No. 16602, Department of Environmental Sciences. ETH Zurich, Zurich, p. 160
Kueffer C (2006b) Integrative ecological research: case-specific validation of ecological knowledge for environmental problem solving. Gaia 15:115–120
Kueffer C, Hirsch Hadorn G (2008) How to achieve effectiveness in problem-oriented landscape research – the example of research on biotic invasions. Living Rev Landscape Res 2: 2. http://www.livingreviews.org/lrlr-2008-2.
Kueffer C, Klingler G, Zirfass K, Schumacher E, Edwards P, Güsewell S (2008) Invasive trees show only weak potential to impact nutrient dynamics in phosphorus-poor tropical forests in the Seychelles. Funct Ecol onlineEarly. Funct Ecol 22:359–366. doi: 10.1111/j.1365-2435.2007.01373.x.
Kueffer C, Larcher P, Paulsen T, Bose L, Guisan A, Schaffner U (2007a) Invasive Pflanzen in der Schweiz: Identifizieren von Lücken und Problemen beim Wissensaustausch. [Invasive plants in Switzerland: identification of gaps and problems hindering knowledge transfer between

research and management]. Swiss Academies of Arts and Sciences & Swiss National Centre of Competence in Research (NCCR) *Plant Survival*, Berne, Switzerland

Kueffer C, Schumacher E, Fleischmann K, Edwards PJ, Dietz H (2007b) Strong belowground competition shapes tree regeneration in invasive *Cinnamomum verum* forests. J Ecol 95:273–282

Levine JM, Adler PB, Yelenik SG (2004) A meta-analysis of biotic resistance to exotic plant invasions. Ecol Lett 7:975–989

Lonsdale WM (1999) Global patterns of plant invasions and the concept of invasibility. Ecology 80:1522–1536

Lososova Z et al. (2006) Patterns of plant traits in annual vegetation of man-made habitats in central Europe. Perspect Plant Ecol Evol Syst 8:69–81

Lugo AE (2004) The outcome of alien tree invasions in Puerto Rico. Front Ecol Environ 2:265–273

Malmstrom CM, McCullough AJ, Johnson HA, Newton LA, Borer ET (2005) Invasive annual grasses indirectly increase virus incidence in California native perennial bunchgrasses. Oecologia 145:153–164

Maskell LC, Firbank LG, Thompson K, Bullock JM, Smart SM (2006) Interactions between non-native plant species and the floristic composition of common habitats. J Ecol 94:1052–1060

Mauremootoo J, Payendee R (2002) Against the odds: restoring the endemic flora of Rodrigues. Plant Talk 28:26–28

McLachlan JS, Hellmann JJ, Schwartz MW (2007) A framework for debate of assisted migration in an era of climate change. Conserv Biol 21:297–302

Millennium Ecosystem Assessment (2005) Ecosystems and human well-being: biodiversity synthesis. World Resources Institute, Washington, DC

Minton MS, Verling E, Miller AW, Ruiz GM (2005) Reducing propagule supply and coastal invasions via ships: effects of emerging strategies. Front Ecol Environ 6:304–308

Mooney HA, Cleland EE (2001) The evolutionary impact of invasive species. PNAS 98:5446–5451

Mooney HA, Mack RN, McNeely JA, Neville LE, Schei PJ, Waage JK (eds) (2005) Invasive alien species. A new synthesis. Island Press, Washington, London

Morrison LW (2002) Long-term impacts of an arthropod-community invasion by the imported fire ant, *Solenopsis invicta*. Ecology 83:2337–2345

Morse LE, Randall JM, Benton N, Hiebert R, Lu S (2004) An invasive species assessment protocol: evaluating non-native plants for their impact on biodiversity. Version 1. NatureServe, Arlington, Virginia, USA

Müller-Schärer H, Schaffner U, Steinger T (2004) Evolution in invasive plants: implications for biological control. Trends Ecol Evol 19:417–422.

National Academies of Sciences (2002) Predicting invasions of nonindigenous plants and plant pests. National Academy Press, Washington DC

Nicholls H (2006) Restoring nature's backbone. PLoS Biol 4:e202. doi: 210.1371/journal.pbio.0040202

Orians GH et al. (1986) Ecological knowledge and environmental problem solving. National Academy Press, Washington, DC

Parker IM et al. (1999) Impact: towards a framework for understanding the ecological effects of invaders. Biol Invasions 1:3–19

Pauchard A et al. (in press) Ain't no mountain high enough: Plant invasions reaching high elevations. Front Ecol Environ

Peacor SD, Allesina S, Riolo RL, Pascual M (2006) Phenotypic plasticity opposes species invasions by altering fitness surface. PLoS Biol 4:2112–2120

Petersen C, Huntley B (2005) Mainstreaming biodiversity in production landscapes. Working paper 20. Global Environment Facility, Washington DC

Pohl C, Hirsch Hadorn G (2007) Principles for designing transdisciplinary research. oekom, Munich, Germany

Prach K, Pyšek P, Smilauer P (1999) Prediction of vegetation succession in human-disturbed habitats using an expert system. Restor Ecol 7:15–23
Pyšek P, Jarošík V (2005) Residence time determines the distribution of alien plants. In: Inderjit (ed) Invasive plants: ecological and agricultural aspects. Birkhäuser, Basel, Switzerland, pp. 77–95
Pyšek P, Jarošík V, Chytry M, Kropá. Z, Tichy L, Wild J (2005) Alien plants in temperate weed communities: prehistoric and recent invaders occupy different habitats. Ecology 86:772–785
Pyšek P, Jarošík V, Kucera K (2002) Patterns of invasion in temperate nature reserves. Biol Conserv 104:13–24
Reichard SH, White P (2001) Horticulture as a pathway of invasive plant introductions in the United States. BioScience 51:103–113
Rejmanek M (1996) A theory of seed plants invasiveness: the first sketch. Biol Conserv 78:171–181
Richardson DM, Binggeli P, Schroth G (2004) Plant invasions - problems and solutions in agroforestry. In: Schroth GGF, Harvey CA, Gascon C, Vasconcelos H, and Izac AM. (ed) Agroforestry and biodiversity conservation in tropical landscapes. Island Press, Washington. pp. 371–396
Richardson DM, Pysek P (2006) Plant invasions: merging the concepts of species invasiveness and community invasibility. Progr Phys Geogr 30:409–431
Richardson DM, Pysek P, Rejmánek M, Barbour MG, Panetta FD, West CJ (2000) Naturalization and invasion of alien plants: concepts and definitions. Div Distr 6:93–107
Richardson DM, Cowling RM, Le Maitre DC (1990) Assessing the risk of invasive success in *Pinus* and *Banksia* in South African mountain fynbos. J Veg Sci 1:629–642
Robbins P (2001) Tracking invasive land covers in India, or why our landscapes have never been modern. Ann Assoc Am Geogr 91:637–659
Rodriguez LF (2006) Can invasive species facilitate native species? Evidence of how, when, and why these impacts occur. Biol Invasions 8:927–939
Rose M, Hermanutz L (2004) Are boreal ecosystems susceptible to alien plant invasion? Evidence from protected areas. Oecologia 139:467–477
Ruiz GM (ed) (2003) Invasive species. Vectors and management strategies. Island Press, Washington, London
Safford RJ, Jones CG (1998) Strategies for land-bird conservation on Mauritius. Conserv Biol 12:169–176
Sagoff M (2005) Do non-native species threaten the natural environment? J Agric Environ Ethics 18:215–236
Sax DF, Kinlan BP, Smith KF (2005) A conceptual framework for comparing species assemblages in native and exotic habitats. Oikos 108:457–464
Schlaepfer MA, Sherman PW, Blossey B, Runge MC (2005) Introduced species as evolutionary traps. Ecol Lett 8:241–246
Schneider L, Geoghegan J (2006) Land abandonment in an agricultural frontier after a plant invasion: the case of Bracken Fern in Southern Yucatán, Mexico. Agric Res Econ Rev 35:167–177
Schumacher E (2007) Variation in growth responses among and within native and invasive juvenile trees in Seychelles. Diss. No. 16988, Department of Environmental Sciences. ETH Zurich, Zurich
Schumacher E, Kueffer C, Tobler M, Gmür V, Edwards PJ, Dietz H (2008) Influence of drought and shade on seedling growth of native and invasive trees in the Seychelles. Biotropica 40: 543–549. doi: 10.1111/j.1744–7429.2008.00407.x
Scott JM, Goble DD, Wiens JA, Wilcove DS, Bean M, Male T (2005) Recovery of imperiled species under the Endangered Species Act: the need for a new approach. Front Ecol Environ 7:383–389
Simberloff D, Von Holle B (1999) Positive interactions of nonindigenous species: invasional meltdown? Biol Invasions 1:21–32
Smith RG, Maxwell BD, Menalled FD, Rew LJ (2006) Lessons from agriculture may improve the management of invasive plants in wildland systems. Front Ecol Environ 4:428–434

Stinson KA et al. (2006) Invasive plant suppresses the growth of native tree seedlings by disrupting belowground mutualisms. PLoS Biol 4:e140. doi: 110.1371/journal.pbio.0040140
Strayer DL, Eviner VT, Jeschke JM, Pace ML (2006) Understanding the long-term effects of species invasions. Trends Ecol Evol 21:645–651
Suding KN, Gross KL, Houseman GR (2004) Alternative states and positive feedbacks in restoration ecology. Trends Ecol Evol 19:46–53
Theoharides KA, Dukes JS (2007) Plant invasion across space and time: factors affecting nonindigenous species success during four stages of invasion. New Phytol 176:256–273
Vitousek PM (1990) Biological invasions and ecosystem processes: towards an integration of population biology and ecosystem studies. Oikos 57:7–13
Vitousek PM, Walker LR, Whiteaker LD, Mueller-Dombois D, Matson PA (1987) Biological invasion by *Myrica faya* alters ecosystem development in Hawaii. Science 238:802–804
Wilkinson DM (2004) The parable of Green Mountain: Ascension Island, ecosystem construction and ecological fitting. J Biogeogr 31:1–4
Williamson M (1996) Biological invasions. Chapman & Hall, London
Young TP, Petersen DA, Clary JJ (2005) The ecology of restoration: historical links, emerging issues and unexplored realms. Ecol Lett 8:662–673
Zavaleta E, Hobbs RJ, Mooney HA (2001) Viewing invasive species removal in a whole-ecosystem context. Trends Ecol Evol 16:454–459

Chapter 6
Temporal Management of Invasive Species

Catherine S. Jarnevich and Thomas J. Stohlgren

Abstract Successful management of invasive species requires using spatial models of current distributions and forecasts of spread with explicit consideration of the effects of time on the invasion. Forecasts must also include components contributing to the spread rate such as invasion stage and Allee effects. There are several different analysis techniques available for spatial models and forecasting, and the appropriate technique will depend on the particular research or management question. Many of the best forecasting examples with time as a parameter exist for insect species, but the same techniques are useful in forecasting the spread of plant species. Often, data availability is a limiting factor in doing this, so we need to change the data being collected. Inclusion of this temporal information in prioritization of resources for control/eradication efforts will help them be effective and efficient.

Keywords Forecasting • Invasions • Spread rates

6.1 Introduction

Invasion ecology must progress from a reactive science to a proactive science (Lodge et al. 2006). That's because the likelihood of eradicating or containing an infestation decreases as the size of the infestation increases (Rejmanek and Pitcairn 2002)

An important component of prevention and control of invasive species is forecasting where invasions are most likely to initially occur, and where they are likely to spread once local populations are established in a country, region, or locally. Many "predictions" to this point have involved "hinde-casting" past invasions or spatially extrapolating patterns of nonnative species, based on environmental

C.S. Jarnevich (✉) and T.J. Stohlgren
Fort Collins Science Center, U.S. Geological Survey, Fort Collins, CO 80526, USA
jarnevichc@usgs.gov

attributes or their relationship to native species (Jarnevich et al. 2006; Stohlgren et al. 2006; Stohlgren et al. 2003). Additionally, modeling approaches commonly attempt to predict the potential distribution of individual invasive species using snap-shot-in-time datasets, meaning a dataset collected over a short time for a specific location (Elith et al. 2006). Although useful, these models generally provide no estimate of when a species may arrive at a particular location – only that it may at some unknown future point. Predictions for individual species' potential abundance or rates of spread are less common, often due to lack of data. As technology has advanced over the past couple of decades, spatial models have become more sophisticated and accurate. These models have proven valuable in the management of invasive species. The increased availability of geographic positioning system (GPS) technology has also helped data collection. However, these forecasting tools are still limited by gaps in the field data being collected and synthesized to calibrate or independently validate their predictions. There are also still limitations in our ability to model natural systems where nonnative species establish and spread.

In a recent survey of existing invasive species databases in the United States, Crall et al. (2006) found that 38% of the 254 databases discovered contained data covering 10 or more years. This survey covered databases, so groups collecting field data stored in a less technologically advanced system were not included. Thus, this number ignores many collections, but is probably somewhat reflective of reality, with less than half of the datasets holding long-term data. Additionally, 82 of the datasets covered an area equal to a county or smaller. Even the datasets that do exist may not always be readily available and in the same format. Data integration would help solve some of these limitations. To effectively manage invasions, we need information across broad spatial extents and over long time periods as invasions occur over these scales.

Partly due to the lack of these data, predictions including a specific temporal component are much rarer in the literature than spatial predictions (i.e., a species' potential habitat). A literature search in Web of Science including the terms spatial, modeling, and invasive revealed many articles, while one including temporal or time, modeling, and invasive revealed a dearth of articles published in peer reviewed journals.

Invasive species management involves many concepts and careful consideration of analysis techniques. There are several important points to keep in mind when creating predictions of species spread in addition to specific individual species traits. Range expansion of a species will be a function of the number and spatial arrangements of introductions, time since invasion, propagule pressure (frequency of propagules), a vector for dispersal, seeds being dispersed to a favorable location, hybridization, and many other factors. Forecasting invasions including richness, distribution, and abundance of invasive species with a temporal component as opposed to species distribution models that predict potential distribution regardless of time can be accomplished with several different methods. The most appropriate analysis method for forecasting a particular invasion may vary depending on the spatial resolution, the species, and the stage of invasion. Additionally, managing species effectively through time involves assessing the long-term potential distribution and abundance of an invasive species, including the long-term results of any control efforts on invasive species and the effects of changing climates, land use, and trade and transportation through time. In this chapter, we will review the important components involved in predicting spread rates, review the current compared with the needed state of data collection and data analysis techniques, and examine strategies to effectively control invasive species in the long term.

6.2 Forecasting Considerations

Forecasts of invasive species spread rates can aid management activities at many different levels. At large scales such as national or state/province equivalents, they may assist high level managers with the distribution of limited resources for prevention and control efforts. Forecasts may also provide "watch lists" to management units of species that may invade their area. A specified year or time period of expected arrival associated with these predictions narrows down the number of species for resource managers and their helpers to be especially vigilant for at any given time. Additionally, species that are predicted to spread more rapidly than others may require a much quicker response for containment or eradication than a species that spreads very slowly with low propagule production over decades. For example, a plant species producing many small, windblown seeds may quickly infest large areas. Individual plant characteristics like this example may have a great effect on spread rates along with the more general factors discussed in more detail below. Understanding the spread of a species over time through a local area can aid in setting priorities for control and eradication efforts among various species and for neighboring small infestations of a particular species. When generating these invasion forecasts, there are several different concepts that should be considered.

6.2.1 Time Since Establishment

A general factor that should be included with any invasive species spread model is the time since invasion, or at least how far along the invasion process the species has proceeded at the particular location being examined. In the case of multiple introductions, the invasion stage may vary at different focal points and may have an accelerated response compared with a single introduction. These multiple introductions are similar to anthropogenic long distance dispersal, which can cause an accelerated expansion rate A species early in the invasion process may not have had the opportunity to establish in all locations with environmental conditions within its "ecological envelope" (e.g., locations with suitable habitat). A species that is nearing the final stages of invasion, having maximized its potential range, may be treated more like a native species in development and interpretation of models of potential distribution and abundance as it will have a relatively more stable distribution, having had time for propagules to reach all suitable habitat. However, despite the abundance of model comparison papers, we were unable to find any that have analyzed this topic, although Wilson et al. (2007) do discuss the importance of residence time as a consideration when modeling potential distribution. They created simple logistic models of rate of spread including this term.

This temporal context for invasive species is important in assessing current and potential impacts and spread. A general trend in nonnative species invasions is a lag time between establishment and explosive spread, where populations in the new range follow the logistic growth curve (Hobbs and Humphries 1995; Sakai et al. 2001; Strayer et al. 2006). Although some of this lag may be contributed to low detection rates when a species has not yet been identified as a problem, there are definitive examples of species exhibiting lag times whose establishment and subsequent spread has been well studied (e.g., Liebhold and Tobin 2006). The potential length of a lag period is an important consideration when trying to predict the spread of a species and its invasion potential. The duration of the lag time may vary considerably with a species' reproductive strategy and propagule pressure (Lockwood et al. 2005), its adaptability to a new environment, the identity, and availability of vectors of spread (Barney 2006), or hybridization with other nonnative or native species (Ellstrand, Schierenbeck 2000; Ellstrand and Schierenbeck 2006). Cheatgrass (*Bromus tectorum*), a species now very widespread in the Western US, had a 30 year lag time before spreading rapidly (Mack 1981). In contrast, Brotherson and Field (1987) suggest that salt cedar (*Tamarix* sp.) had a lag time of 100 years before becoming well established in riparian ecosystems throughout the United States. These lag times are very different, and may not even cover the full range that exists. Given these disparate times, integrating potential lag times and rates of spread for invasive species into predictive models is not an easy task.

With low population levels, short-range movements are much more likely than long-distance dispersal, at least for gypsy moths (*Lymantria dispar*) in the early stages of invasion (Liebhold and Tobin 2006). Given that long distance dispersal drives the invasion process (Hastings et al. 2005; Nehrbass et al. 2007), this finding reinforces the lag effect. The almost 10 year lag time identified for gypsy moths informs management decisions by indicating that time could be taken in eradication of local infestations to ensure that the entire area is treated. As population increases, Allee effects become less important and long-distance movement becomes much more likely.

Spread rates will also not be constant through time, as demonstrated by the gypsy moth lag effect, which resulted from interannual stochasticity in population growth rates and Allee effects (Liebhold and Tobin 2006). Spread rates may be affected by differences in climate between years (Neubert et al. 2000); perfect conditions one year for a particular species will result in quicker spread and greater abundance than poor conditions (e.g., drought, flooding, extreme cold). Additional introductions of an already established and spreading species may also affect the rate of spread by either new foci appearing on the landscape or a greater number of propagules available for dispersal at an existing foci.

Allee effects are also important to consider individually and are defined as "a positive relationship between any component of individual fitness and either numbers or density of conspecifics" (Stephens et al. 1999). In Washington state, fecundity of the nonnative smooth cordgrass (*Spartina alterniflora*) was greater in high-density areas than in low-density areas, resulting in a mean spread rate of 31% compared with 19% (Davis et al. 2004; Taylor et al. 2004). This example allee effect in the low-density areas is a weak effect, where the spread rate slows but still occurs, but is still an important factor in considering forecasts of the spread of smooth cordgrass.

6.2.2 Spatial Considerations

Understanding the relationship of spread rates to spatial scale, both resolution and extent, is another important aspect of forecasting. An invading species may not have the same rate of spread at different leading edges of its expansion, and rates may change through time as they are moderated by other factors like climate. For example, the spread of hemlock woolly adelgid (*Adelges tsugae*) in the Eastern United States had different rates through time at different leading edges and at the same location through time (Evans and Gregoire 2007). Different modeling approaches and different datasets to predict woolly adelgid spread all yielded varying rates, but were consistent in predicting spread rates to differ geographically. The adelgid spread more quickly in the southern US than at the northern edge, probably because of colder temperatures in the north. Additionally, at a smaller spatial scale, counties on the leading edge were invaded more slowly if the leading edge arrived during a colder winter, further evidence that interannual climatic variation may also affect the rate of spread.

Spread rates may also differ depending on the spatial resolutions. Large scale (like continental) spread rates result from long distance dispersal events which are often human mediated, like the spread of zebra mussels (*Dreissena polymorpha*) appearing in western US waterways unconnected to the heavily infested ones in the eastern US. However, at the local scale such as a preserve or meadow allee effects can be important, where spread rates differ as a result of population dynamics. These rates of spread would be measured in units of meters per year, while the long distance dispersal events measurements differ by orders of magnitude.

6.2.3 Vectors of Invasion

Including vectors or pathways of spread in predictive models may also be important in forecasting invasions (Leung et al. 2006). Spread of an invasive species by non-anthropogenic means may be predictable, though difficult; spread by humans will be harder to predict as humans may have less predictable pattern in their movement than natural processes. The number of introductions of a species can vary and alter the rate of expansion. Human pathways of introduction such as horticulture for ornamental species and ballast water contaminants may cause introductions more often, affecting propagule pressure, and across a wider range of locations than accidentally introduced species. These variants across the temporal scale can be important influences on the rate of spread, but difficult to monitor or predict.

Modeling pathways of spread can be difficult, even if a pathway does not rely on the unpredictable movements of humans. For example, Myers et al. (2004) examined the importance of deer in the Eastern United States for long distance dispersal of plant seeds, and postulated that deer may be an important vector in the rapid spread of species in that region. This mode of transportation adds another level of complexity to modeling the spread of species, as it necessitates prediction of the feeding habits, spatial movement and rate of movement of deer along with the other factors related to the invasive plant species' spread. These types of vectors are not necessarily readily apparent when determining the ability of a species to spread. An organism like the zebra mussel, which depends mainly on human transport between bodies of water, may be modeled more easily with techniques like gravity models that are appropriate for human pathways (Leung et al. 2006). Similar techniques may prove appropriate for aquatic weeds.

6.3 Forecasting Invasions

There have been many studies to spatially model invasions. However, upon review, there are very few papers that examine the spread of invasive species through time where time is an independent variable in the model. There are different techniques that can be used to analyze invasive species spread both spatially and temporally (Table 6.1). These can include simple literature review and data collation techniques, monitoring field site locations, using statistical techniques, and classifying remotely-sensed imagery.

6.3.1 Data Synthesis

Although plot-based field data that may contain detailed information for a certain location is typically limited spatially and temporally, there are broad scale occurrence datasets based on observations of organisms that might be used to examine species' broad scale distributions and patterns of spread. These datasets include museum and herbarium records, where the location is often very general (e.g., county level) and the only information about the organism that can be gleaned is that it was present at a particular location at some point in time (e.g., generally a particular year). We typically lack detailed information on species abundance over time linked to specific coordinates, so analyzing historical datasets like these may

Table 6.1 Techniques involved in forecasting invasions including applications to: early detection, predicting species spread, data analyses, spatial scales, and taxonomic completeness

	Early detection	Species spread	Analysis types	Spatial scale	Taxonomic completeness
Data synthesis	None	Past invasion rates	Past patterns of invasion; can validate current methods	Limited by data that exists; large spatial extent, coarse spatial resolution	Any species with historic occurrence records
Field data	Detect completely new species in an area	Field measurements	Early detection, validation locations, current distribution in small area	Limited spatial extent; can resample through time; fine resolution	Single or few species
Statistical techniques	Prioritize survey locations; Generate watch lists; relies on field data and legacy data	Use field data to predict spread rates, distribution, abundance, etc.	Generate watch lists, predict habitat suitability, or potential abundance	Relies heavily on field data, but can be large extent and high resolution	Single to multispecies dependent on technique
Remote sensing	Can detect infestations of known species to a certain resolution; requires some field data	Temporal analyses of imagery can determine spread rates	Current distributions	Trade off between extent and resolution, can be cost prohibitive	Intensive for single species; may not work for some species; plant centric

provide insight into the requirements for the establishment and subsequent spread of invasive species that can be applied to current invasions. Perhaps by examining these types of datasets, we may be able to answer such questions as how species initially establish and what conditions must occur for them to be able to rapidly expand their populations.

Even without specific quantifiers for rates of spread, these data can be useful to examine the invasion time scale. Crayfish data from herbarium records and the literature for Wisconsin, USA, could be divided into three time bins: pre-invasion of the nonnative species, postinvasion (coexistence), and extant years when native populations had declined and the nonnative species was abundant (Olden et al. 2006). Analysis of this data revealed impacts on the crayfish community caused by the invading species, including changes in native species abundance and the structure of communities. Here, rusty crayfish have taken the role of most dominant species from the two of the native crayfish fauna. These impact data may inform management decisions in other locations where the invader establishes. It may also be possible to use this historic data to parameterize models to predict future invasions.

We extracted a time series county level dataset of invasive plants from the INADERS on-line database (Rice 2006) to use for temporal analyses. This dataset helped validate a simple forecasting model for use on species initially arriving in the US. We also used this dataset to examine patterns of invasion through time in the Pacific Northwest of the US (Stohlgren et al. 2008). These data showed that invasions were continuing to increase through time. Both examples using existing, coarse scale temporal data could be useful in informing management of issues related to invasions by nonnative species.

6.3.2 Field Data

For invasive plant species, weed mapping is a very common technique used by many resource management agencies. This technique involves a person in the field with either a map and writing implement or a GPS unit that can be used to capture coordinates for point locations or polygons of patches by physically traveling around an area searching for a single species or a small suite of species. These data, if captured through time with samples taken at the same locations, can be used to track changes in species presence or absence or changes in abundance.

However, ecologists often lack "absence" data, meaning records of areas searched for a particular species where it was not found. Knowing where a species was not found is important for several different reasons. From a temporal standpoint, we want to know when a species first arrives at a particular location, which we are unable to do if we do not know it was absent last year and is present this year. Without these data, spread rates are indeterminable. Additionally, many different statistical modeling techniques require absence locations along with presence locations. Characterizing unsuitable habitat for a species may be as important as knowing where suitable habitat exists.

We also lack monitoring data, where we return to the same locations to map species each year. This oversight makes it difficult to capture changes in species distributions and abundance through time. It has the same temporal implications as unrecorded absence data, where we do not know initial establishment because of lack of time series data. These issues could be solved by adding recording what locations were sampled to mapping protocols. For example, the North American Weed Mapping Association has developed a weed mapping standard that is widely used but only includes instructions for mapping locations where a species is present (North American Weed Management Assocation 2002). A suggestion to modify the standards to address the oversight of recording all sampled locations is to add a step to the protocol for data collectors to record a location every 10 min, regardless of whether the target species is present. If not present, the location should be recorded as an absence location. This addition would create a record of what areas have been surveyed for the species, providing a better temporal view of a species' distribution. Additionally, returning to at least a subset of the sampled present and absent locations each year would provide information on local spread, such as how quickly a particular patch of an invader is growing. Adding a "professional" layer of plot data to a larger set of weed mapping data as demonstrated by Barnett et al (2007) can also help with these issues.

Typically weed mapping is only focused on a single or a small subset of species. However, when people are out searching for species, they may find a previously unreported species that has established in the area of which managers were unaware. Although it is unlikely that these types of data collectors would distinguish a cryptic invader, they might notice a showy species or one that is suddenly abundant in the area. Additionally, they could be provided with a short list of species that have a high probability of invading the area based on distribution predictions. Data collected by weed mapping over time can then be used to forecast invasions.

6.3.3 Statistical Techniques

Using statistics to forecast the distribution of a particular species and its rate of spread involves developing a relationship between data collected in the field and other predictor variables available across the full extent of the area of interest such as climate data or satellite imagery. Statistical methods can be used with weed mapping data or with research plots that capture information on a suite of species rather than a single or a few individual species.

Gilbert et al. (2005) offer one of the few examples of a published modeling effort that actually predicts the spread of a species through time over a large area, and their products provide useful information for the control of the species. The researchers followed the spread of the horse chestnut leafminer (*Cameraria ohridella)* from its initial invasion from 2002 to 2004 in the United Kingdom and were able to use this information to develop models to predict the further spread of the species in the next four subsequent years (2005, 2006, 2007, and 2008). So, they produced a map of the United Kingdom showing the distribution in each of the seven years (Fig. 6.1).

However, models like this one are not easy to develop. A major limitation of these statistical techniques is the data needed to parameterize them. More detailed predictions such as those including rates of spread require very specific data. Modeling the spread of an organism that moves quickly like an insect or disease may be hampered by the inability to gather needed datasets or by not having the time to fully understand the organism's ecology and dynamics. In contrast, the challenge may be quite different when attempting to model rates of spread of a plant species that may take decades to move across a landscape. When looking at this

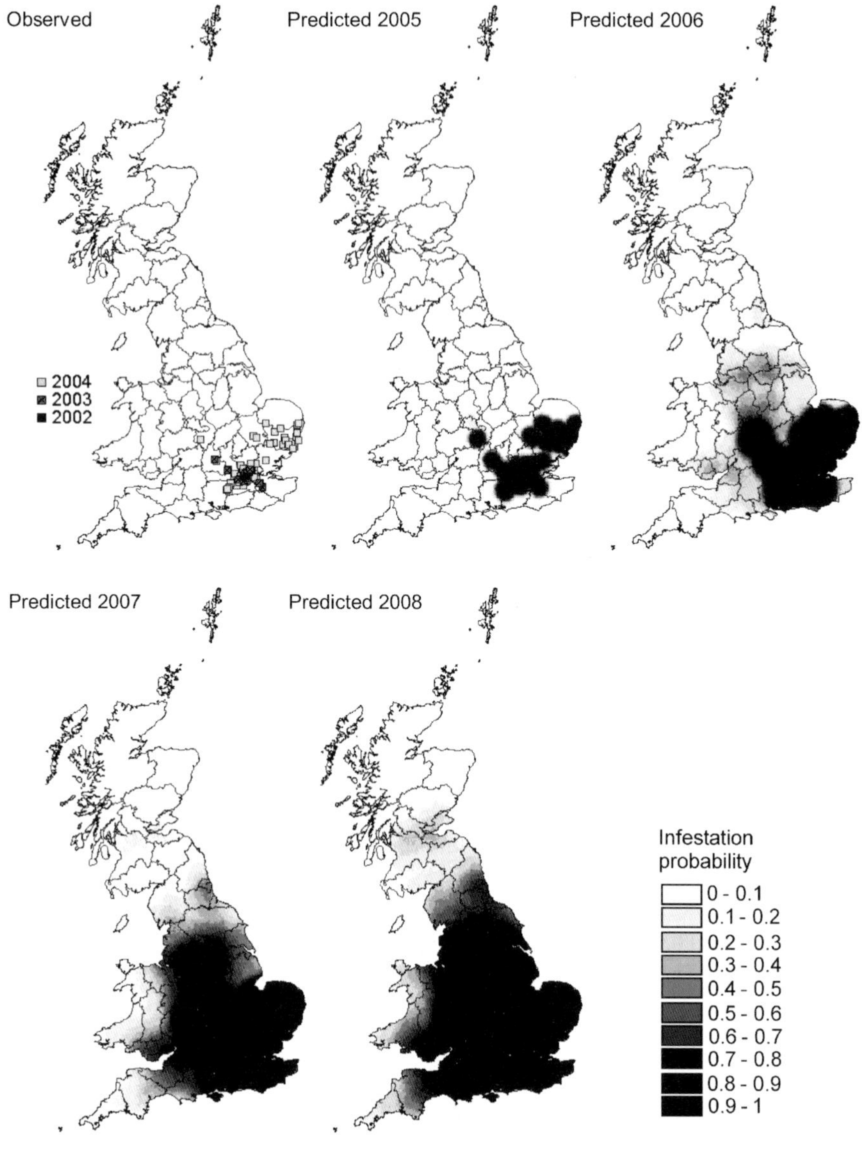

Fig. 6.1 Spread of the leaf miner in England, with the model calibrated from previous invasions in mainland Europe. This figure illustrates the type of modeling we desire for all invasive species. Figure used with permission from Gilbert et al. (2005)

longer period of time, the possibility of additional introduction sites, stochastic events that affect spread rates like a year with more wind storms than average or a drought, and successful local control or eradication efforts may render the models invalid.

With many species it is difficult to detect a very small initial infestation in an area the size of the United Kingdom and even more difficult when a larger area is viewed. Additionally, for a taxonomic group like plants, a small initial infestation like that described for the insect would be more easily eradicated, making the temporal spread model unnecessary. This type of modeling effort is also quite data intensive.

The leafminer model was possible because detailed temporal data from other invaded countries existed that could be used to calibrate a spread model, and detailed data for the initial infestation in the new location was also available. Even the data collection methods relied heavily on prior knowledge of the species. Generally, these data would not be available for a species unless the species was a problem elsewhere or if it was a well-studied species in its native range, relegating the usefulness of available statistical techniques to a limited number of species with readily available data. In a paper describing a spatially explicit population dynamics models for the spread of the grey squirrel (*Sciurus carolinensis*) in a region of Italy, the authors acknowledge they were only successful because detailed data on population parameters for the squirrel were already available from previous research (Tattoni et al. 2006). Using these models, the authors were able to provide guidance to management activities by applying different control and removal scenarios. Again, while these models may be very useful to managers, this type of detailed population dynamic information will not be readily available for most invasive species.

The above examples rely heavily on knowledge of the biology of the species. We recently developed a very simple early warning method for the spread of plant species in United States counties. One of the goals of the method was to develop a technique that could generally be applied to invasive species as soon as they move into an area, without requiring time to be spent collecting detailed information specific to that species. There are general climatic factors that affect the distribution of all species, such as temperature and precipitation. However, these models do not include a temporal component. They predict the potential distribution of a species, but do not attempt to predict when a species will actually reach its maximum distribution. A combination of these types of methods may be necessary depending on the availability of species specific data.

6.3.4 Remote Sensing

Another method to capture the spread of a species involves remotely-sensed imagery. Here, the spectral signature of a species is captured at different locations in the field and then, using one of several different algorithms, the signature is used to try to identify the species in the image across its extent. Images containing spectral information from the reflectance of sunlight can be captured at the same location through time, and the spread of species can then be determined by examining sequential images from the same location. However, this methodology only pertains to showy, open-grown plant species. Cryptic, rare, or understory species are more difficult to monitor with remote sensing. The usage of these techniques is very species specific, and image acquisition must be timed with phenological characteristics of the target species in relation to surrounding vegetation so that it can be distinguished from other species. For example, detection of spotted knapweed (*Centaurea maculosa*) and babysbreath (*Gypsophila paniculata*) in Idaho require the understory vegetation to be fully matured and bleached, and spotted knapweed must retain some of a previous year's stems for detection Lass et al. (2005).

Sensor technology and detection algorithms are improving, but this technique is still limited to certain species. Pure pixels of the invasive species, meaning an area on the ground corresponding to a grid cell in the image that is completely covered in the species, are often required as training data to classify it in an image. For many species it may be difficult to find pixels that are unmixed (e.g., some spotted knapweed and some sagebrush rather than only spotted knapweed). It is much more difficult to detect understory species in a dense forest or submerged aquatic vegetation than a species on the prairie. For example, Underwood et al. (2006) mapped two aquatic species, Brazilian waterweed (*Egeria densa*), which is a submerged aquatic species, and water hyacinth (*Eichhornia crassipes*), an emergent, floating species at two different spatial extents. At the local scale (average size 51 ha), average accuracy was highest for Brazilian pepperweed (93% compared with 73%), but dropped drastically when the extent was increased to 177,000 ha (29%) while the accuracy for water hyacinth decreased only slightly (65%). The cost of developing these classifications for each individual species renders developing a different model for every 51 ha unreasonable in most cases.

As with statistical techniques, there are several different algorithms that can be used to classify a remotely-sensed image. Several different papers compare methods and promote some as better than others (Elith et al. 2006; Higgins and Richardson 1996; Hirzel et al. 2001), but as with statistical techniques, different methodologies probably work better for different species in different localities.

How small can an infestation be and still be detected? We do have time series data, but are we able to tell when a species first appears in an area, or does it have to reach a threshold level of abundance? If it does, is that level of infestation large enough so that eradication at the new location would be impossible or incredibly expensive? These questions still await answers. While it is true that covering large area such as a state or province for field data collection over multiple years would be difficult, remotely-sensed imagery cannot supply all the necessary information to answer questions for management. For example, it would be difficult to use products where cheatgrass, a small annual grass, was modeled at 30–60 m resolution (Bradley and Mustard 2006) for early detection of new infestations of cheatgrass in central Nevada at a local level.

Detecting species at low levels of abundance even over small areas with high resolution data is still difficult. Mundt et al. (2005) classified hoary cress (*Cardaria draba*) in 3 m spatial resolution hyperspectral images for an area 1.75 by 22 km in southwestern Idaho, evaluating their classified surface using Incremental Cover Evaluation. With this method, accuracy was calculated for different levels of cover (e.g., greater than 10%) to determine what percent cover of hoary cress was required for it to be accurately detected. They were able to detect it with 0–10% cover, but only correctly identified 5 of 19 locations. Thirty percent cover was required for the accuracy to be useful to management applications (82% accuracy). Similarly, Glenn et al. (2005) detected small infestations of leafy spurge (*Eupohorbia esula*) in the Swan Valley in Idaho using 3.5-m resolution hyperspectral data, with the species detectable at 10% cover but for repeatability the discrimination threshold was around 40% cover. So, even in this relatively small area, detecting new infestations while they are small is still problematic. The costs associated with high resolution data over areas greater than a few hectares currently makes it prohibitive to use remote sensing for early detection (Shaw 2005), even if it could be used to detect species when they have very low cover.

6.3.5 Statistics with Remote Sensing

Several statistical methods can incorporate remotely-sensed imagery layers or derived layers like Normalized Difference Vegetation Index (NDVI) as parameters in models. These types of models could be used to determine areas most likely to be invaded by a particular species. These models could then direct field surveys to areas with a high probability of invasion for early detection efforts or to areas with high levels of uncertainty.

Remotely-sensed imagery products may also capture temporal variability important in predicting invasions. For example, NASA researchers created three derived products from MODIS satellite data that captured annual variability in NDVI, including the range in greenness throughout a year, the mean NDVI value, and the average date of green up for a pixel (Morisette et al. 2006). These products improved models for habitat suitability.

6.3.6 The Future

None of the techniques we found addressed all of the considerations we discussed in the previous section (Table 6.2). Each technique has it pros and cons, and may be useful to answer different questions. Many of the techniques are time and labor intensive, and often require extensive previous knowledge of a species. We need to become better at making this information easily available to other researchers and at making our models more sophisticated, incorporating considerations of time since invasion, invasion stage, issues related to scale, and other important factors.

Table 6.2 Essential factors, in addition to species-specific characteristics, to consider when developing any forecast for the spread of invasive species.

Type	Description
Invasion stage	The stage in the invasion process – initial introduction, establishment, spread, naturalized; lag effects are important in relation to the stage
Residence time	Time since initial introduction, which interacts with other factors to influence spread through lag times
Propagule pressure	Number of individuals originally introduced and the rate of subsequent introductions to a particular location
Vectors or pathways and barriers to invasion	Mechanisms of dispersal for spread or barriers that prevent spread; mechanisms responsible for moving the species long distances are especially important
Environmental stochasticity	Differences in the environment through time can alter spread rates (a component of lag times)
Long-distance dispersal	Dispersal events driving the range expansion of a species
Allee effects	A relationship between population growth and density of individuals that is species specific and can influence rate of spread (a component of lag times)

All of these factors are rarely if ever incorporated into forecasting models, but are necessary components for a well-developed forecast

6.4 Managing Invasive Species Spread in Long Term

6.4.1 Lessons for Control Prioritization

Forecasts over time could potentially help make control and eradication efforts more cost effective by informing prioritization of management efforts based on potential impact of various locations. Spatial considerations such as distance from source populations are important in determining where control efforts should be carried out. At a local scale, forecasts could be used to prioritize patches of invasion on the basis of predicted patch size and therefore impact (Fig. 6.2). For example, there are two small patches of an invasive species in a management unit, only one of which can be affordably eradicated. A local scale spread model for the species predicts that one will persist at very low levels and not spread to the surrounding area. The other is predicted to become very dense and act as a source for other populations. The second patch therefore would be the one on which to focus control efforts. Another consideration is the original source of an infestation. If new propagules are continually arriving, control efforts would be more effective if the source was also controlled. Without the input from temporal predictions, the resource manager would have to randomly choose or divide resources between locations, potentially not being able to effectively control the spread of the species in their area as well as they would with temporal predictions.

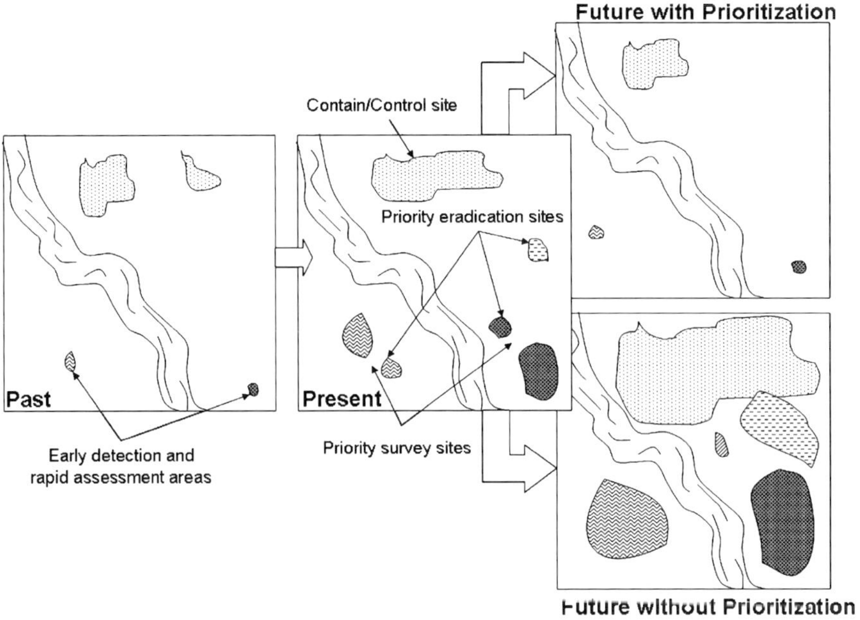

Fig. 6.2 Prioritization of control is required for efforts to be successful through time. For example, if the river and wind move top to bottom in the diagram, the patches at the top will act as sources to the bottom. Without control efforts at that source location, any efforts to control the small infestations would fail

The same concept could be used to prioritize national control and eradication efforts for particular species. Those species predicted to rapidly spread across a country, a state, or a county require a more rapid response than those whose spread may be on the order of decades.

Additionally, different conclusions could be drawn about a species with only a single point in time observation, highlighting the need for long-term monitoring. The impacts of an invader may change through time, and this will only be captured with sampling at multiple points along the invasion time scale. For example, studies of the red imported fire ant (*Solenopsis invicta*) at two different points of time in the invasion phase in the same location in central Texas, USA revealed very different results (Morrison 2002). Sampling first occurred in the 1980s during the initial phase of invasion. As the ant species spread, it became dominant and its presence was correlated with decreases in native ant populations. However, when the same location was sampled 12 years later, native ant populations had returned to preinvasion levels. The fire ant was still the most abundant ant species, but the relative abundance of the species was drastically different from 12 years before. Examining either the initial study or the later study singly, very different conclusions would be drawn. Another example is of zebra mussels (*Dreissena polymorpha*) in the Hudson River Estuary, USA. Native bivalve populations decreased by 60–100% over the

5-year period following the initial invasion of zebra mussels in 1991 (Strayer and Malcom 2007). However, monitoring of the four most common native bivalves since 2000 showed stabilization or recovery of their populations, despite the enduring presence of zebra mussels. Continued monitoring through the stages of invasion like the examples above is necessary to determine what long-term impact invaders may have on native populations as single point in time observations are not accurate to determine the long-term trajectory of ecological communities. These conclusions about long-term impacts are useful in prioritizing control efforts to focus on species that may have greater long-term effects. A species that impacts native populations throughout the entire invasion process may have a higher priority for early detection and removal in new areas than ones whose impacts decrease over time.

6.4.2 Long-Term Viewpoint

Once control efforts have been carried out, it is necessary to monitor the location and effects. If control of a particular species occurs, and the organisms that occupy the space after control are different invasive species rather than native vegetation, the question must be asked whether the control efforts were effective. Understanding what state control efforts are trying to restore a location to is important. This temporal view can aid in setting priorities for resources to areas where control efforts will have the desired effect through time rather than only over the short term, like a single growing season.

Another temporal aspect of control prioritization relates to the spatial configuration of infestations. The frequency and number of individuals introduced is highly correlated with establishment success (Colautti et al. 2006; D'Antonio et al. 2001; Kolar and Lodge 2001). It only follows, then, that this factor should be included in determining locations for control efforts. For example, if there is an adjacent population that could easily provide propagule supplies that is not controlled, then a temporal view of control efforts suggests that a different approach should be taken: controlling source populations before focusing on other populations, or controlling both at the same time to eradicate the smaller infestation while decreasing the propagule supply from the well-established population.

One common management practice is for government agencies to re-seed areas after fires. Managers reseeded three areas in the Grand Staircase Escalante National Monument in Utah burned in July 1996, July 1997, and August 1998 (which partly overlapped the 1996 burn) with native and nonnative seeds, two different native only seed mixes, and no seeding, respectively. Sampling both burned and adjacent unburned sites in 2000 revealed differences in response to reseeding at different locations (Evangelista et al. 2004). In one case, the treatment of native only seeding in a burned area resulted in greater similarity to undisturbed sites in relation to native and nonnative plant species richness and cover. A different location with this same treatment (native only seed and burned), however, was more similar to the untreated burned area and generally remained high in nonnative species richness.

It appeared that site factors such as soil nutrients and native plant species cover were the most important factors in the post-burn landscape. Thus, this example highlights the importance of knowing the site-specific characteristics where the treatments have succeeded and where they have failed (e.g., what works where).

Biological control programs like the one in Theodore Roosevelt National Park to control leafy spurge (*Euphorbia esula*) are another common management practice. The Park, like many other management groups, has also invested a lot of money and time into herbicide applications. Effectiveness of the two different control methods over a period of three years showed that the benefits of herbicide treatments were short term compared with those from biological control and that herbicide treatments appeared to have a negative effect on the biological control organisms (Larson et al. 2007). These data inform management strategies in the Park. All of these examples indicate the importance of long-term monitoring of management actions to determine the long-term success or failure to achieve desired results. The results of the long-term monitoring can then be used to guide future control efforts.

6.4.3 Ecosystem Changes

Ecosystems are not static, especially with current anthropogenic impacts. These impacts vary widely from those at the global to regional scale, ranging from large scale effects of climate change and nitrogen deposition to local scale impacts such as construction of a new road. A large scale effect can be predicted for smooth brome (*Bromus inermis*) across the central grasslands of North America (Vinton and Goergen 2006). Analysis of this species spread indicated a competitive advantage over native grass species due to its ability to efficiently recycle litter rich in nutrients. Thus, nitrogen deposition may play a critical role in the future for the persistence of this invasive species.

Predicting the role climate change may have on the potential distribution and abundance of invasive species is an important consideration. A newly introduced species that prospers under new climatic conditions may add additional hardships to native species trying to adapt and survive under the new climate regime. These types of models are difficult to produce because of the lack of accepted models for future climate. Creating models to predict future climate and for invasive species spread through time share problems, such as the inability to validate models – at least until time has passed and changes have occurred. Using a variety of climate change scenarios, models for the potential distribution of an invader could be created for current climate and for climate conditions at certain times in the future, like in 10 years, 20 years, 50 years, etc. Adding spread rates would only improve these models. Comparison of the times series could illustrate populations of the invader that are decreasing due to projected unsuitable climate in the future, stable populations due to projected suitable climate conditions currently and in the future, and increasing populations due to increased climatic suitability as climate changes. This information could be useful to guide early detection/rapid response activities to areas of projected increasing population and to prioritize areas for control to areas of projected stable populations.

6.5 Summary

In forecasting the spread of invasive species, there are many different tools available and many different factors to consider. A combined approach of the different techniques presented here along with a few improvements in the gathering of consistent data through time should provide managers with information they need to perform their jobs more effectively and efficiently. There are several different factors that could potentially influence the rate of spread of a particular species (Table 6.2). These factors should be important components of any models of the distribution or abundance of a particular species in space or time, and should be considered in the collection of field data. Much of this data is difficult to obtain, such as time since invasion, number and frequency of introductions, population factors (such as allee effects) and propagule supply. However, with increased awareness of the importance of these data and technological advances, we should be able to improve our modeling techniques to address at least some of these important factors.

Current techniques are also very species focused, and often all relate to common, dominant invasive species, not necessarily on what might be the next big invader that will be important to detect early in its invasion. We need to become better at applying the techniques we develop to different species, and at focusing on species that may not yet be an extensive problem, but will be in the future if we do not do anything to control the spread.

References

Barnett, DT, TJ Stohlgren, CS Jarnevich et al (2007) The art and science of weed mapping. Environ Monit Assess 132: 235–252.

Barney, JN (2006) North American history of two invasive plant species: phytogeographic distribution, dispersal vectors, and multiple introductions. Biol Invasions 8: 703–717.

Bradley, BA, JF Mustard (2006) Characterizing the landscape dynamics of an invasive plant and risk of invasion using remote sensing. Ecol Appl 16: 1132–1147.

Brotherson, JD, D Field (1987) Tamarix: Impacts of a successful weed. Rangelands 9:110–112.

Colautti, RI, IA Grigorovich, HJ MacIsaac (2006) Propagule pressure: A null model for biological invasions. Biol Invasions 8: 1023–1037.

Crall, AW, LA Meyerson, TJ Stohlgren, et al (2006) Show me the numbers: what data currently exist for non-native species in the USA? Front Ecol Environ 4: 414–418.

D'Antonio, CM, JM Levine, M Thomsen (2001) Ecosystem resistance to invasion and the role of propagule supply: A California perspective. Journal of Mediterranean Ecology 2: 233–245.

Davis, HG, CM Taylor, JC Civille et al (2004) An Allee effect at the front of a plant invasion: Spartina in a Pacific estuary. J Ecol 92: 321–327.

Elith, J, CH Graham, RP Anderson et al (2006) Novel methods improve prediction of species' distributions from occurrence data. Ecography 29: 129–151.
Ellstrand, N, K Schierenbeck (2000) Hybridization as a stimulus for the evolution of invasiveness in plants? Proc Natl Acad Sci USA 97: 7043–7050.
Ellstrand, NC, KA Schierenbeck (2006) Hybridization as a stimulus for the evolution of invasiveness in plants? Euphytica 148: 35–46.
Evangelista, P, TJ Stohlgren, D Guenther et al (2004) Vegetation response to fire and postburn seeding treatments in juniper woodlands of the Grand Staircase-Escalante National Monument, Utah. West N Am Nat 64: 293–305.
Evans, AM, TG Gregoire (2007) A geographically variable model of hemlock woolly adelgid spread. Biol Invasions 9: 369–382.
Gilbert, M, S Guichard, J Freise et al (2005) Forecasting *Cameraria ohridella* invasion dynamics in recently invaded countries: from validation to prediction. J Appl Ecol 42: 805–813.
Glenn, NF, JT Mundt, KT Weber et al (2005) Hyperspectral data processing for repeat detection of small infestations of leafy spurge. Remote Sens Environ 95: 399–412.
Hastings, A, K Cuddington, KF Davies et al (2005) The spatial spread of invasions: new developments in theory and evidence. Ecol Lett 8: 91–101.
Higgins, SI, DM Richardson (1996) A review of models of alien plant spread. Ecol Model 87: 249–265.
Hirzel, AH, V Helfer, F Metral (2001) Assessing habitat-suitability models with a virtual species. Ecol Model 145: 111–121.
Hobbs, RJ, SE Humphries (1995) An integrated approach to the ecology and management of plant invasions. Conserv Biol 9: 761–770.
Jarnevich, CS, TJ Stohlgren, D Barnett et al (2006) Filling in the gaps: modelling native species richness and invasions using spatially incomplete data. Divers Distrib 12: 511–520.
Kolar, CS, DM Lodge (2001) Progress in invasion biology: predicting invaders. Trends Ecol Evol 16: 199–204.
Larson, DL, JB Grace, PA Rabie et al (2007) Short-term disruption of a leafy spurge (Euphorbia esula) biocontrol program following herbicide application. Biol Control 40: 1–8.
Lass, LW, TS Prather, NF Glenn et al (2005) A review of remote sensing of invasive weeds and example of the early detection of spotted knapweed (Centaurea maculosa) and babysbreath (Gypsophila paniculata) with a hyperspectral sensor. Weed Sci 53: 242–251.
Leung, B, JM Bossenbroek, DM Lodge (2006) Boats, pathways, and aquatic biological invasions: Estimating dispersal potential with gravity models. Biol Invasions 8: 241–254.
Liebhold, AM, PC Tobin (2006) Growth of newly established alien populations: comparison of North American gypsy moth colonies with invasion theory. Popul Ecol 48: 253–262.
Lockwood, JL, P Cassey, T Blackburn (2005) The role of propagule pressure in explaining species invasions. Trends Ecol Evol 20: 223–228.
Lodge, D, S Williams, HJ MacIsaac, et al (2006) Biological invasions: recommendations for U.S. policy and management. Ecol Appl 16: 2035–2054.
Mack, RN (1981) Invasion of *Bromus tectorum L* into Western North America - and ecological chronical. Agro-Ecosystems 7: 145–165.
Morisette, JT, CS Jarnevich, A Ullah et al (2006) A tamarisk habitat suitability map for the continental United States. Front Ecol Environ 4: 11–17.
Morrison, LW (2002) Long-term impacts of an arthropod-community invasion by the imported fire ant, Solenopsis invicta. Ecology 83: 2337–2345.
Mundt, JT, NF Glenn, KT Weber et al (2005) Discrimination of hoary cress and determination of its detection limits via hyperspectral image processing and accuracy assessment techniques. Remote Sens Environ 96: 509–517.
Myers, JA, M Vellend, S Gardescu et al (2004) Seed dispersal by white-tailed deer: implications for long-distance dispersal, invasion, and migration of plants in eastern North America. Oecologia 139: 35–44.

Nehrbass, N, E Winkler, J Mullerova et al (2007) A simulation model of plant invasion: long-distance dispersal determines the pattern of spread. Biol Invasions 9: 383–395.

Neubert, MG, M Kot, MA Lewis (2000) Invasion speeds in fluctuating environments (vol B267, pg 1603, 2000). Proceedings of the Royal Society of London Series B-Biological Sciences 267: 2568–2569.

North American Weed Management Assocation (2002) North American invasive plant mapping standards. http://www.nawma.org.

Olden, JD, JM McCarthy, JT Maxted et al (2006) The rapid spread of rusty crayfish (Orconectes rusticus) with observations on native crayfish declines in Wisconsin (USA) over the past 130 years. Biol Invasions 8: 1621–1628.

Rejmanek, M, MJ Pitcairn (2002) When is eradication of exotic pest plants a realistic goal? Pages 249–253 In: C. R. Veitch and M. N. Clout (ed). Turning the tide: the eradication of invasive species. IUCN SSC Invasive Species Specialist Group, IUCN, Gland, Switzerland and Cambridge, UK.

Rice, PM (2006) *INVADERS Database System (http://invader.dbs.umt.edu)*. Division of Biological Sciences, University of Montana, Missoula, MT 59812–4824.

Sakai, A, F Allendorf, J Holt et al (2001) The population biology of invasive species. Annu Rev Ecol Syst 32: 305–332.

Shaw, DR (2005) Translation of remote sensing data into weed management decisions. Weed Sci 53: 264–273.

Stephens, PA, WJ Sutherland, RP Freckleton (1999) What is the Allee effect? Oikos 87: 185–190.

Stohlgren, T, DT Barnett, C Flather et al (2006) Species richness and patterns of invasion in plants, birds, and fishes in the United States. Biol Invasions 8: 427–447.

Stohlgren, TJ, DT Barnett, CS Jarnevich et al (2008) The myth of plant species saturation. Ecology Letters In Press.

Stohlgren, TJ, DT Barnett, JT Kartesz (2003) The rich get richer: patterns of plant invasions in the United States. Front Ecol Environ 1: 11–14.

Strayer, DL, VT Eviner, JM Jeschke et al (2006) Understanding the long-term effects of species invasions. Trends Ecol Evol 21: 645–651.

Strayer, DL, HM Malcom (2007) Effects of zebra mussels (Dreissena polymorpha) on native bivalves: the beginning of the end or the end of the beginning? J N Am Benthol Soc 26: 111–122.

Tattoni, C, DG Preatoni, PWW Lurz et al (2006) Modelling the expansion of a grey squirrel population: Implications for squirrel control. Biol Invasions 8: 1605–1619.

Taylor, CM, HG Davis, JC Civille et al (2004) Consequences of an Allee effect in the invasion of a pacific estuary by Spartina alterniflora. Ecology 85: 3254–3266.

Underwood, EC, MJ Mulitsch, JA Greenberg et al (2006) Mapping invasive aquatic vegetation in the Sacramento-San Joaquin Delta using hyperspectral imagery. Environ Monit Assess 121: 47–64.

Vinton, MA, EM Goergen (2006) Plant-soil feedbacks contribute to the persistence of Bromus inermis in tallgrass prairie. Ecosystems 9: 967–976.

Wilson, JRU, DM Richardson, M Rouget et al (2007) Residence time and potential range: crucial considerations in modelling plant invasions. Divers Distrib 13: 11–22.

Chapter 7
Applying Ecological Concepts to the Management of Widespread Grass Invasions

Carla M. D'Antonio, Jeanne C. Chambers, Rhonda Loh, and J. Tim Tunison

Abstract The management of plant invasions has typically focused on the removal of invading populations or control of existing widespread species to unspecified but lower levels. Invasive plant management typically has not involved active restoration of background vegetation to reduce the likelihood of invader reestablishment. Here, we argue that land managers could benefit from the ecological principles of *biotic resistance* and *ecological resilience* in their efforts to control invading plants and restore native species. We discuss two similar but contrasting case studies of grass invasion that demonstrate how these principles can be applied to control and management. In seasonally dry Hawaiian woodlands, management of invasive fire-promoting grasses has focused on seeding native species that are resilient to fire disturbance and can coexist with grasses. Resistance to grass invasions appears to be weak in unburned native habitats. Thus, the focus of management efforts has been to increase resilience of the native vegetation to inevitable disturbance. We contrast this with the Great Basin of the western USA where the annual Mediterranean grass, *Bromus tectorum*, also has promoted an increase in fire frequency in shrublands and woodlands. Here, a three-tiered approach has been employed in which preventative management in the form of fire or fire surrogates is used in the initial stages of invasion to increase the resilience and resistance of the native herbaceous vegetation. In transitional stages where *B. tectorum* is well established but not dominant, mechanical or herbicide treatments are used to open up dense and senescing shrub canopies, thereby increasing vigor of native perennial herbaceous species through competitive release. The released competitors (perennial grasses) are then assumed to provide resistance to *B. tectorum* invasion. Following complete *B. tectorum* dominance, the focus of management is intensive seeding of native species to create resistant plant communities that reduce the likelihood of reinvasion.

Keywords Biotic resistance • Cheatgrass; Ecological resilience • Ecological restoration • Exotic grasses • Fire • Grass/fire cycle • Great Basin • Hawaii • State and transition models

---

C.M. D'Antonio (✉), J.C. Chambers, R. Loh, and J.T. Tunison
Environmental Studies Program, University of California, Santa Barbara, CA 93106, USA
dantonio@es.ucsb.edu

7.1 Introduction

While the control of some plant invasions may be as simple as removing a small founding population, more often control is part of a larger vision for the ecological management and eventual restoration of a natural area (D'Antonio and Meyerson 2002; D'Antonio and Chambers 2006). Ideally, attributes of a "fully restored ecosystem" as described by the International Society for Ecological Restoration (http://www.ser.org/) are applicable to the management of widespread invasions in the context of ecosystem restoration rather than simple species removals. One of these attributes states that the restored ecosystem should be "self-sustaining and able to persist under existing environmental conditions." Using this as a guiding principle, long-term control of nonnative species should involve the creation of systems that are resilient to future disturbance and resistant to reinvasion. Thus, two ecological concepts that are particularly critical to long-term management of plant invasions are *biotic resistance* and *ecological resilience*.

In this chapter we explore and contrast two case studies in which the concepts of resistance and resilience play different roles in the management of specific plant invasions. We focus on two examples of invasion by fire-promoting grasses, one in Hawai'i and the other in the Great Basin region of the Western USA. These two systems were chosen because together they provide a range of insights into how the ecological principles of resistance and resilience can be applied to the management of persistent or recalcitrant plant invaders. We focus on grass invaders because they are widespread and difficult to control and because they often dramatically alter ecosystem structure and functioning. Thus, efforts to manage them or the communities of which they are now a part are critical to maintaining native biodiversity and ecosystem functioning in some regions.

7.1.1 Ecological Resistance and Resilience

Since the rise in interest in the field of biological invasions, community ecologists have explored the role of ecological resistance as an ecosystem property influencing invasion success (e.g., Rejmanek 1989; D'Antonio 1993; Maron and Vila 2001; Levine et al. 2004). Despite the extensive recent history of research in this area (reviewed by Levine et al. (2004) and Bruno et al. (2005)), land managers have been slow to embrace the concept into their approaches to *weed* control. Instead, weed invasions in natural areas have been controlled largely by chemical and mechanical means, or through prescribed burning and grazing. Such approaches do not explicitly engage ecological principles but focus instead on a *top-down* approach where the manager is acting as a predator removing the species from the system (McEvoy and Coombs 1999). This approach has been successful at eradication of invaders in many discrete isolated areas. Typically, however, little attention is given to whether the post-removal community has higher resistance to further weed invasion and increased resilience to future disturbance.

Resistance in an ecological sense refers to the ability of a community to withstand encroachment by nonnative species. An overriding conclusion from the literature on ecological resistance is that resistance is a probabilistic phenomenon (D'Antonio et al. 2001b), varying across community types for a given invader species (e.g., D'Antonio 1993) and among invader species within the same community type. Likewise, the relative importance of abiotic vs. biotic mechanisms of resistance varies across landscapes and among years within the same landscape. Thus, some communities are inherently more susceptible to invasion than others while communities on the resistant end of the spectrum can be highly susceptible to invasion after disturbance or a stress that opens up space in the community (Davis et al. 2000). Despite the variation in resistance over space and time, experimental studies have demonstrated that background vegetation plays an important role in suppressing reinvasion of target weeds after initial top–down control (McEvoy and Coombs 1999). D'Antonio and Thomsen (2004) have argued that ecological resistance should be a more important part of invasive plant management than it is currently.

In contrast to ecological resistance, *resilience* describes the ability of a community to return toward its predisturbance, and presumably preinvasion, state after a disturbance. It is a restoration goal because high resilience in a restored site means less hands-on input is required by managers to keep vegetation within target bounds. With relevance to invasive species, high resilience of native or desired species will potentially reduce the need for immediate (postdisturbance) and future control of invaders and for reseeding or replanting of native species after disturbances. Hence, high resilience increases the likelihood of resistance to invasion, and both should be considered as management goals.

7.1.2 *Fire-Enhancing Grass Invasions*

The invasion of natural areas by introduced grasses is a widespread phenomenon (e.g., D'Antonio and Vitousek 1992), and grasses are disproportionately represented on virtually all published lists of high-impact natural area invaders (Daehler 1998). Grasses can be difficult to control and manage for several reasons: (1) They frequently have excellent mechanisms of resilience reestablishing from buds, roots, or seedbanks after disturbance. (2) Many disperse readily across the landscape either through effective passive dispersal or through attachment to animals. (3) They are not easily controlled through classical biological control. (4) They can transform ecosystems through their interactions with fire regimes (D'Antonio and Vitousek 1992; D'Antonio 2000; Brooks et al. 2004). (5) They often play conflicting roles in landscapes because within the same region they can provide forage for livestock while also promoting fire and/or loss of native species. As a result consensus for control is difficult to achieve.

In the Hawaiian Islands, perennial grasses from several other continents became widely established by the mid 1900s (Smith 1985). Within the dry and mesic parts of the islands they have been found to enhance the occurrence of fire (Mueller-Dombois 1981; Tunison et al. 2001). This has occurred largely to the detriment of native species (Hughes et al. 1991), although there is some variation in the severity of fire impacts across environmental gradients (D'Antonio et al. 2000). Some of the perennial grasses are resilient to fire because of rapid resprouting from basal root crowns (Smith 1985; D'Antonio et al. 2001a). Others, however, are killed by fire but regenerate rapidly from seed (Tunison et al. 1994, 1995; D'Antonio et al. 2000, 2001a).

Nonnative, fire-enhancing grasses have also invaded xeric and semiarid portions of the mainland USA with large-scale ecological consequences. The most widespread of these is *Bromus tectorum* or cheatgrass, an annual species from the eastern Mediterranean region. *B. tectorum* was widespread throughout the intermountain western USA by the 1930s (Mack 1986) and was associated with widespread increases in fire frequency and size and declines in native species a few decades later (e.g., Whisenant 1990). This species increases fire frequency and size by increasing the homogeneity or horizontal continuity of fuels and the rate of fire spread across what is otherwise a patchy shrub-steppe ecosystem that experiences summer dry lightning (Link et al. 2006). Because of rapid spring growth and early maturation, *B. tectorum* plants typically produce seeds prior to the fire season. Populations recover rapidly after fire via dormant seeds not killed by fire and very high seed output due to increased resource availability in the years following fire. Many of the native shrub species in these ecosystems are killed by fire, and in areas with depleted herbaceous understories, *B. tectorum* can rapidly dominate the ecosystem (e.g., Whisenant 1990).

7.1.3 Chapter Overview

In this chapter, we use two case studies to emphasize a way to incorporate ecological concepts into the management of persistent plant invaders. The example of grass invasions in Hawaii that we describe is a case study of a situation in which ecological resistance plays little role in the planning and implementation of control and revegetation of grass-invaded, fire-prone ecosystems. In Hawaii Volcanoes National Park's drier lowlands and mid-elevation habitats, exotic grasses have invaded virtually every place they could have invaded, and degradation and invader seed sources are widespread. Management is focusing on creating resilient native plant assemblages that can coexist with the invasive grasses particularly after fire has already occurred. Technically this is a restorative activity rather than ecosystem restoration (sensu Jordan 2003). Nonetheless, the goal is to create communities with some of the desired attributes of a restored ecosystem (http://www.ser.org/). By contrast with the Hawai'i example, grass invasions in the Great Basin of the USA are not complete and some potentially resilient native communities still exist. In sites that have not converted to complete *B. tectorum* dominance, management is focusing on maintaining native resilience while increasing resistance to invasion. In vast areas of the Great Basin, however, native communities have been replaced with *B. tectorum* grasslands. In these areas management is focusing on restoration of native communities that will be resistant to further invasion by the introduced grasses.

7.2 Case Study I: Hawaiian Dry and Submontane Seasonal Environments

7.2.1 *Study System*

The Hawaiian Islands are characterized by large environmental gradients driven by the prevailing trade winds and the volcanic shield topography. Dry forests and shrublands exist on the leeward side of the high islands while wet forests occur on the windward sides of the high islands. In contrast to the strongly varying microclimates, soil chemistry across the islands is relatively constant with all soils ultimately deriving from basaltic lava or ash. The soils are typically nitrogen limited when young, colimited by nitrogen (N) and phosphorus (P) at intermediate ages (10,000–100,000 years), and P limited on the older surfaces (Vitousek 2004).

Grasses from other regions of the world have invaded virtually all of the many microclimates in Hawaii from dry coastal terraces receiving <20-cm precipitation to sites with >3 m of rainfall annually. A detailed list of grass invaders is not available but most species are listed in the flora of Hawaii (Wagner et al. 1999).

7.2.2 *Grass Invasion Impacts*

While introduced perennial grasses are widespread throughout the Hawaiian Islands, those that have invaded dry shrubland and open woodland habitats appear to be causing the most dramatic changes because they change the continuity and density of fuels (Mueller-Dombois 1981; Smith 1985; Smith and Tunison 1992). The most ecologically significant grass invaders on Hawai'i Island include species from both the new and old world. Three of these are (1) *Pennisetum setaceum* (fountain grass), a perennial bunchgrass from Africa, (2) *Melinis minutiflora* (molasses grass), a mat-forming grass from Africa, and (3) *Schizachyrium condensatum* (bushy beardgrass), a perennial bunchgrass from Central America. *P. setaceum* has invaded the leeward side of Hawaii Island from sea level up to 3,000 m. It competes with native vegetation (Cabin et al. 2000, 2002) and regrows after fire. In mid-elevation environments that have experienced fire, native species richness has declined due to the lack of ability of native species to compete with *P. setaceum* and their relatively slow regrowth after fire (Shaw et al. 1997). Although *P. setaceum* is an ecologically important invader and a large new project has been initiated to assess ways to reduce its impact (S. Cordell, Institute of Pacific Island Forestry, personal communications), we will not discuss that work here. We will instead focus on management of habitats invaded by *M. minutiflora* and *S. condensatum*.

M. minutiflora was introduced to the Hawaiian Islands as livestock forage (Parsons 1972). It invades mesic shrubland and open woodland ecosystems where it promotes an increase in fire occurrence and intensity (Tunison et al. 2001). Individual *M. minutiflora* are typically killed by fire, but the abundant seedbank is a source of resilience and high, rapid postfire seed production of first-year plants results in rapid recovery of *M. minutiflora* populations (D'Antonio et al. 2001a). *M. minutiflora* co-occurs with *S. condensatum* in many natural areas. This latter species is from Central and South America where it is not typically associated with fire-prone ecosystems. It readily invades submontane forests, competes with native vegetation before fire (D'Antonio et al. 1998), and resprouts rapidly after fire (D'Antonio et al. 2001b). Its relative, *Andropogon virginicus* (broomsedge), from the southeastern USA, also occurs across these environments and is also associated with increased fire frequency (Tunison et al. 2001).

Within Hawaii Volcanoes National Park, fires are associated primarily with volcanic activity and humans. Fire regimes began to change after the spread of *A. virginicus*, *S. condensatum*, and *M. minutiflora* within park boundaries in the 1960s. Tunison et al. (2001) documented that fire frequency has increased by threefold and fire size by sixtyfold since the establishment of these grasses. D'Antonio et al. (2000) analyzed impacts of grass-fueled fires on native species diversity and cover in 19 sites from the coastal lowlands to the upper submontane seasonal zone of the Park and found the strongest impacts in the submontane zone where the prefire dominants are intolerant of fire. These dominants include the native tree, *Metrosideros polymorpha*, and the native shrub, *Leptecophylla tameiameia* (formerly *Styphelia tameiameia*, Wagner et al. 1999). In the coastal lowlands *A. virginicus* and *S. condensatum*-fueled fires regenerate toward at least some native species although native diversity is reduced (Tunison et al. 1994). *M. minutiflora*-fueled fires in both the coastal lowlands and submontane seasonal zone greatly reduce native species.

By sampling burned vs. unburned forests across the same elevation and rainfall within the submontane zone, Friefelder et al. (1998) documented that the homogenous structure of the grass canopy in burned sites resulted in an approximately threefold increase in wind speeds than were found above the canopy of unburned forests (with grasses). This resulted in modeled fire spread rates that were 20 times higher than those in unburned forest with exotic grasses in the understory. Mack et al. (2001) and Mack and D'Antonio (2003a, b) documented extensive changes in productivity, microclimate, and nitrogen cycling in burned compared with unburned woodland. They found that by greatly altering species composition including the elimination of native woody species, fire created an ecosystem that has much lower primary production and is much leakier for nitrogen. For example, Mack et al. (2001) report that net primary production is reduced by 55% in burned sites compared with nearby unburned counterparts, but that annual net nitrogen mineralization rates are an order of magnitude higher in burned sites. The lack of primary production during periods of high soil N mineralization resulted in periods with high potential for loss of nitrogen through leaching or trace gas losses (Mack, unpublished), a result corroborated by lower recovery of an N15 tracer added to plots within each site (Mack et al. 2001).

In addition to these ecosystem changes, the elimination of native species due to fire changes the substrates upon which nitrogen fixation, and thus ecosystem N accretion, occurs. Native plant species with symbiotic nitrogen fixation are rare in these unburned woodlands. Instead nitrogen fixation occurs via asymbiotic fixation in association with the litter of native species (Ley and D'Antonio 1998; Mack et al. 2001). The primary substrates for fixation are leaf litter of the dominant unburned tree species *M. polymorpha* and the organic (O) layer of the soil in unburned sites. Because *M. polymorpha* is eliminated by fire and the O layer is greatly altered and no longer fixes N, the capacity of these systems to fix N decreases by two orders of magnitude. This could significantly alter ecosystem development since these sites are on young volcanic soils low in nitrogen (Mack and D'Antonio 2003a).

The almost complete replacement of native woody species with invasive grasses in the submontane seasonal zone should have a profound influence on wildlife composition. However, no studies have been done to document such impacts. Furthermore, native birds are rare in intact native forests below 1,200 m because of the prevalence of introduced avian malaria. Nonetheless, we have observed two species of native birds, the Amakihi (*Hemignathus virens*) and the Apapane (*Himatione sanguinea*) in our unburned sites. By contrast we have not observed any native bird species in the burned sites.

7.2.3 *Difficulty of Control and Management*

By the early 1970s, feral goats had extensively browsed the woody vegetation in the coastal lowlands and submontane woodlands of this region. After their removal in the 1970s, grasses invaded virtually all of the coastal lowland and submontane shrublands and forest with at least some soil above the lava bedrock (T. Tunison, personal observation). By the 1990s, roughly 80% of the submontane seasonal zone and lowland shrub/grasslands had experienced at least one grass-fueled fire (Tunison et al. 2001). The primary management strategy for the ecosystem was fire suppression achieved through restricting human access to sites during the dry season and immediate fire suppression when ignitions occurred. Chemical and mechanical control of grasses was only being done for *P. setaceum*, which is restricted in its distribution within the park. *M. minutiflora, S. condensatum*, and *A. virginicus* were considered too widespread for any sort of chemical or mechanical control. Use of grazing animals, none of which would be native to this environment, was not considered a viable management strategy.

Restoration of the prefire native community was not considered to be a viable management strategy because most of the prefire dominants are not resilient to fire. Surveys of sites that had burned once but not burned again for 20 years demonstrated that only one of the prefire native shrubs, *Dodonaea viscosa*, regenerates over time in these ecosystems (Hughes et al. 1991). A resurvey of these sites in October 2007, unburned since 1973, confirms the lack of recovery of native species (D'Antonio et al., unpublished). The prefire dominant tree, *M. polymorpha*, grows very slowly as a seedling and it along with *L. tameiameia*, the dominant shrub, cannot tolerate competition from the grasses (Hughes and Vitousek 1992). In addition, they would readily be killed by the next, inevitable fire. For these reasons, it was decided that any restoration efforts toward these species would be futile.

Surveys of burned sites in the coastal lowlands (Tunison et. al. 1994) and our own and other observations (Tunison et al. 1995) suggest that there are fire-tolerant species in the Hawaiian flora that can be found in these ecosystems. Their scarcity throughout these sites is considered in part to be a function of previously high goat browsing, which is known to have reduced many woody species (Mueller-Dombois and Spatz 1975; Loope and Scowcroft 1985). Only one of these native species, *D. viscosa*, was common in burned sites (Hughes et al. 1991; D'Antonio et al. 2000). However, along the roads leading to these sites, burned individuals of planted trees showed regeneration from seed and stumps (Tunison et al. 1995, and personal observation). Also, the occurrence of other native dry forest and shrubland species in fire-prone habitats higher on Mauna Loa and Mauna Kea volcanoes suggested that there are fire-tolerant species in the Hawaiian flora that could grow at these sites. We hypothesized that to persist successfully in these habitats, a native species would have to tolerate growing with the dense exotic grasses that form an almost continuous canopy within 2–3 years after fire, and they must have mechanisms of resilience to regenerate rapidly after the inevitable fire. The success of the introduced grasses in invading intact woodland (D'Antonio et al. 2001a) suggested that no matter what plant communities exist on the sites, they are likely to offer little resistance to grass invasion. Thus, we designed a revegetation program that focused on developing fire-resilient native species assemblages that can coexist with grasses and persist with the new disturbance regime. It was unlikely that resistance could be strong enough to significantly dampen grass invasion in the near term especially since *S. condensatum* regenerates rapidly from root crowns after fire and can reduce the growth of native species (D'Antonio et al. 1998).

Our approach to the management of these sites is appropriately termed ecosystem rehabilitation (Bradshaw 1997). Ecosystem restoration implies the return of the composition and functioning of a system to the predisturbance state (Jordan 2003). Efforts to manage plant invasions in these sites focus on the process of resilience to restore ecosystem functions to the sites, with little understanding of whether resilience was a feature of pre exotic-grass ecosystems. It is possible that we are restoring aspects of composition that occurred prior to the extensive goat grazing that occurred before botanical records of the area were kept, but we do not know this.

7.2.4 Current Approaches to Management

In January 1993, we conducted our first experimental burn within Hawaii Volcanoes National Park with a goal of reestablishing more native species by reburning an already burned area heavily dominated by *M. minutiflora* and *S. condensatum*. The site had burned originally in 1972 and had few native species (Hughes et al. 1991; D'Antonio, unpublished). We started by introducing seeds of two species that were fire-tolerant and perhaps fire-enhanced, *D. viscosa* and *Sophora chrysophylla*. The latter is a small, N-fixing tree known to support native bird species. We seeded plots both before fire and after fire to evaluate whether fire would stimulate or inhibit their germination. We also seeded adjacent plots of identical age (previous woodland) that we did not burn in order to evaluate whether restoration could be achieved in a previously burned, exotic-grass-dominated site without reburning it. We followed seedling emergence and survival over the next 4 years. This initial fire was a low-intensity fire consuming only 50% of the aboveground grass biomass (Mack et al. 2001). The grass layer regenerated very quickly both from seed (*M. minutiflora*) and from resprouting individuals (*S. condensatum*) greatly limiting the window of time for native species to establish. Nonetheless, we found that both *S. chrysophylla* and *D. viscosa* germinated and grew within the burned plots (Fig. 7.1). The former species also germinated and grew in the unburned plots but individuals did not reach as large of a size. In the burned plots *S. chrysophylla* individuals reached flowering age within 5 years. Fourteen years later some of these individuals were very large (Fig. 7.2) and had seedlings of their own species growing nearby, suggesting that further recruitment had occurred. *D. viscosa* established at a lower rate than *S. chrysophylla* perhaps because its seeds benefit from higher intensity fire events (Tunison et al. 1995). Despite this, *D. viscosa* establishment was higher in the burned than in the control plots (Fig. 7.1). Some individuals of this species persisted throughout the monitoring time and they too reached reproductive maturity and were still present after 14 years (Fig. 7.2).

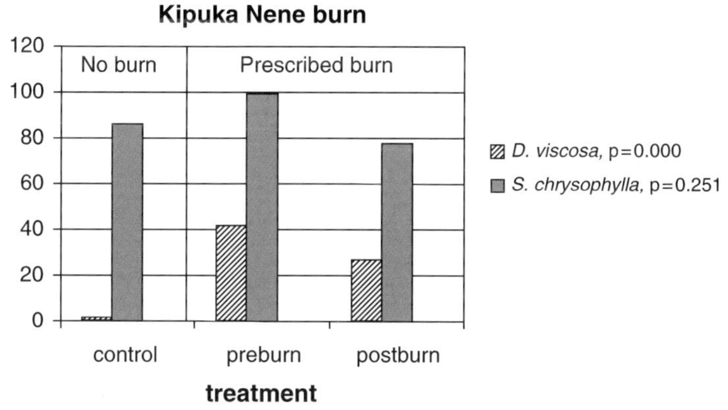

Fig. 7.1 Example of results of revegetation burns in Hawaii Volcanos National Park. Shown here are results of seeding in the first controlled burn conducted for revegetation of grass-invaded, burned seasonal submontane habitats conducted by Tunison, D'Antonio, and Loh in January 1993. Data are individuals of either *Dodonaea viscosa or Sophora chrysophylla* that were >10 cm in height after 4 years. N = 5 per treatment. Error bars not shown but P values represent significant treatment affects in a one-way ANOVA for each species. "Unburned" plots were grass dominated and were like the burned plots that had burned originally in 1973

Fig. 7.2 Example of a mature S. *chrysophylla* tree 14 years after being experimentally seeded into a grass-dominated burned Hawaiian submontane seasonal habitat. A fruiting individual of a second seeded species, *D. viscosa* (a shrub), is at the base immediately to the left of the tree. The matrix in which they are growing is the African grass *Melinis minutiflora* that has dominated these sites since invasion and the first fire in the early 1970s (photograph by C. D'Antonio)

Our next experimental burns occurred in July 1995 in two patches of vegetation burned previously in 1972 that were also dominated by *M. minutiflora, S. condensatum*, and *D. viscosa*. In contrast to the previous burn, these experimental burns were of very high intensity, consuming 98% of aboveground biomass and 20% of root and soil organic matter in the upper 20 cm (Mack et al. 2001). Plots were again seeded with *D. viscosa* and *S. chrysophylla* both before and after fire and in unburned control plots. Establishment was generally higher in postburn compared with preburn seeding and declined after initial germination due to periodic drought (Loh et al., unpublished report). Nonetheless, the successful germination and early growth of *D. viscosa* and *S. chrysophylla* in the burned plots compared with the unburned control plots suggested that high-intensity reburning of grass-dominated sites could promote native species establishment.

Over the next 5 years, the National Park Resources Management Division under direction of J.T. Tunison and R. Loh conducted five more experimental burns. Six additional species were planted in the submontane zone to evaluate regeneration and persistence with the ever-present invasive grasses; several of them showed potential for postfire revegetation and fire tolerance (Table 7.1). Because of the funding limitations the results have been monitored only sporadically but they suggest that several species of native plants are suitable for postfire rehabilitation and will persist under the exotic grass-dominated site conditions. In addition to these

Table 7.1 Hawaiian dry forest plants that were utilized in postfire revegetation prescribed burn experiments in Hawaii Volcanoes National Park, seasonal submontane woodland in and shrubland fires

Plant	Hawaiian name	Family	Life form
Bidens hawaiiensis	Ko'oko'olau	Asteraceae	Shrub
Dodonaea viscosa	A'ali'i	Sapotaceae	Shrub
Myoporum sandwicensis	Naio	Myporaceae	Small tree
Osteomeles anthyllidifolia	U'lei	Rosaceae	Shrub
Santalum paniulatum	Iliahi	Santalaceae	Tree, hemiparasite
Scaevola Kilauea	Naupaka	Goodeniaceae	Shrub
Sophora chrysophylla	Mamane	Fabaceae	Small tree
Sida fallax	Ilima	Malvaceae	Sub shrub

experimental burns, an accidental fire within the zone dominated by *S. condensatum* and *A. virginicus* occurred in 2000. Resources Management staff at Hawaii Volcanoes National Park used the approach recommended by this experimental burning program and actively seeded the burn with most of the suggested native species and other available as yet untested species. They also transplanted seedlings of species evaluated in the pilot burns (Loh et al. 2007). Approximately 900 acres were seeded and replanted with a total of 30 native species. Many of these had been studied in our experimental burns or in laboratory heating trials (Loh et al. 2007). Over 2.7 million seeds and 18,000 individual plants were placed into the burn. By 2004 eleven native species had reached reproductive maturity. These are still present as of October 2007.

7.2.5 Ongoing Challenges and Unanswered Questions

At least three of the native species tested showed a strong ability to establish after fire, and seeds of these species have been collected and stockpiled for postfire seeding when further wildfires occur. Several of the remaining species that responded at least somewhat positively to fire are harder to collect native seed from. Germination of three additional species tested only in heat trials in the lab was heat-stimulated (Loh et al., unpublished), suggesting that they could be useful, but all three are uncommon making it difficult to collect and store seed. Stockpiling of seed for future postfire seeding has therefore been limited both in species composition and amount. While postfire rehabilitation via native seeding is used in other portions of the western USA such as the Great Basin, the supply of native seed tends to be limiting and rarely can more than a small percent of burned areas be reseeded. Seeding rates are typically low and contribute to poor success. This is likely to be an enduring challenge in Hawaii Volcanoes National Park where seed sources for some species are sparse and technology for native seed production is poorly developed. Thus, although initial trials suggest that postfire seeding can lead to the successful establishment of native species, in reality restoration will be limited by availability of appropriate seed and will therefore likely be of limited species composition. Since 2006, R. Loh and staff have been establishing native plant seed *orchards* in various portions of the park to serve as a source of material for future restoration efforts. Their efforts will provide critical information on native plant propagation and possibilities for large-scale seed production.

Are the communities that are being created actually resilient to fire? This was a critical element of the initial argument for this approach to restoration/revegetation. Testing this, however, will require reburning of the *revegetated* sites. This has only been done for two small sites and it was observed that *S. chrysophylla* individuals regrew from root sprouts and *D. viscosa* individuals regenerated from the seedbank. We do not yet know how the other species would respond or over what range of fuel conditions populations of desired native species will show resilience.

On the leeward side of Hawaii, S. Cordell and others (Institute of Pacific Island Forestry, Hilo Hawaii, personal communications) are experimenting with *green stripping* as part of large-scale ecosystem rehabilitation in fire-prone shrublands and woodlands. Their goal is to prevent the spread of *P. setaceum* fueled wildfires into remnant patches of native dry forest that harbor rare species. Their approach focuses on finding species with fuel characteristics that will suppress grass growth and reduce fire spread rates. Although such an approach seems unfeasible in the vast stretches of grass-invaded submontane forests and shrublands in Hawaii Volcanoes National Park, it is being considered in areas where fire could spread from grass-invaded portions of the Park into a nearby residential subdivision. In addition, active fire suppression will help to slow the further loss of the prefire forests and shrublands. Special ecological areas with rare or unique species have been identified by Park personnel. Protecting these areas from fire is a high priority. This may be done through fire suppression, manual fuel reduction, or potentially green stripping if it can be effectively done.

Could the reestablishment of native species such as we have tried eventually lead to the establishment of some biotic resistance to exotic grass invasion? This is possible for *M. minutiflora* because its growth is sensitive to shading (D'Antonio et al. 2001a). However, *S. condensatum* and *A. virginicus* both tolerate a high level of shade so it would take a very dense shrub or tree canopy to reduce their invasion rate. Nonetheless, high densities of native shrubs could decrease the fine fuel biomass that accumulates with exotic grass invasion, potentially reducing fire intensity. Despite this, the long-term persistence of native species in these areas will rely more on creating a community of species resilient to fire than any ecological resistance that these communities might provide.

Are we restoring important ecosystem functions to these sites? Reestablishment of a heterogeneous plant canopy is a potential benefit of managing these grass invasions by rehabilitating these ecosystems with native woody species. As mentioned previously, Freifelder et al. (1998) showed that the homogenous exotic grass canopy promoted high wind speeds and therefore faster fire spread rates. The breakup of this homogenous canopy through the establishment of taller woody species could therefore help to slow the spread of wind-driven fires through the region. An additional benefit of rehabilitation with woody species is the possible use of these species by native birds. No research has yet been done on this question.

7.3 Case Study II: *B. tectorum* in the Sagebrush Steppe

7.3.1 Study System

The intermountain area of the western USA is an arid to semiarid region in which most of the precipitation arrives during the winter or early spring. High spatial and temporal variability in precipitation both among years and within growing seasons is a defining characteristic. Nutrient availability, especially nitrogen, is typically low, but increases with elevation (Alexander et al. 1993; Dahlgren et al. 1997), and closely tracks moisture availability (Evans and Ehleringer 1994). High topographic variability results in strong gradients in both soil water and nutrient availability. These gradients determine the distribution of species and ecological types within the region. Sagebrush (*Artemisia* species) is the most abundant shrub with the subspecies *A. tridentata wyomingensis* dominating areas with effective moisture of 20–25 cm (8–12 in.) and *A. tridentata vaseyana* dominating areas with higher effective moisture of 30–41 cm (12–16 in.) (West 1983). For sites in moderate to high ecological condition the associated species are predominantly perennial grasses with lesser amounts of annual and perennial forbs.

Settlement of the region around 1860 by European Americans resulted in major changes in vegetation structure and composition of sagebrush communities and increased their susceptibility to invasion by exotic species (Mack 1986; Knapp 1996). Initially, widespread overgrazing by cattle and sheep led to decreases in native perennial grasses and forbs (Miller and Eddleman 2001). The decrease in the herbaceous species reduced the necessary fine fuels for carrying natural fires and altered competitive relations in favor of woody species. As a result, *Artemisia* species generally increased in dominance.

B. tectorum, cheatgrass, was accidentally introduced into the region at several different locations in the late 1800s (Mack 1986). The annual grass rapidly spread into the depleted sagebrush ecosystems, especially the warmer and drier *A. tridentata wyomingensis* shrubland types (Mack 1986). The fine, more continuous fuels contributed by *B. tectorum* resulted in more frequent and larger fires (Whisenant 1990; Knapp 1996). In many parts of the region an annual grass–fire cycle now exists in which fire return intervals have decreased from about 60–110 years to as little as 3–5 years (Whisenant 1990). It has been estimated that areas dominated by *B. tectorum* covered a minimum of 20,000 km^2 or 5% in the 1990s (Bradley and Mustard 2005) with an additional 150,000 km^2 at high risk of conversion (Suring et al. 2005).

7.3.2 Grass Invasion Impacts

Invasion by *B. tectorum* rapidly alters many ecosystem properties. *B. tectorum* dominance can alter nutrient cycling and soil microbial communities even in the absence of fire (Evans et al. 2001; Hawkes et al. 2006). Fire-induced community changes following invasion can lead to reduced soil water recharge and reduced soil moisture patchiness (Obrist et al. 2004). In addition, although invasive annual grasses can stabilize topsoil, loss of vegetative cover following fires or other disturbances increases overland flow and surface erosion resulting in the loss of soil nutrients, siltation of streams and rivers, and increased susceptibility to flooding (Knapp 1996). At regional scales repeated fire and progressive increases in annual grasses can result in conversion of shrublands and woodlands from carbon sinks to carbon sources (Bradley et al. 2006). Large-scale change in land cover from diverse shrublands to homogenous grasslands potentially can influence the region's albedo affecting evapotranspiration and, ultimately, moisture transfer, convective activity, and rainfall (Millenium Ecosystem Assessment 2005). The net effect could be an increase in aridity of the region.

Conversion of sagebrush communities to annual grasses results in increased landscape homogeneity and decreased patch diversity. A growing number of sagebrush-obligate species are at risk due to habitat loss (Knick et al. 2003), and approximately 20% of the ecosystem's native flora and fauna are already considered imperiled (Center for Science, Economics and Environment 2002). More frequent fires associated with cheatgrass invasion are resulting in increased costs for land management agencies due to increased fire suppression and land rehabilitation costs (US Department of Interior, BLM 1999, 2000). Local communities benefit from money spent for fire suppression, but can suffer from wildfires due to loss of livestock forage and property, health, and safety risks caused by smoke and particulate matter, and reduced recreational value and income.

7.3.3 Difficulty of Control and Management

Control of *B. tectorum* has been difficult because of its persistent seedbank, rapid response to disturbance, highly plastic seed production, and ability to compete with native species. Removal of herbaceous perennials in sagebrush communities can cause increases in both soil water and nitrate availability, conditions that promote *B. tectorum* growth (Chambers et al. 2007). Fire causes the death of fire-intolerant shrubs (Young and Evans 1978), and can result in greater soil water availability (Chambers and Linnerooth 2001) and dramatically increase soil mineral nitrogen (Stubbs and Pyke 2005; Rau et al. 2007), which can be up to 12 times higher in the postburn compared with preburn community (Blank et al. 1994, 1996). *B. tectorum* has the capacity for high growth rates (Arredondo et al. 1998) and can rapidly respond to increased availability of nitrogen (Lowe et al. 2002, Monaco et al. 2003) and soil water (Link et al. 1990, 1995). Biomass and seed production of *B. tectorum* can increase 2–3 times following removal of perennial grasses and forbs, 2–6 times after fire, but 10–30 times following both removal and burning (Chambers et al. 2007). Field and modeling studies show that *B. tectorum* populations have an 80–90% risk of exploding to densities near 10,000 plants m^{-2} within 10 years after fire (Young and Evans 1978; Pyke 1995).

7.3.4 Determinants of Resistance and Resilience

Evidence exists that the resilience of sagebrush communities and their resistance to *B. tectorum* are greatest on sites with relative high percentages of native perennial grasses and forbs. Long-term observations show that an inverse relationship exists between *B. tectorum* and total perennial herbaceous cover of sagebrush-steppe recovering from livestock grazing (Anderson and Inouye 2001) and of sagebrush semidesert responding to wildfire and livestock grazing (West and York 2002). Mechanistic research indicates that following overgrazing or fire, susceptibility to invasion by *B. tectorum* is lowest on sites with relatively high cover of perennial herbaceous species (Chambers et al. 2007). Under these conditions native perennials typically increase following fire, thus limiting growth and reproduction of *B. tectorum* (Chambers et al. 2007). Native species with similar growth forms and phenology, like *Elymus elymoides* (squirreltail), have the capacity to preclude or limit the establishment and reproduction of *B. tectorum* (Stevens 1997; Booth et al. 2003; Humphrey and Schupp 2004). Sites with low abundances of perennial grasses and forbs typically have reduced resilience following perturbations and, thus, are less resistant to invasion or increases in *B. tectorum*. The seedbanks of perennial herbaceous species, especially grasses, are typically small (Hassan and West 1986). Also, the seedlings of native perennial grasses are generally poor competitors with *B. tectorum* because the annual grass can germinate earlier in the fall and under colder winter temperatures (Aguirre and Johnson 1991). *B. tectorum* exhibits greater root elongation at low soil temperatures (Harris 1967) and is capable of competitive displacement of the root systems of native plants (Melgoza and Nowak 1991).

The ability to control *B. tectorum* or increase the resistance of sagebrush communities to its invasion varies in these topographically diverse ecosystems. The current distribution of *B. tectorum* indicates that while the species is abundant and widespread at lower elevations, invasion of high elevation *A. tridentata* systems has been minimal (Suring et al. 2005). *B. tectorum* exhibits relatively high germination at cold temperatures (Evans and Young 1972) and has considerable ecotypic variation in optimal night/day germination temperatures (Meyer et al. 1997; Bair et al. 2006). However, ecophysiological limitations due to cold temperatures can restrict its growth and, consequently, reproduction within *A. tridentata vaseyana* communities during short and cool growing seasons and in higher elevation mountain brush communities in general (Chambers et al. 2007). Precipitation, via its effects on available soil water, appears to be the primary control on *B. tectorum* invasibility when temperature is not a factor. High variability in available soil water at lower elevations may result in lower average native perennial cover and increased windows of opportunity for growth and reproduction of *B. tectorum* when available soil water is above a certain level (Chambers et al. 2007).

7.3.5 Current Management Approaches

The type of management approach or restoration activity used depends on the stage of invasion and the environmental characteristics of the affected communities (see Whisenant 1999; Chambers 2005; D'Antonio and Chambers 2006). Site prioritization depends on management goals and the need to maintain or improve habitat for a growing number of animal species obligate to sagebrush ecosystems, such as sage grouse (*Centrocerus urophasianus*) (Hemstrom et al. 2002). For the purposes of discussing management approaches for areas exhibiting *B. tectorum* invasion, we describe three ecological states (Figs. 7.3 and 7.4). In the first state, resilience is high for most *Artemisia* community types. The existing vegetation is managed to maintain or increase resilience to disturbance and resistance to *B. tectorum* invasion. Shrubs or trees may be increasing in abundance, but native herbaceous perennials are still a significant component of the community. *B. tectorum* may be present, but has relatively low abundance. Preventative management can be used to increase resistance by reintroducing disturbance in the form of fire or fire surrogate treatments (Wright and Chambers 2002; Chambers 2005; Miller et al. 2005).

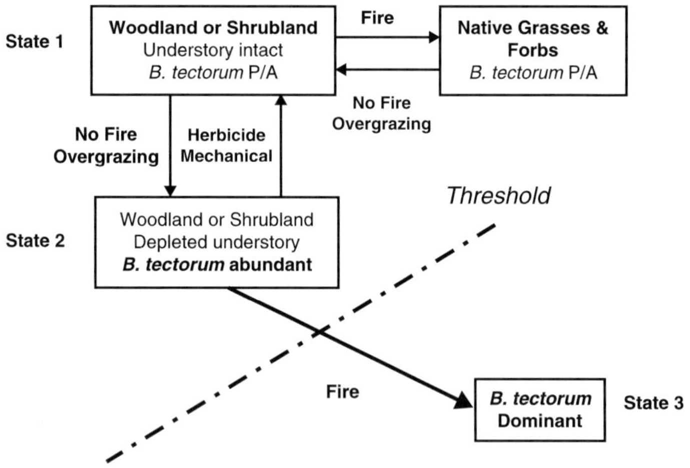

Fig. 7.3 State and transition model for Great Basin sagebrush steppe in western USA. Boxes represent ecosystem states. P/A indicates present or absent but not abundant. Arrows represent processes promoting transitions among states

Fig. 7.4 Examples of sites in different stages of *B. tectorum* invasion in Great Basin sagebrush steppe. Left panel = first state where A. *tridentata* shrubland has understory of native perennial grasses and minimal invasion. Center panel = transitional site with A. *tridentata* canopy but understory of *B. tectorum*. Right panel = state where shrubland has been fully converted to B. *tectorum* dominance (photographs: courtesy of Bob Blank, USDA-ARS, Reno Nevada)

Management objectives include increasing native grasses and forbs through competitive release from shrubs and trees, and reducing woody fuel loads to minimize the risk of high-severity fires. Treatments target *A. tridentata* or pinyon and juniper trees and may include prescribed fire or application of the herbicide tebuthiuron (Monson et al. 2004). A separate objective may be to rejuvenate shrub stands characterized by old age individuals and a lack of seedling recruitment. In this case, treatment often involves brush beating or mowing of *A. tridentata* to decrease shrub density, promote shrub seedling recruitment, and increase native herbaceous species through competitive release (Monson et al. 2004). Ideally, areas selected for these treatments have sufficient native perennial herbaceous species that reseeding is not required.

In the next state which we refer to as *transitional*, the community has low resilience and is at risk of crossing a biotic threshold following fire or other disturbance that could result in a new ecological state dominated by *B. tectorum*. Herbaceous perennials may be distributed throughout the community, but are present in relatively low percentages. Native shrubs are a significant part of the community, but *B. tectorum* is present and moderately abundant (Fig. 7.4, center). Management objectives are to decrease woody fuel loads, rejuvenate *A. tridentata* stands, and increase perennial herbaceous species through competitive release. Treatments to decrease woody vegetation typically do not involve fire because of high risk of *B. tectorum* dominance following fire. Instead, hand or mechanical treatments or application of the herbicide tebuthiuron may be used to decrease woody species abundance (Monson et al. 2004). In addition, preemergent herbicides may be used to decrease *B. tectorum* germination (Vallentine et al. 2004). Sites in this state are often revegetated immediately after wildfire because they lack sufficient perennial herbaceous species to provide the resilience for regeneration prior to increases by *B. tectorum*.

In the final state, a biotic threshold has been crossed due to fire and invasion. This new state is dominated by *B. tectorum* (Fig. 7.4, right). An increase in contiguous fine fuels due to *B. tectorum* dominance often results in higher fire frequencies (Whisenant 1990; Link et al. 2006). Low to very low percentages of herbaceous perennial species are present, and fire-intolerant native shrubs, including *Artemisia* species, are largely absent. This is the current condition of thousands of hectares of land in the Great Basin. Control of *B. tectorum* and aggressive revegetation are necessary to restore the native community. Integrated management strategies are being tried in which pretreatments are used to reduce the seedbank of *B. tectorum* followed by revegetation to establish the desired community (Sheley and Krueger-Mangold 2003; Vallentine 2004). Collaborative research and management projects are being implemented across the region to develop solutions for restoring these ecosystems. These are described in the next section.

7.3.6 Ongoing Challenges and Unanswered Questions

Managers face several challenges in defining and implementing the appropriate restoration treatments for sagebrush ecosystems exhibiting *B. tectorum* invasion. The first of these is accurately defining the state of invasion and the potential for recovery following the different types of available treatments. Developing state and transition models that illustrate the vegetation states and successional stages for the various sagebrush ecological types (community types) by the US Natural Resources Conservation Service (NRCS) and Bureau of Land Management (BLM) has helped managers to understand the possible trajectories for these systems. However, the specific conditions (site characteristics and vegetation structure and composition) that result in transitions or threshold crossings following disturbance or management actions have rarely been examined (but see Wright and Chambers 2002; Chambers et al. 2004). Currently, a regional, multiagency project funded by the Joint Fire Sciences Program, "Sage Step," is investigating the thresholds of recovery for sagebrush communities threatened by *B. tectorum* invasion and *Pinus monophylla* (single-needle pinyon pine) and *Juniperus osteosperma* and *J. occidentalis* (Utah and western juniper) encroachment (sagestep.org). The project is examining use of fire and mechanical removal in areas exhibiting *P. monophylla* and *Juniperus* encroachment, and fire, brush mowing, and herbicides in *B. tectorum*-invaded sagebrush sites. Study locations are positioned across the Great Basin to evaluate the generality of these treatments. At each location a gradient of *B. tectorum* invasion or woody plant dominance is being evaluated to determine the thresholds beyond which different treatments are not effective in promoting a resistant understory. Results will help managers select management tools and areas for treatment.

A significant challenge that managers face is the revegetation of sites that are transitional or already dominated by *B. tectorum*. This requires control of *B. tectorum* populations and establishment of native species communities that are resistant to *B. tectorum*. A proven method of controlling *B. tectorum* is the use of herbicides such as glyphosate, which result in high levels of *B. tectorum* mortality when properly applied (Vallentine 2004), but this can be expensive over large areas. Grazing by livestock has been suggested as a means of controlling *B. tectorum* seed production, but field trials show that the annual grass has highly plastic growth and produces seeds even after repeated short clipping (Hempy-Mayer and Pyke 2008). Herbage removal is therefore not effective in eliminating *B. tectorum*, and repeated removal of *B. tectorum* biomass by livestock can harm resident natives. The use of a head smut pathogen (*Ustilago bullata*), which often causes epidemic levels of head smut disease in Intermountain populations of *B. tectorum* is being explored as a biocontrol agent (Meyer et al. 2000, 2005). The pathogen also, however, infects native grasses. Research is underway to determine conditions under which it might be useful as a control agent.

Because invaders are highly responsive to nitrogen, a restoration approach that has been tried in many locations is to utilize methods that decrease N availability. Carbon amendments have been shown to decrease the growth, reproduction, and cover of some invasive species (e.g., Reever-Morgan and Seastedt 1999; Alpert and Maron 2000; Paschke et al. 2000), but they can also affect growth of native species (Monaco et al. 2003; Corbin and D'Antonio 2004). Nonetheless, lower competitive pressure from exotics may compensate for reduced nutrients. Despite some successes, this approach has not been shown to have long-term efficacy for weed control because carbon amendments can be expensive, difficult to use over large areas, and often have only short-term effects (Mazzola et al. 2008).

A recently completed USDA Initiative for Future Agriculture and Food Systems (IFAFS) project on "Integrated Strategies Toward Weed Control on Western Rangelands" evaluated several different approaches for restoring *B. tectorum*-dominated *A. tridentata wyomingensis* communities at eight locations across Oregon, Idaho, Utah, and Nevada (Nowak et al. 2006). The first approach involved identifying native grass and forb species with high probabilities of establishment and strong competitive abilities. Introduced grasses, especially crested wheatgrass (*Agropyron cristatum*, *A. desertorum*, *A. fragile*), are used extensively in the western USA to increase forage production of degraded rangelands and revegetate post-fire landscapes (Lesica and DeLuca 1996; Richards et al. 1998). However, the rapid loss of native sagebrush ecosystems including sagebrush-obligate wildlife species has emphasized the need to focus revegetation efforts on recreating native communities (Wisdom et al. 2005). Also, recent research shows that areas seeded with introduced *Agropyron* species are no more resistant to *B. tectorum* following fire than intact native communities (Chambers et al. 2007). The IFAFS project compared the performance of almost 20 native grass and forb species and two *Agropyron* accessions (Nowak et al. 2006). One accession, *A. desertorum* "CD II," outperformed native grasses in 30% of comparisons, while the other (*A. desertorum* "Valvilov") did not perform better. Several native accessions performed well at multiple locations, are commercially available, and provide viable alternatives to introduced species.

An approach for building community resistance is active seeding of functionally diverse species that will maximize resource uptake by the entire community once established. Revegetation mixtures that include grasses, forbs, and shrubs with varying life forms and rooting depths should facilitate resource extraction through the soil profile, while species with different phenologies should maximize use of available soil resources throughout the growing season. Thus, resistance would be maximized both in the short and longer terms.

The challenges of integrating these approaches are illustrated by the IFAFS project. It evaluated the effectiveness of control of *B. tectorum* using glyphosate, a short-lived herbicide, to reduce population abundance, followed by immobilization of soil nitrogen through sugar (sucrose) applications (Nowak et al. 2006) simultaneous with seeding a diverse mix of native grasses, forbs, and shrubs. Although sucrose addition decreased available N and initial *B. tectorum* biomass and seed production, by the second growing season the effect had disappeared (Mazzola et al. 2008). Sucrose addition also reduced growth of native plants and may have resulted in increased overwinter mortality of seeded natives (Mazzola 2008). Competitive effects of the seeded native species on *B. tectorum* reinvasion were slight although expected to increase as the native species mature. The native species were seed limited relative to *B. tectorum*. Higher seeding rates (600 vs. 150 or 300 plants m^{-2}) resulted in higher establishment of natives as long as *B. tectorum* densities were relatively low (<300–500 plants m^{-2}) (Mazzola 2008).

Recreating sagebrush communities with the functional diversity necessary to support sagebrush-obligate wildlife species and to resist *B. tectorum* invasion is therefore a challenging management goal. Although establishment of several native grass accessions was relatively high in the IFAFS project, establishment of native forbs and shrubs was low and seed availability was limited. For systems dominated by *B. tectorum*, it may be possible first to seed with competitive native grass accessions and then seed with a more diverse species mixture. Although native grasses with broad amplitudes are commercially available, the volume of seed needed to reseed burned areas at a reasonable rate is very high. Native species are generally seed limited relative to *B. tectorum*, and typical seeding rates for native species are probably inadequate (Mazzola 2008). Also seed increase programs and seed zones for native forb and shrub species are just beginning to be developed and for most species, seed supplies are far too limited for large-scale restoration efforts. These limitations are being addressed, in part, by the Great Basin Native Plant Selection and Increase Project, a collaborative effort of the BLM, Great Basin Restoration Initiative US Forest Service, Grasslands, Shrublands and Deserts Project, and other regional agencies and universities (http://www.fs.fed.us/rm/boise/research/shrub/greatbasin.shtml).

The project seeks to increase seed supplies of native plant species, particularly forbs. Its components include plant selection (source-identified seed sources, methods of propagation), seed and seeding technology, and seed production (federal and state nurseries, NRCS, private growers).

Sagebrush ecosystems are changing at a rapid rate. The 2005, 2006, and 2007 fire seasons had among the largest areas burned on record with 437,060, 542,683, and 360,170 ha (1,080,000, 1,341,000, and 890,000 acres), respectively, burned in the state of Nevada alone (http://www.forestry.nv.gov/docs/2007_accomplishment_report.pdf). A high percentage of sagebrush communities that burned will be invaded by or converted to *B. tectorum* (Hemstrom et al. 2002). It has been suggested that managers use a triage process involving "sorting through the sagebrush communities to allocate resources to maximize the number, size, type, and distribution of communities that survive" (Wisdom et al. 2005). The process includes (1) determining which communities are resilient and which are not, i.e., determining their ecological state, (2)developing a systematic process of prioritizing sites, across the entire region, for management activities, (3) utilizing appropriate management techniques to maintain sagebrush communities with a high degree of resilience, and (4) restoring some transitional or converted communities to serve goals of enhancing intact communities that will be resistant to *B. tectorum* invasion.

7.4 Concluding Thoughts

The two examples discussed here provide alternative views on how to promote native species in the face of grass invasions. In both cases the invaders are widespread and persistent, but managers are exploring ways to promote native species by focusing on maintaining or restoring resilience to the native assemblage, or establishing resistant plant assemblages that will reduce the intensity of reinvasion. The Hawaiian example is unusual in that it advocates promotion of a different type of plant community than is known to have existed on the invaded sites – a form of rehabilitation. Such an approach may be the only means for promoting more desirable species in the face of persistent, disturbance-promoting invaders where the new disturbance, fire, is not part of the historical successional framework of the sites and has resulted in an alternative persistent state. In the Great Basin, by contrast, fire was part of the preinvasion (pre-European) disturbance regime and successional framework (Fig. 7.3) and if perennial herbaceous species are still present in the native community, it is often possible to restore resilience and resistance prior to degradation by the invader. In the Great Basin, native grasses and forbs that respond favorably to increased resources following disturbance or management treatments can decrease invasion by *B. tectorum* and, thus, it is also possible to promote resistance. However, in both case studies the supply of native seed for enhancing resistance or resilience (or both) is an important factor limiting management options. The concepts of resistance and resilience are fundamental ecological processes that can assist greatly in "weed control" to help managers work toward goals such as creating "self-sustaining ecosystems able to persist under existing environmental conditions" (SER Primer, http://www.ser.org/). While practicalities may limit implementation of practices based on these concepts, they nonetheless can provide a scientific framework for the development of programs to guide future management efforts.

Acknowledgments The authors would like to thank the Hawaii Volcanoes National Park Resources Management staff and Fire division for their assistance with the revegetation program. J. Chambers and C. D'Antonio thank the IFAFS and Sage Step groups for stimulating discussions of state and transition frameworks for Great Basin vegetation. We also thank an anonymous reviewer and Inderjit for comments on an earlier draft of this manuscript and Robert Blank for photographs.

References

Aguirre L, Johnson DA (1991) Influence of temperature and cheatgrass competition on seedling development of two bunchgrasses. J Range Manage 44:347–354

Alexander EB, Mallory JI, Colwell ML (1993) Soil-elevation relationships on a volcanic plateau in the southern Cascade Range, northern California, USA. Catena 20:113–128

Alpert P, Maron JL (2000) Carbon addition as a countermeasure against biological invasion by plants. Biol Invasions 2:33–40

Anderson JE, Inouye RS (2001) Landscape-scale changes in plant species abundance and biodiversity of a sagebrush steppe over 45 years. Ecol Monogr 71:531–556

Arredondo JT, Jones TA, Johnson DA (1998) Seedling growth of Intermountain perennial and weedy annual grasses. J Range Manage 51:584–589

Bair NB, Meyer SE, Allen PS (2006) A hydrothermal after-ripening time model for seed dormancy loss in *Bromus tectorum*. Seed Sci Res 16:17–28

Blank RR, Allen FL, Young JA (1994) Extractable anions in soils following wildfire in a sagebrush-grass community. Soil Sci Soc Am J 58:564–570

Blank RR, Allen FL, Young JA (1996) Influence of simulated burning of soil litter from low sagebrush, squirreltail, cheatgrass, and medusahead sites on water-soluble anions and cations. Int J Wildland Fire 6:137–143

Booth MS, Caldwell MM, Stark JM (2003) Overlapping resource use in three Great Basin species: implications for community invisibility and vegetation dynamics. J Ecol 91:36–48

Bradley BA, Mustard JF (2005) Identifying land cover variability distinct from land cover change: cheatgrass in the Great Basin. Remote Sens Environ 94:204–213

Bradley BA, Houghton RA, Mustard JF, et al (2006) Invasive grass reduces aboveground carbon stocks in shrublands of the Western US. Glob Change Biol 12:1815–1822

Bradshaw A (1997) What do we mean by restoration? In: Urbanska KM, Webb NR, Edwards PJ (eds) Restoration ecology and sustainable development, Cambridge University press, Cambridge, pp. 8–16

Brooks ML, D'Antonio CM, Richardson DM, et al (2004) Effects of invasive alien plants on fire regimes. Bioscience 54:677–688

Bruno JF, Fridley JD, Bromberg KD, Bertness MD (2005) Insights into biotic interactions from studies of species invasions. In: Sax DF, STachowicz JJ, Gaines SD (eds). Species invasions: insights into ecology, evolution and biogeography, Sinauer, Sunderland MA.

Cabin RJ, Weller SG, Lorence DH, et al (2000) Effects of long-term ungulate exclusion and recent alien species control on the preservation and restoration of a Hawaiian tropical dry forest. Conserv Biol 14:439–453

Cabin RJ, Weller SG, Lorence DH, et al (2002) Effects of light, alien grass, and native species additions on Hawaiian dry forest restoration. Ecol Appl 12:1595–1610

Center for Science, Economics and Environment (2002) The state of the nation's ecosystems: measuring the lands, waters and living resources of the United States. Cambridge University Press, Cambridge, UK. Available at http://www.heinzctr.org/ecosystem/index.html. Accessed 17 July 2007

Chambers JC (2005) Fire related restoration issues in woodland and rangeland ecosystems. In: Taylor L, Zelnik J, Cadwallader S, Hughes B (compilers), Mixed Fire Regimes: Ecology and Management. Symposium Proceedings, Spokane, WA. AFE MIXC03, pp 149–160

Chambers JC, Linnerooth AR (2001) Restoring sagebrush dominated riparian corridors using alternative state and threshold concepts: environmental and seedling establishment response. Appl Veg Sci 4:157–166

Chambers JC, Miller JR, Germanoski D, et al (2004) Process based approaches for managing and restoring riparian ecosystems. In: Chambers JC, Miller JR (eds) Great Basin Riparian Ecosystems – Ecology, Management and Restoration, Island Press, Covelo, CA, pp 196–231

Chambers JC, Roundy BA, Blank RR, et al (2007) What makes Great Basin sagebrush ecosystems invasible by *Bromus tectorum*? Ecol Monogr 77:117–145

Corbin J, D'Antonio CM (2004) Can sawdust addition increase competitiveness of native grasses? A case study from California. Restor Ecol 12:36–43

Daehler C (1998) The taxonomic distribution of invasive angiosperm plants: ecological insights and comparison to agricultural weeds. Biol Conserv 84:167–180

D'Antonio CM (1993) Mechanisms controlling invasion of coastal plant communities by the alien succulent, *Carpobrotus edulis*. Ecology 74:83–95

D'Antonio CM (2000) Fire, plant invasions and global changes. In: Mooney HA, Hobbs RJ (eds) Invasive species in a changing world, Island Press, Covela, CA. pp 65–94

D'Antonio CM, Chambers J (2006) Using ecological theory to manage or restore ecosystems affected by invasive plant species. In: Falk D, Palmer M, Zedler J (eds) Foundations of Restoration Ecology, Island Press, Covelo, CA, pp 260–279

D'Antonio CM, Meyerson L (2002) Exotic species and restoration: synthesis and research needs. Restor Ecol 10:703–713

D'Antonio CM, Thomsen M (2004) Ecological resistance in theory and practice. Weed Technology 18:1572–1577

D'Antonio CM, Vitousek PM (1992) Biological invasions by exotic grasses, the grass–fire cycle and global change. Annu Rev Ecol Syst 23:63–88

D'Antonio CM, Hughes RF, Mack MM et al (1998) Response of native species to the removal of non-native grasses in a Hawaiian woodland. J Veg Sci 9:699–712

D'Antonio CM, Tunison JR, Loh R (2000) Variation in impact of exotic grass fueled fires on species composition across an elevation gradient in Hawai'i. Austral Ecol 25:507–522

D'Antonio CM, Hughes RF, Vitousek PM (2001a) Factors influencing dynamics of invasive C4 grasses in a Hawaiian woodland: role of resource competition and priority effects. Ecology 82:89–104

D'Antonio CM, Levine JM, Thomsen M (2001b) Propagule supply and resistance to invasion: a California botanical perspective. J Medit Ecol 2:233–245

Dahlgren RA, Boettinger JL, Huntington GL, et al (1997) Soil development along an elevational transect in the western Sierra Nevada. Geoderma 78:207–236

Davis MA, Grime JP, Thompson K (2000) Fluctuating resources in plant communities: a general theory of invasibility. J Ecol 88:528–534

Evans RA, Young JA (1972) Microsite requirements for establishment of annual rangeland weeds. Weed Science 20:350–356

Evans RD, Ehleringer JR (1994) Water and nitrogen dynamics in arid woodland. Oecologia 99:233–242

Evans RD, Rimer R, Sperry L, Belnap J (2001) Exotic plant invasion alters nitrogen dynamics in an arid grassland. Ecol Appl 11:1301–1310

Freifelder R, Vitousek PM, D'Antonio CM (1998) Microclimate effects of fire-induced forest/ grassland conversion in seasonally dry Hawaiian woodlands. Biotropica 30:286–297
Harris G (1967) Some competitive relationships between *Agropyron spicatum* and *Bromus tectorum*. Ecol Monogr 37:89–111
Hassan MA, West NE (1986) Dynamics of soil seed pools in burned and unburned sagebrush semi-deserts. Ecology 67:269–272
Hawkes C, Belnap J, D'Antonio CM, Firestone MK (2006) Arbuscular mycorrhizal assemblages in native plant roots change in the presence of invasive exotic grasses. Plant Soil 281:369–380
Hempy-Mayer K, Pyke DA (2008) Defoliation effects on *Bromus tectorum* seed production – implications for grazing. Rangeland Ecol Manage 61:116–123
Hemstrom MA, Wisdom MJ, Hann WJ, et al (2002) Sagebrush-steppe vegetation dynamics and restoration potential in the Interior Columbia Basin, USA. Conserv Biol 16:1243–1255
Hughes RF, Vitousek PM (1993) Barriers to shrub reestablishment following fire in the seasonal submontane zone of Hawai'i. Oecologia 93:557–563
Hughes RF, Vitousek PM, Tunison JT (1991) Alien grass invasion and fire in the seasonal submontane zone of Hawai'i. Ecology 72:743–746
Humphrey LD, Schupp EW (2004) Competition as a barrier to establishment of a native perennial grass (*Elymus elymoides*) in alien annual grass (*Bromus tectorum*) communities. J Arid Environ 58:405–422
Jordan WR III (2003) The Sunflower forest: ecological restoration and the new communion with nature. University of California Press, Berkeley, CA
Knapp PA (1996) Cheatgass (*Bromus tectorum* L.) dominance in the Great Basin Desert. Glob Environ Chang 6:37–52
Knick ST, Dobkin DS, Rotenberry JT, et al (2003) Teetering on the edge or too late? Conservation and research issues for avifauna of sagebrush habitats. Condor 105:611–634
Lesica P, DeLuca T (1996) Long-term harmful effects of crested wheatgrass on Great Plains grassland ecosystems. J Soil Water Conserv 51:408–409
Levine JM, Adler PB, Yelenik SG (2004) A meta-analysis of biotic resistance to exotic plant invasions. Ecol Lett 7:975–989
Ley R, D'Antonio CM (1998) Exotic grasses alter rates of nitrogen fixation in seasonally dry Hawaiian woodlands. Oecologia 113:179–187
Link SO, Gee GW, Downs JL (1990) The effect of water stress on phenological and ecophysiological characteristics of cheatgrass and Sandberg's bluegrass. J Range Manage 43:506–513
Link SO, Bolton H Jr, Theide ME, et al (1995) Responses of downy brome to nitrogen and water. J Range Manage 48:290–297
Link SO, Keeler CW, Hill RW, et al (2006) *Bromus tectorum* cover mapping and fire risk. Int J Wildland Fire 15:113–119
Loh R, McDaniel S, Benitez D, et al (2007) Rehabilitation of seasonally dry ohi'a woodland and mesic koa forest following the Broomsedge Fire, Hawai'i Volcanoes National Park. Technical report no. 147. Pacific Cooperative Studies Unit, University of Hawai'i, Honolulu, HI, 21 p
Loope LL, Scowcroft PG (1985) Vegetation responses within exclosures in Hawai'i: a review. In: Stone CP, Scott P (eds) Hawai'i's terrestrial ecosystems: preservation and management. University of Hawai'I Cooperative National Park Resources Studies Unit, University of Hawaii Press, Honolulu, HI
Lowe PN, Lauenroth WK, Burke IC (2002) Effects of nitrogen availability on the growth of native grasses and exotic weeds. J Range Manage 55:94–98
Mack MC, D'Antonio CM (2003a) Exotic grasses alter controls over soil nitrogen dynamics in a Hawaiian woodland. Ecol Appl 13:154–166
Mack MC, D'Antonio CM (2003b) Direct and indirect effects of introduced C4 grasses on decomposition in a Hawaiian woodland. Ecosystems 6:503–523
Mack MC, D'Antonio CM, Ley R (2001) Pathways through which exotic grasses alter N cycling in a seasonally dry Hawaiian woodland. Ecol Appl 11:1323–1335

Mack RN (1986) Alien plant invasion into the intermountain west: a case history. In: Mooney HA, Drake JA (eds) Ecology of biological invasions of North America and Hawaii. Springer, New York, pp 191–213

Maron JL, Vila M (2001) When do herbivores affect plant invasion: evidence for the natural enemies and biotic resistance hypotheses. Oikos 95:361–373

Mazzola MB (2008) Spatio-temporal heterogeneity and habitat invasibility in sage steppe ecosystems. PhD Dissertation, University of Nevada, Reno, NV

Mazzola MB, Allcock KG, Chambers JC et al (2008) Effects of nitrogen availability and cheatgrass competition on the establishment of Vavilov Siberian wheatgrass. Rangeland Ecol Manag 61:475–484

McEvoy PB, Coombs EM (1999) Biological control of plant invaders: regional patterns, field experiments and structured population models. Ecol Appl 9:387–401

Melgoza G, Nowak RS (1991) Competition between cheatgrass and two native species after fire: implications from observations and measurements of root distributions. J Range Manage 44:27–33

Meyer SE, Allen PS, Beckstead J (1997) Seed germination regulation in *Bromus tectorum* (Poaceae) and its ecological significance. Oikos 78:475–485

Meyer SE, Nelson DL, Clement S, et al (2000) Exploring the potential for biocontrol of cheatgrass with the head smut pathogen. In: Entwistle PG, DeBolt AM, Kaltenecker JH, Steenholf K (compilers) Proceedings: Sagebrush Steppe Ecosystem Symposium; 23–25 June 1999, Boise, ID. BLM/ID/PT-001001 + 1150. US Department of the Interior, Bureau of Land Management, Idaho State Office, Boise, ID

Meyer SE, Nelson DL, Clement S, et al (2005) Genetic variation in *Ustilago bullata*: molecular genetic markers and virulence on *Bromus tectorum* host lines. Int J Plant Sci 166:105–115

Millennium Ecosystem Assessment. Sustainable Rangeland Roundtable. Available at http://www.millenniumassessment.org/en/index.aspx. Accessed 17 July 2007

Miller RF, Eddleman LE (2001) Spatial and temporal changes of sage grouse habitat in the sagebrush biome. Oregon State University, Agricultural Experiment Station Bulletin 151, Corvallis, OR

Miller RF, Bates JD, Svejcar AJ et al (2005) Biology, ecology, and management of western juniper (*Juniperus occidentalis*). Agricultural Experiment Station, Oregon State University, Corvallis, OR

Monaco TA, Johnson DA, Norton JM, et al (2003) Contrasting responses of Intermountain West grasses to soil nitrogen. J Range Manage 56:289–290

Monson SB, Stevens R, Shaw NL (eds) (2004) Restoring western ranges and wildlands. Gen. Tech. Rep. RMRS-GTR-136. 3 volumes. US Department of Agriculture, Forest Service, Rocky Mountain Research Station, Fort Collins, CO

Mueller-Dombois D (1981) Fire in tropical ecosystems. In: Mooney HA, Bonnicksen TA, Christensen NL et al (eds) Fire regimes and ecosystem properties. Gen. Tech. Report WO-26. USDA Forest Service, Washington, DC, pp 502–520

Mueller-Dombois D, Spatz G (1975) The influence of feral goats on the lowland vegetation in Hawaii Volcanoes National Park. Phytocoenologica 3:1–29

Nevada Division of Forestry Western Region (2007) Fire management report. Available at http://www.forestry.nv.gov/docs/2007_accomplishment_report.pdf. Accessed 23 May 2008

Nowak R, Glimp H, Doescher P et al (2006) Integrated restoration strategies towards weed control on western rangelands. Final Report. CREES Agreement No. 2001–52103–11322 (Submitted 14 December, 2006)

Obrist D, Yakir D, Arnone J (2004) Temporal and spatial patterns of soil water following wildfire-induced changes in plant communities in the Great Basin in Nevada, USA. Plant Soil 262:1–12

Parsons J (1972) Spread of African pasture grasses to the American tropics. J Range Manage 25:12–17

Paschke MW, McLendon T, Redente EF (2000) Nitrogen availability and old-field succession in a short-grass steppe. Ecosystems 3:144–158

Pyke DA (1995) Population diversity with special reference to rangeland plants. In: West NE (ed) Biodiversity of rangelands. Natural resources and environmental issues, Vol. IV. College of Natural Resources, Utah State University, Logan, UT, pp 21–32

Rau BM, Blank RR, Chambers JC, et al (2007) Prescribed fire in a Great Basin sagebrush ecosystem: dynamics of soil extractable nitrogen and phosphorus. J Arid Environ 71:362–375

Reever-Morgan KJ, Seastedt TR (1999) Effects of soil nitrogen reduction on nonnative plants in restored grasslands. Restor Ecol 7:51–55

Rejmanek M (1989) Invasibility of plant communities. In: Drake JA, Mooney HA, di Castri F, et al (eds) Biological invasions: a global perspective. Wiley, New York

Richards RT, Chambers JC, Ross C (1998) Use of native plants on federal lands: policy and practice. J Range Manage 51:625–632

Shaw RB, Castillo JM, Laven RD (1997) Impact of wildfire on vegetation and rare plants within the Kipuka Kalawamauna endangered plant habitat area, Pohakaloa training Area, Hawai'i. In: Greenlee JM (ed) Proceedings of the first conference on fire effects on rare and endangered species and habitats conference, Interrnational association of wildland fire. Couer d'Alene, ID, pp 253–264

Sheley RL, Krueger-Mangold J (2003) Principles for restoring invasive plant-infested rangeland. Weed Sci 51:260–265

Smith CW (1985) Impact of alien plants on Hawaii's native biota. In: Stone CP, Scott JM (eds) Hawaii's terrestrial ecosystems: preservation and management. Cooperative National Park Resources Studies Unit, University of Hawaii, Honolulu, HI, pp 180–243

Smith CW, Tunison JT (1992) Fire and alien plants in Hawai'i: research and management implications for Hawaii's native ecosystems. In: Stone CP, Smith CW, Tunison JT (eds) Alien plant invasions in native ecosystems of Hawaii: management and research. Cooperative National Park Resources Studies Unit, University of Hawaii, Honolulu, HI, pp 394–408

Stevens AR (1997) Squirreltail (*Elymus elmoides*) establishment and competition with cheatgrass (*Bromus tectorum*). PhD Dissertation, Brigham Young University, Provo, UT

Stubbs MM, Pyke DA (2005) Available nitrogen: a time-based study of manipulated resource islands. Plant Soil 270:123–133

Suring LH, Wisdom MJ, Tausch RJ, et al (2005) Modeling threats to sagebrush and other shrubland communities. In: Wisdom MJ, Rowland MM, Suring LH (eds) Habitat threats in the sagebrush ecosystems: methods of regional assessment and applications in the Great Basin. Alliance Communications Group, Lawrence, KS, pp 114–149

Tunison JT, Leialoha JAK, Loh RK, et al (1994) Fire effects in the coastal lowlands, Hawaii Volcanoes National Park. Tech. Report 88. Cooperative National Park Resources Studies Unit, University of Hawaii, Honolulu, HI

Tunison JT, Loh RK, Leialoha JAK (1995) Fire effects in the submontane seasonal zone, Hawaii Volcanoes National Park. Tech. Rep. 97. Cooperative National Park Resources Studies Unit, University of Hawaii, Honolulu, HI

Tunsion JT, D'Antonio CM, Loh R (2001) Fire, grass invasions and revegetation of burned areas in Hawaii Volcanoes National Park. In: Galley KE, Wilson TP (eds) Proceedings of the Invasive Species Workshop: The role of fire in the controls and spread of invasive species. Tall Timbers Research Station Publication No. 11, Allen Press, Lawrence, KS, pp 122–131

US Department of Interior, Bureau of Land Management (2000) The Great Basin – healing the land. Available at http://www.blm.gov/pgdata/etc/medialib/blm/nifc/gbri/documents.Par.61257.File.dat/Healing.pdf. Accessed on 5 May 2008

Vallentine JF (2004) Herbicides for plant control. In: Monson SB, Stevens R, Shaw NL (compilers) Restoring western ranges and wildlands. Gen. Tech. Rep. RMRS-GTR-136, 3 volumes. US Department of Agriculture, Forest Service, Rocky Mountain Research Station, Fort Collins, CO, pp 89–100

Vitousek PM (2004) Nutrient cycling and limitation: Hawaii as a model system. Princeton University Press, Princeton, NJ

Wagner WL, Herbst DR, Sohmer SH (1999) Manual of flowering plants of Hawaii, 2nd ed. B.P. Bishop Museum, Honolulu, HI

West NE (1983) Great Basin-Colorado Plateau sagebrush semi-desert. In: West NE (ed) Temperate deserts and semi-deserts. Elsevier, Amsterdam, pp 331–349

West NE, York TP (2002) Vegetation responses to wildfire on grazed and ungrazed sagebrush semi-desert. J Range Manage 55:171–181

Whisenant SG (1990) Changing fire frequencies on Idaho's Snake River plains: ecological and management implications. In: Proceedings from the symposium on cheatgrass invasion, shrub dieoff and other aspects of shrub biology and management. USFS Gen. Tech. Rep. INT-276, pp 4–10

Whisenant SG (1999) Repairing damaged wildlands: a process-oriented, landscape-scale approach. Cambridge University Press, Cambridge, UK

Wisdom MJ, Rowland MM, Suring LH (2005) Habitat threats in the sagebrush ecosystem: methods of regional assessment and applications in the Great Basin. Alliance Communication Group, Lawrence, KS

Wright JM, Chambers JC (2002) Restoring sagebrush dominated riparian corridors using alternative state and threshold concepts: biomass and species response. Appl Veg Sci 5:237–246

Young JA, Evans RA (1978) Population dynamics after wildfires in sagebrush grasslands. J Range Manage 31:283–289

Chapter 8
Weed Invasions in Western Canada Cropping Systems

K. Neil Harker, Robert E. Blackshaw, Hugh J. Beckie, and John T. O'Donovan

Abstract Agricultural ecosystem weeds can be invasive species. On the Canadian Prairies, the vast majority of weeds that annually invade crops and interfere with crop production are self-sustaining, non-native species that have spread over large areas. Weeds have vulnerabilities that can be exploited by combining optimal agronomic practices in addition to herbicide application. Some of these practices include competitive cultivars, relatively high crop seed rates, and strategic fertilizer placement. Prairie weeds have been reduced in numbers by the consistent use of herbicides, but herbicide resistance is now a major challenge. Weed resistance and other shifting crop production or environmental factors may significantly alter invasive weed dynamics and crop–weed interactions, and will undoubtedly pose future challenges to crop producers. These and other challenges can be countered by reducing soil disturbance and diversifying cropping systems that combine several synergistic components. There is a need for the implementation of true integrated crop, weed, and pest management systems that are multi-disciplinary. The common alien weed species that persistently invade Prairie cropland each year will not be subdued over the long term in the absence of such systems.

Keywords Alien species • Direct seeding • Herbicide resistance • Integrated weed management • Tillage

8.1 Introduction

In Canada and many other countries, increasing attention is being focused on assessing the present and future risks and consequences of invasive alien (non-indigenous) plant species. Invading alien species can lower crop yields, cause export market loss and/or commodity devaluation, reduce the quality of crop- and rangelands, and lead to the

K.N. Harker (✉), R.E. Blackshaw, H.J. Beckie, and J.T. O'Donovan
Agriculture and Agri-Food Canada, Lacombe Research Centre, 6000 C & E Trail, Lacombe, AB, Canada T4L 1W1
harkerk@agr.gc.ca

Inderjit (ed.), *Management of Invasive Weeds*,

expenditure of billions of dollars in chemical and biological control measures (Pimentel et al. 2000; Swanton et al. 1993). Besides direct economic costs, invasive plant species can reduce the stability of ecosystems and threaten biodiversity (White et al. 1993).

Globally, most emphasis on invasive plants has been directed towards species that impact relatively undisturbed ecosystems. The species considered as threats to natural Canadian Prairie (western provinces of Alberta, Saskatchewan, Manitoba) ecosystems are leafy spurge (*Euphorbia esula* L.), crested wheatgrass [*Agropyron cristatum* (L.) Gaertn.], and smooth brome (*Bromus inermis* Leyss.) (Haber 2002). All three species are minor, however, when their relative abundance in Prairie cropping systems is considered; the most abundant of these species, smooth brome, ranks 84th in relative abundance (Thomas and Leeson 2007).

Prior to introduction, most of our agricultural weeds had long been associated with European crops that are now widely grown in Canada (Clements et al. 2004). Thus, some considered these weeds to be non-invasive (Thomas and Leeson 2007). However, according to Pyšek et al. (2004), naturalized (sustain self-replacing populations for at least 10 years) alien species that spread are invasive. Invasive plants are naturalized non-native plants (aliens) that, without human assistance, can spread over large areas and replace native plants (Pyšek et al. 2004). Weeds are plants that grow where they are not wanted; because they can be self-sustaining without or in spite of direct human intervention, they have the potential to spread over large areas and thus can also be classified as invasive (Pyšek et al. 2004). The interpretation that agricultural ecosystem weeds can be invasive species is consistent with definitions from a 1999 United States Presidential Executive Order (13112) and the Invasive Species Advisory Council (DiTomaso 2008).

Eighty percent of agricultural weeds in Canada are classified as aliens (Government of Canada 2004). These weeds annually invade agricultural land and challenge farmers, extension personnel, and researchers in all major agricultural cropping regions. Of the 36 weed species considered to be *abundant* on the Canadian Prairies, 89% are alien species (Thomas and Leeson 2007). Therefore, the majority of Prairie crop yield losses due to weed competition are caused by alien species. Moreover, 99% of herbicide expenditures in western Canada are used to control alien species (Leeson et al. 2006). Therefore, repeated annual incursions of native or alien weed cohorts on cropland are invasions that we can ill afford to ignore (Harker et al. 2005; Thomas and Leeson 2007). Here, we review factors such as tillage intensity, cropping diversity, and herbicide resistance that influence repeated annual weed invasions in Canadian Prairie cropping systems. Additionally, mitigation strategies and tactics to manage such invasions are outlined. All of the weeds discussed later are invasive aliens of cultivated cropping systems.

8.2 Historical Perspective

Before herbicides became available, farmers employed cultural practices such as crop rotation, delayed seeding, green manure crops, and tillage to manage weeds. The introduction of the first selective herbicide, 2,4-D, in 1946, which was soon followed by others, led to dramatic improvements in weed control. Today in western Canada, there are 30 unique herbicide active ingredients registered for use in wheat (*Triticum aestivum* L.) alone (Brook 2007). Over the last few decades, there is strong evidence to suggest that weed management practices employed on the Prairies have substantially reduced weed populations. Post-management weed surveys initiated in the 1970s indicated that by the early 2000s, the number of weed species detected in average field samples had been reduced from 7 to 5, while mean plant density had been reduced from 100 to 31 plants m^{-2} (Leeson et al. 2005b). The remainder includes species that are tolerant of our predominant management methods, and therefore pose a significant challenge to management systems that have not changed substantially in 50 years.

Unfortunately, our first and often only response to weed infestations is to treat them with herbicides. We have seldom examined the causes of the perpetual presence of weeds (Buhler 1999). Herbicidal successes have led to overuse and the neglect of many useful cultural weed management techniques. With herbicides as the dominant weed management tool, resistance is now common in some of our major weeds, such as wild oat (*Avena fatua* L.) (Beckie et al. 1999) and kochia [*Kochia scoparia* (L.) Schrad.] (Beckie et al. 2001). The combination of weed resistance to herbicides, herbicide costs, and concerns regarding the non-target effects of herbicides has led to a resurgence of interest in non-chemical weed control methods, or at least more integrated crop management (ICM) systems. Such systems, including some of their components, are detailed later.

8.3 The Crop–Weed Association

The high level of land *disturbance* that accompanies crop production facilitates the success of weed communities. Froud-Williams (1988) states that: 'Arable land is characterized by regular, recurrent, and often highly predictable disturbance. The consequence of this disturbance is that weeds of cultivated land represent the most ephemeral of plant communities, completing their life cycles within a relatively short time and producing copious quantities of dormant seed of potentially long life span.' Indeed, the three habitat traits that favour weed invasion in natural environments are almost always present in modern agroecosystems; they are as follows: (1) disturbance (tillage, seeding and harvest operations, grazing), (2) low species richness (crop monocultures), and (3) high resource availability (bare soil/crop stubble, sunlight, fertilizer) (Booth et al. 2003). Rapid response to high resource availability is a physiological attribute of most early succession plants, especially weeds (Bazzaz 1979).

8.4 Tillage Intensity

Since the beginning of the twentieth century, the widespread use of tillage in agricultural systems has selected for weeds that thrive in tilled soils (Mohler 2001). Many weeds have seeds that are relatively small; the fitness of such weeds is enhanced indirectly by tillage, which removes other species and improves the light, nutrient, soil contact, and temperature environment for seedling establishment. Over the past 30 years, the adoption of direct-seeding and no-tillage (no-till) cropping systems on the Prairies has steadily increased. Direct seeding refers to planting a crop directly into the previous crop's stubble; soil disturbance can vary from low to high levels depending on the planting equipment. No tillage is a subset of direct seeding with <30% soil disturbance. Such systems protect the soil from wind and water erosion, conserve soil moisture, and dramatically reduce labour and fuel costs. Moreover, no-till systems conserve crop residues and soil organic matter, increasing biological life in soils and contribute to long-term sustainability by enhancing nutrient cycling (Soon and Clayton 2002), biological diversity (Lupwayi et al. 1998, 1999), and disease suppression. After a few decades of direct seeding, we have come to understand that the primary value of tillage was for weed management when adequate herbicides and other no-till technology were not available.

In direct-seeding systems where crop residues are left on the soil surface, many weeds with small seeds are disadvantaged. This conclusion is supported by studies conducted in the relatively moist subhumid Parkland region of the Prairies (O'Donovan and McAndrew 2000). Spring seedling populations of green foxtail [*Setaria viridis* (L.) P. Beauv.], stinkweed (*Thlaspi arvense* L.), wild buckwheat (*Polygonum convolvulus* L.), and lamb's-quarters (*Chenopodium album* L.) decreased as tillage intensity was reduced and were lowest under no- (zero) till (Fig. 8.1). Reduction in these small-seeded weed populations in the no-till system occurred despite large amounts of weed seeds sometimes present at the soil surface. Most of these seeds did not germinate. Liebman and Mohler (2001) state that:

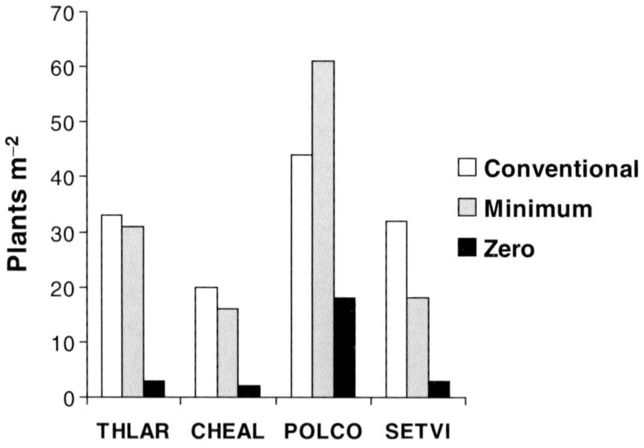

Fig. 8.1 Effect of tillage system on spring weed seedling populations after 4 years of continuous barley in central Alberta. THLAR, CHEAL, POLCO, and SETVI are five-letter codes for stinkweed (*Thlaspi arvense L.*), lamb's quarters (*Chenopodium album L.*), wild buckwheat (*Polygonum convolvulus* L.), and green foxtail [*Setaria viridis* (L.) P. Beauv.], respectively. Adapted from O'Donovan and McAndrew (2000)

'...detrimental effects of crop residue are greater for small-seeded species than larger-seeded species. Because seeds of most major crops are one to three orders of magnitude larger than the weeds with which they regularly compete, residue management offers important opportunities for weed suppression.' Small-seeded weed seedlings do not have the reserves necessary to support sustained growth and to establish themselves in the hostile (dark, dry, physical impedance) conditions that surface crop residues create.

In no-till systems, ungerminated weed seeds left on the soil surface have higher mortality rates (Blackshaw 2005; Roberts and Feast 1972, 1973). This effect is likely due, in part, to vertebrate and invertebrate seed predators, which have easier access to weed seeds left on the soil surface and are also afforded shelter from their own predators. Weed seed predators occur in much greater numbers in no-till fields than in conventional-tillage or organic fields relying on tillage (Menalled et al. 2000). Some insect predators and bacteria and fungi pathogens survive best in undisturbed plant residues that are common in direct-seeding and no-till systems (Derksen et al. 1996). Increased soil microbial diversity can directly or indirectly increase weed seed mortality, reduce weed emergence and growth, and increase crop competitiveness with weeds. However, it is also conceivable that greater microbial diversity could favour weed seed survival and seedling growth. Holmes and Froud-Williams (2005) determined that for weeds such as wild oat, lamb's-quarters, and Canada thistle [*Cirsium arvense* (L.) Scop.], non-avian seed predators removed more seeds than birds. Weed seeds left on the soil surface also experience greater mortality for at least two additional reasons: physiological aging (respiration-related exhaustion of reserves) and germination at soil positions and times of year that are not suitable for seedling emergence or survival (Mohler 2001). It is also likely that potential allelochemicals from crop residues would be more concentrated and inhibitory to weed seeds germinating at or near the soil surface. Thus, the surface microenvironment of conservation-tillage systems can be rather inhospitable for weed seeds and seedlings, as illustrated in the inter-row area of a barley (*Hordeum vulgare* L.) crop seeded on canola (*Brassica napus* L.) stubble (Fig. 8.2).

Although tillage intensity influences weed community density and composition, weed species may not consistently respond the same way to varying levels of soil disturbance (Blackshaw 2005). Therefore, classifying weeds into functional groups based on response to tillage intensity is difficult. Weed adaptation to specific soil types and environments will probably have greater influence on weed communities than tillage. Consequently, predicting trends in abundance of our important invasive alien weeds in various agroecoregions across the Prairies in response to increasing adoption of no-tillage cropping systems is problematic. However, 'farmers should not be deterred from adopting zero tillage production practices because of concerns of increased weed control problems but rather monitor their fields, and be aware of potential changes in weed communities and how they may be effectively managed' (Blackshaw 2005). Thus, weed monitoring is the first step to understanding weed community changes and devising management tactics. For example, in a no-till system, a combination of a selective in-crop herbicide with a pre-harvest application of glyphosate reduced a relatively high Canada thistle infestation to a very low

Fig. 8.2 Inter-row soil surface environment in a juvenile barley crop direct-seeded into canola stubble (Lacombe, AB). Weed seedling germination and survival is often more successful in low surface-residue environments vs. the high-surface residue conditions depicted here

level after 4 years (O'Donovan et al. 2001). Overall, farmers in western Canada often report lower weed densities after 5–10 years of no-till crop production.

8.5 Weed Vulnerabilities to Resource Availability

Fortunately, both native and alien weeds have exploitable vulnerabilities. Weeds may suffer greater relative declines in growth than crops when nutrients are limited because they are adapted to exploit high resource levels (Mohler 2001). Their generally small seed size is often associated with higher relative growth rates and greater biomass partitioning to *thin* leaves (Mohler 2001). However, successful weed phenotypes are selected by almost infinite, random environments that they encounter in agroecosystems. As a consequence of this selection, the relatively greater genetic diversity of weeds compared with most crop species can make them more responsive or *plastic* to varied resource availability (Dekker 1997). Nevertheless, whether weed or crop, some genotypes are simply more plastic than others and the capacity for phenotypic plasticity is genetically controlled (Via et al. 1995).

In resource-limited or resource-abundant environments where *artificial* resources such as fertilizers are applied, weeds may have the advantage when crop managers are not careful to somehow sequester the applied resources for the crops. Accordingly, wise crop producers place nutrients where crops have better access than weeds (Blackshaw et al. 2002, 2005; Kirkland and Beckie 1998). For some species such as green foxtail and stinkweed, a combination of strategic nitrogen placement and no till can dramatically reduce weed populations over time (O'Donovan et al. 1997).

Weeds are also relatively susceptible to the negative effects of shade (Mohler 2001). In addition, some weed species require a *light* signal for germination, and are inhibited by the red-light depleted radiation that filters through crop leaves (Górski 1975; King 1975; Silvertown 1980). Astute managers are aware of these vulnerabilities and strive to promote rapid, uniform crop emergence and ground cover to preempt resources potentially available to weeds. The soil moisture necessary for rapid crop germination and emergence to facilitate preemptive resource acquisition is more likely to be found in direct-seeding and no-till systems than in tilled systems.

8.6 Diversified Cropping Systems Promote Crop Health

The most effective and sustainable form of weed management involves diversified cropping systems. Weeds fortunate enough to grow in simple, repeated cropping systems will continue to have little difficulty adapting and thriving in those systems (Harker and Clayton 2003). Buhler (1999) suggests that the majority of all current cropping systems are highly simplified, allowing the best adapted weed species to proliferate.

The weed seed population of the soil is greatly influenced by the type of crop grown. "Soil conditions being similar, the composition of the flora under continuous wheat and barley is very much the same, but the relative composition of the constituent species varies greatly, some being favoured by the wheat crop and others by the barley" (Brenchley and Warington 1933). "Continuous production of a single crop and short sequences of crops with similar management practices promote the increase of weed species adapted to conditions similar to those used for producing the crops. In contrast, over the course of a diverse rotation employing crops with different planting and harvest dates, different growth habits and residue characteristics, and different tillage and weed management practices, weeds can be challenged with a wide range of stresses and mortality risks, and will be given few consistent opportunities for unchecked growth and reproduction" (Liebman and Staver 2001).

The successful utilization of crop diversity in weed management systems is mostly governed by the life cycle of the most dominant weed(s) and the life cycle of the rotational crops. Many crop rotations involve substantial crop species diversity, but lack crop life cycle diversity. For example, if summer annual wild oat is the dominant weed species, crop producers should not solely grow summer annual crops in their rotation. Conversely, downy brome (*Bromus tectorum* L.), a winter annual, is easily managed in winter wheat when the crop rotation includes canola – a summer annual crop in Canada (Blackshaw 1994).

Using crop rotations with varying crop life cycles is not the only way of introducing diversity into a cropping system. Diversity can also be introduced by varying crop sowing date (Clayton et al. 2004), or by varying the date of crop harvest (Harker et al. 2003b). Almost any other method that introduces diversity into cropping systems can reduce the prevalence and impact of a relatively few dominant weed species. Therefore, crop diversity can be temporal (varied planting and harvest dates), spatial where different species are grown in the same space but at different times (variable life cycles), or by growing different crop species at the same time and in the same space (intercropping).

8.7 System Approach Vs. Single Components

It is important to recognize that any single practice will not be adequate for long-term weed management. Successful crop managers combine a variety of weed management practices both in time and in space. For example, combining early herbicide application with a competitive canola cultivar and higher-than-normal seeding rate not only reduced weed biomass and variability, but also improved crop yield (Harker et al. 2003a). When combining several optimal practices, a major goal should be the promotion of crop health. Healthy and early developing crop canopies limit weed invasion (Harker et al. 2005), because they exploit weed vulnerability to low light and altered light quality (Mohler 2001) and pre-empt nutrient and soil moisture resource availability. Overall, annual weed invasions can be managed when two or more related cropping principles are used in combination: in this case, rotational diversity and crop health.

Beck (2006) reminds us that 'Successful crop production, regardless of the methods used, is a careful piecing together of numerous components into a system. Simply replacing one component with another is seldom successful'. Focusing on crop health and competitiveness will lead producers to adopt a tool kit of management practices that includes sanitation to limit weed seed spread, low-disturbance seeding (maintaining crop residues), higher crop seed rates, relatively narrow row spacing, optimum fertilizer placement, and diverse crop rotations. For example, the simple practice of increasing crop seed rates consistently reduces weed competition and improves herbicide performance (Blackshaw et al. 2006; Mohler 2001), and is a very effective form of biological weed control (Blackshaw et al. 2000; Harker et al. 2005). Seeding crops at relatively high rates can also increase weed economic threshold values, and thus reduce the need for herbicide application (O'Donovan et al. 2005). Poor fertility can reduce crop health to the degree that all of the tools employed for pest management are negated. Similarly, disease and insect management are also important for weed management because of their impact on crop health and competitiveness. It may be that the best weed management approach for a weakly competitive crop such as pea (*Pisum sativum* L.) is to ensure that optimal practices are combined to limit weed seed production in the crop grown before pea.

ICM exploits synergies that are possible when technologies are combined with natural resources for sustainable and profitable crop production. Crop management solutions can be so urgent that *quick-fix* remedies seem imperative. In response, research and extension personnel have often stressed single tools (usually a technological input such as a herbicide). However, crop management challenges are the culmination of numerous decisions over a long time period. Highly simplified cropping systems with heavy dependence on herbicide tools are susceptible to the selection of herbicide-resistant weed biotypes (described later).

8.8 Herbicide Resistance

8.8.1 Abundance

The relatively recent phenomenon of herbicide resistance can potentially influence weed invasions in agroecosystems. Herbicide-resistant weed populations are found in over 10% of cultivated land (ca. 5 million ha) in the Canadian Prairies (Beckie et al. 2008). Incidence and economic impact of herbicide resistance are much greater in grass than in broadleaf weed species. Wild oat resistant to acetyl-coA carboxylase (ACCase) or acetolactate synthase (ALS) inhibitors account for 80% of these infestations. Green foxtail resistant to ACCase inhibitors is common in the central and eastern Prairies. Cases of ALS inhibitor resistance in a number of broadleaf weed species, such as wild mustard (*Sinapis arvensis* L.), stinkweed, chickweed [*Stellaria media* (L.) Vill.], and spiny annual sowthistle [*Sonchus asper* (L.) Hill], are less common than the two grass weeds but are increasing steadily (Beckie et al. 2008).

Because green foxtail and wild oat rank first and second, respectively, in relative abundance among weed species in the Prairies (Leeson et al. 2005b), herbicides are frequently used to control them. Since 1996, ACCase inhibitors have been applied to nearly 50% of cropped land annually, resulting in high and sustained selection pressure (Beckie et al. 2008); ALS inhibitors are routinely applied to one-third or more of cropped fields (Leeson et al. 2005a). Where the mechanism of resistance has been ascertained, most cases of ACCase inhibitor resistance in grass weed species and ALS inhibitor resistance in grass or broadleaf weed species in the Prairies are due to target-site mutation selected by herbicides applied at rates giving high efficacy (>90% control).

As in other species, ALS inhibitor resistance in kochia is due to target-site resistance; some plants can contain multiple target-site mutations (Warwick et al. 2008). In contrast to resistant biotypes of wild oat (Beckie et al. 2005) or other species, the incidence or spread of ALS inhibitor-resistant kochia is increasing the fastest.

8.8.2 Has Herbicide Resistance in Kochia Accelerated Its Invasiveness? A Case Study

Kochia, an invasive alien, is highly variable in morphology, growth, and development as influenced by environment (Bell et al. 1972; Eberlein and Fore 1984). The biological characteristics of kochia that enhance its spread or invasive capability include tumbling plant architecture, a facultative outcrossing breeding system, evolution of resistant biotypes to ALS inhibitor herbicides (Morrison and Devine 1994; Heap 2008), drought tolerance (Erickson and Moxon 1947), tolerance to saline and alkaline/high pH soils (Erickson and Moxon 1947), and tolerance to predation by grasshoppers (*Melanoplus* and *Camnula* spp.) and other insects (Olfert et al. 1990).

Kochia is one of the top ten most abundant agricultural weeds in the Prairies; it is undergoing rapid range expansion in western Canada (Leeson et al. 2005b). Since the 1970s and 1980s, kochia has expanded northward into cooler, wetter regions from traditional areas in the southern semiarid grassland region of the Prairies (Fig. 8.3), and has recently been reported in the most northern agricultural areas (Maurice, personal communication). It is also increasing in abundance, as exemplified by its increased relative abundance ranking (increase of 14 positions) from the 1970s to 2000s (Leeson et al. 2005b). In contrast to kochia, the distribution and abundance of Russian thistle (*Salsola tragus* L.) has remained static or declined over the past 40 years (Leeson et al. 2005b) even though it apparently is biologically and ecologically similar to kochia (Crompton and Bassett 1985).

Limitations to the northern expansion of kochia have been attributed in the past to its inability to flower under long-day conditions because it is a short-day plant,

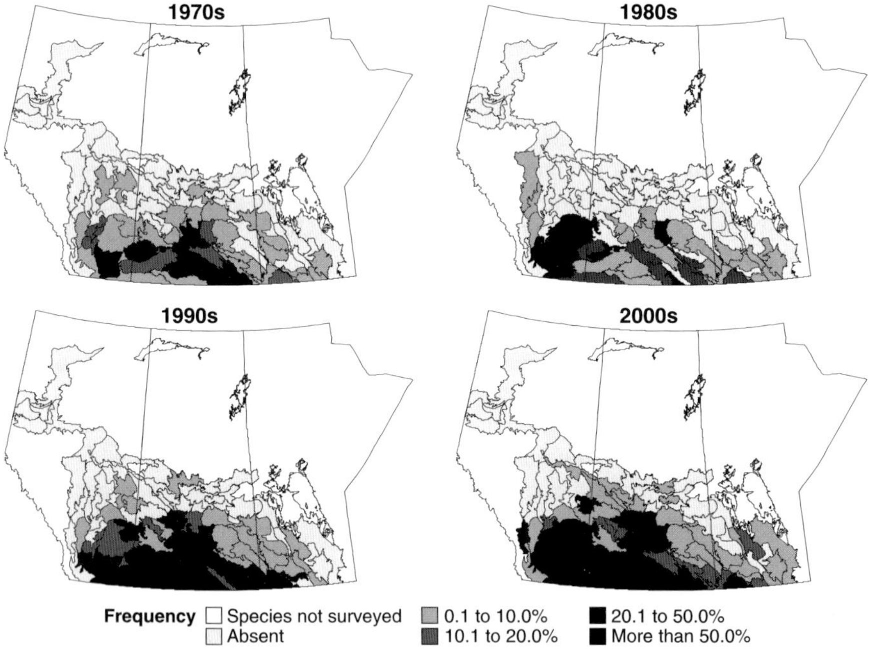

Fig. 8.3 Percent frequency of kochia in surveyed Canadian Prairie province fields (Alberta, Saskatchewan, Manitoba – left to right) from the 1970s to the 2000s. ALS inhibitor resistance in kochia was initially documented in the 1990 surveys. In 2007, a random survey of over 100 fields across all agricultural ecoregions in the three provinces found that over 80% of sites had ALS inhibitor-resistant populations (Beckie, unpublished data). Adapted from Leeson et al. (2005b), p. 349

and/or set viable seed under the shorter frost-free growing season (Bell et al. 1972; Eberlein and Fore 1984). Kochia is a prolific seed producer – up to 30,000 seeds per plant (Stallings et al. 1995). Mature seed is readily germinable, and kochia germinates early in spring when soil temperatures are relatively cool (Nussbaum et al. 1985). It has a sustained emergence period and may emerge after in-crop herbicide application (Mickelson et al. 2004). Although only limited controlled-environment germination research has been conducted on this species (Everitt et al. 1983; Jami Al-Ahmadi and Kafi 2006), data indicate that kochia can germinate over a wide range of temperatures. It is possible that selection of ecotypes differing in germination response or days to flowering may have contributed to its northern spread; however, no information is known of possible Canadian latitudinal/climatic ecotypes.

Climate change is expected to impact weed communities, including kochia. Kochia and Russian thistle utilize the rarer C4 photosynthetic pathway and therefore are ideal indicator species to track climate change effects and impact of the frequent hot, dry conditions of the Prairies in recent years. In the past 40 years, the growing season in the Prairies has increased by 1–4 days; the 1990s were the warmest on record (Gitay et al. 2002). Bioclimatic modelling is useful in assessing the impact of changes in climate on pest population distribution (Olfert and Weiss 2006; Rogers et al. 2007), and is being used to predict the suitability of specific agroecosystems for survival and reproduction of kochia under selected climate-change scenarios (Olfert, unpublished data). Climate change on the Prairies may facilitate invasions of new weedy alien species introduced as a result of contamination of seedlots of more adaptable crop species or varieties.

Kochia was the first weed species reported to evolve herbicide-resistant biotypes in western Canada (Morrison and Devine 1994). Only one Russian thistle resistant biotype was documented in the Prairies in 1989, with no reported cases until 2007 (Beckie, unpublished data). Herbicide-resistant kochia biotypes have spread dramatically since 1988, with reports of over 80% of populations in some areas showing some level of ALS inhibitor resistance (Beckie et al. 2008, unpublished data). Because ALS inhibitor resistance is frequently endowed by a single dominant or semi-dominant nuclear gene, resistance alleles can move via seed or pollen (Mallory-Smith et al. 1993; Stallings et al. 1995). Prolific seed production and rapid turnover of the soil seedbank (Burnside et al. 1981) increase herbicide resistance evolution in kochia. In the USA, ALS inhibitor-resistant kochia has no detectable fitness penalty (Peterson 1999). However, a pleiotropic effect (i.e., one gene affects several traits) of the ALS mutation on germination response was reported. Germination of ALS inhibitor-resistant kochia occurred at cooler soil temperatures and/or more rapidly than susceptible kochia and was presumed to be due to the elevated levels of branched-chain amino acids that result from the mutation (Dyer et al 1993; Thompson et al. 1994). Greater and more rapid germination and emergence of resistant kochia than that of susceptible kochia in cool soil in spring may impact both competitiveness and invasiveness (northerly range expansion).

Genetic diversity is the heritable variation within and among populations of a species, and provides the opportunity for a population to evolve under changing environments and selection pressures. Genetic diversity in kochia as estimated by simple sequence repeat markers was high (Mengistu and Messersmith 2001), with 90% of the variation occurring within populations. Genetic diversity within and among resistant and susceptible populations was similar. This result suggests that despite generations of herbicide selection, kochia maintains high genetic diversity through substantial levels of gene flow within and among populations. High levels of genetic variation, and the accompanying increased ability to respond to selection pressures, natural or artificial, are expected to affect the spread or success of an invasive species (Lee 2002; Sakai et al. 2001). Little is known about genetic diversity of Canadian kochia populations, and high levels of genetic diversity may well be an important factor contributing to their spread.

8.9 True Integrated Pest Management

It has been suggested, and appears to be true, that many integrated pest management (IPM) programs can be more accurately described as "integrated pesticide management" or as the "other IPM" (Ehler 2006). Clearly, to effectively manage invasive weed species over the long term, we have to look beyond herbicides alone. Indeed, true integrated weed management not only requires that tools other than herbicide be employed, but that management practices are multi-disciplinary, and are compatible (O'Donovan et al. 2007).

It is not easy to make pest management strategies compatible. For example, early weed removal is desirable for optimum canola yield, but can cause greater root maggot (*Delia* spp.) (Diptera: Anthomyiidae) damage in canola (Dosdall et al. 2003). In a recent popular press article describing this study, the following result was highlighted: '...up to 70% of root maggots are eaten by predators and 40% of the survivors are killed by parasites' (McMenamin 2006). Therefore, when insect pests are below economic thresholds and sprayed unnecessarily, beneficial insects may also be killed. In the Prairies, economic thresholds for major insect pests are known, updated annually, and rapidly adopted by producers. Similarly, Diadegma wasp (*Diadegma* spp.) parasitize and can almost totally neutralize outbreak of a diamondback moth (*Plutella xylostella* L.) pest. However, '...Diadegma produce more eggs when they can feed on wild mustard or yellow rocket (*Barbarea vulgaris* Ait. f.) early in the year and survive longer on lamb's-quarters and perennial sowthistle (*Sonchus arvensis* L.)' (McMenamin 2006). Thus, weed control can negatively impact management of other crop pests; the relative risks of yield loss due to weed competition or insect damage must be balanced.

8.10 Conclusions

Although some would take issue with the classification of our most common and economically important Prairie weeds as invasive aliens, most of these species are true aliens that have immediate and verifiable impact on crop producer income and farm sustainability. Increasing incidence of herbicide resistance in weeds and other crop production or environmental factors may significantly alter invasive weed dynamics and crop–weed interactions, and will undoubtedly pose future challenges to crop producers. These and other challenges can be countered by reduced soil disturbance and diversified cropping systems that combine synergistic components such as competitive cultivars, relatively high crop seed rate, and judicious fertilizer placement in true IPM systems. The common alien weed species that persistently invade Prairie cropland each year will not be subdued over the long term in the absence of such systems.

References

Bazzaz FA (1979) The physiological ecology of plant succession. Ann Rev Ecol Syst 10:351–371

Beck D (2006) No-till guidelines for the arid and semi-arid prairies. http://www.dakotalakes.com/Publications/Guidelines.PDF. Accessed 8 November 2006

Beckie HJ, Thomas AG, Légère A, Kelner DJ, Van Acker RC, Meers S (1999) Nature, occurrence, and cost of herbicide-resistant wild oat (*Avena fatua*) in small-grain production areas. Weed Technol 13:612–625

Beckie HJ, Hall LM, Tardif FJ (2001) Impact and management of herbicide-resistant weeds in Canada. Proc. Brighton Crop Protection Conference – Weeds. British Crop Protection Council, Farnham, UK, pp.747–754

Beckie HJ, Hall LM, Schuba B (2005) Patch management of herbicide-resistant wild oat (*Avena fatua*). Weed Technol 19:697–705

Beckie HJ, Leeson JY, Thomas AG, Brenzil CA, Hall LM, Holzgang G, Lozinski C, Shirriff S (2008) Weed resistance monitoring in the Canadian Prairies. Weed Technol 22:530–543

Bell AR, Nalewaja JD, Schooler AB (1972) Light period, temperature, and kochia flowering. Weed Sci 20:462–464

Blackshaw RE (1994) Rotation affects downy brome (*Bromus tectorum*) in winter wheat (*Triticum aestivum*). Weed Technol 8:728–732

Blackshaw RE (2005) Tillage intensity affects invasiveness of weed species in agroecosystems. In: Inderjit (ed) Invasive plants: ecological and agricultural aspects, Birkhauser-Verlag AG, Switzerland

Blackshaw RE, Semach GP, O'Donovan JT (2000) Utilization of wheat seed rate to manage redstem filaree (*Erodium cicutarium*) in a zero-tillage cropping system. Weed Technol 14:389–396

Blackshaw RE, Semach G, Janzen HH (2002) Fertilizer application method affects nitrogen uptake in weeds and wheat. Weed Sci 50:634–641

Blackshaw RE, Molnar LJ, Larney FJ (2005) Fertilizer, manure and compost effects on weed growth and competition with winter wheat in western Canada. Crop Prot 24:971–980

Blackshaw RE, O'Donovan JT, Harker KN, Clayton GW, Stougaard RN (2006) Reduced herbicide doses in field crops: a review. Weed Biol Manage 6:10–17

Booth BD, Murphy SD, Swanton CJ (2003) Plant invasions. In: Booth BD, Murphy SD, Swanton CJ (eds) Weed ecology in natural and agricultural systems. CABI, Oxford, UK

Brenchley WE, Warington K (1933) The weed seed populations of arable soil. II. Influence of crop, soil and methods of cultivation upon the relative abundance of viable seeds. J Ecol 21:103–127

Brook H (2007) Crop protection 2007. Alberta Agriculture and Food. Agdex 606-1

Buhler DD (1999) Expanding the context of weed management. In: Buhler DD (ed) Expanding the context of weed management. Haworth Press, New York

Burnside OC, Fenster CR, Evetts LL, Mumm RF (1981) Germination of exhumed weed seed in Nebraska. Weed Sci 29:577–586

Clayton GW, Harker KN, O'Donovan JT, Blackshaw RE, Dosdall LM, Stevenson FC, Ferguson T (2004) Fall and spring seeding date effects on herbicide-tolerant canola (*Brassica napus* L.) cultivars. Can J Plant Sci 84:419–430

Clements DR, DiTommaso A, Jordan N, Booth BD, Cardina J, Doohan D, Mohler CL, Murphy SD, Swanton CJ (2004) Adaptability of plants invading North American cropland. Agric Ecosyst Environ 104:379–398

Crompton CW, Bassett IJ (1985) The biology of Canadian weeds. 65. Salsola pestifer A. Nels. Can J Plant Sci 65:379–388

Dekker J (1997) Weed diversity and weed management. Weed Sci 45:357–363

Derksen DA, Blackshaw RE, Boyetchko SM (1996) Sustainability, conservation tillage and weeds in Canada. Can J Plant Sci 76:651–659

DiTomaso JM (2008) WSSA enters a new ear of invasive weed communication. Invas Plant Sci Manage 1:1–2

Dosdall LM, Clayton GW, Harker KN, O'Donovan JT, Stevenson FC (2003) Weed control and root maggots: making pest management strategies compatible. Weed Sci 51:576–585

Dyer WE, Chee PW, Fay PK (1993) Rapid germination of sulfonylurea-resistant *Kochia scoparia* accessions is associated with elevated seed levels of branched chain amino acids. Weed Sci 41:18–22

Eberlein CV, Fore ZQ (1984) Kochia biology. Weeds Today 15:5–7.

Ehler LE (2006) Integrated pest management (IPM): definition, historical development and implementation, and the other IPM. Pest Manage Sci 62:787–789

Erickson EL, Moxon AL (1947) Forage from kochia. Bull. 384, South Dakota Agricultural Experiment Station, Brookings, SD

Everitt JH, Alaniz MA, Lee JB (1983) Seed germination characteristics of *Kochia scoparia*. J Range Manage 36:646–648

Froud-Williams RJ (1988) Changes in weed flora with different tillage and agronomics management systems. In: Altieri MA, Liebman M (eds) Weed management in agroecosystems: ecological approaches. CRC Press, Boca Raton, FL

Gitay H, Suarez A, Watson RT, Dokken DJ (2002) Climate change and biodiversity. IPCC (Intergovernmental Panel on Climate Change) Technical Paper V. United Nations Environment Programme and World Meteorological Organization. http://www.ipcc.ch/pdf/technical-papers/climate-changes-biodiversity-en.pdf. Accessed 15 April 2008

Górski T (1975) Germination of seeds in the shadow of plants. Physiol Plant 34:342–346

Government of Canada (2004) An invasive alien species strategy for Canada. http://www.cbin.ec.gc.ca/issues/ias_invasives.cfm?lang = e. Accessed 8 May 2007

Haber E (2002) Spread and impact of alien plants across Canadian landscapes. In: Claudi R, Nantel P, Muckle-Jeffs E (eds) Alien invaders in Canada's waters, wetlands, and forests. Canadian Forest Service, National Resources Canada, Ottawa, ON

Harker KN, Clayton GW (2003) Diversified weed management systems. In: Inderjit (ed) Principles and practices in weed management: biology and management. Kluwer Academic, Dordrecht, The Netherlands

Harker KN, Clayton GW, Blackshaw RE, O'Donovan JT, Stevenson FC (2003a) Seeding rate, herbicide timing and competitive hybrids contribute to integrated weed management in canola (Brassica napus). Can J Plant Sci 83:433–440

Harker KN, Kirkland KJ, Baron VS, Clayton GW (2003b) Early-harvest barley (*Hordeum vulgare*) silage reduces wild oat (*Avena fatua*) densities under zero tillage. Weed Technol 17:102–110

Harker KN, Clayton GW, O'Donovan JT (2005) Reducing agroecosystem vulnerability to weed invasion. In: Inderjit (ed) Invasive plants: ecological and agricultural aspects. Birkhauser-Verlag AG, Switzerland

Heap IM (2008) International survey of herbicide resistant weeds. http://www.weedscience.org. Accessed 5 April 2008

Holmes RJ, Froud-Williams RJ (2005) Post-dispersal weed seed predation by avian and non-avian predators. Agric Ecosyst Environ 105:23–27

Jami Al-Ahmadi M, Kafi M (2006) Cardinal temperatures for germination of *Kochia scoparia* (L.). J Environ 68:308–314
King TJ (1975) Inhibition of seed germination under leaf canopies in *Arenaria serpyllifolia*, *Veronica arvensis* and *Cerastum* (sic) holosteoides. New Phytol 75:87–90
Kirkland KJ, Beckie HJ (1998) Contribution of nitrogen fertilizer placement to weed management in spring wheat (*Triticum aestivum*). Weed Technol 12:507–514
Lee CN (2002) Evolutionary genetics of invasive species. Trends Ecol Evol 17:386–391
Leeson JY, Thomas AG, Beckie HJ, Hall LM, Brenzil C, Van Acker RC, Brown KR, Andrews T (2005a) Group 2 herbicide use in the prairie provinces. Proc. 2004 National Meeting of the Canadian Weed Science Society. http://www.cwss-scm.ca. Accessed 10 March 2007
Leeson JY, Thomas AG, Hall LM, Brenzil CA, Andrews T, Brown KR, Van Acker RC (2005b) Prairie weed surveys of cereal, oilseed and pulse crops from the 1970s to the 2000s. Weed Survey Series Publ. 05-1. Agriculture and Agri-Food Canada, Saskatoon Research Centre, Saskatoon. 395 p
Leeson JY, Thomas AG, O'Donovan JT (2006) Economic impact of alien weeds on wheat, barley and canola production. Proc. Canadian Weed Science Society, Victoria, BC. Poster abstract p. 90
Liebman M, Mohler CL (2001) Weeds and the soil environment. In: Liebman M, Mohler CL, Staver CP (eds) Ecological management of agricultural weeds. Cambridge University Press, UK
Liebman M, Staver CP (2001) Crop diversification for weed management. In: Liebman M, Mohler CL, Staver CP (eds) Ecological management of agricultural weeds. Cambridge University Press, UK
Lupwayi NZ, Rice WA, Clayton GW (1998) Soil microbial diversity and community structure under wheat as influenced by tillage and crop rotation. Soil Biol Biochem 30:1733–1741
Lupwayi NZ, Rice WA, Clayton GW (1999) Soil microbial biomass and carbon dioxide flux under wheat as influenced by tillage and crop rotation. Can J Soil Sci 79:273–280
Mallory-Smith CA, Thill DC, Stallings GP (1993) Survey and gene flow in acetolactate synthase resistant kochia and Russian thistle. Proc. Brighton Crop Protection Conference – Weeds. British Crop Protection Council, Farnham, UK, pp. 555–558
McMenamin H (2006) Natural enemies – farmer's unpaid helpers. In: Darbishire P (ed) Top Crop Manager, November 2006 issue. Annex Publishing & Printing, Exeter, ON, pp. 50–54
Menalled F, Landis D, Lee L (2000) Ecology and management of weed seed predators in Michigan agroecosystems. Michigan State University Extension Bulletin E-2716
Mengistu LW, Messersmith CG (2001) Genetic diversity within and among herbicide-resistant and -susceptible populations of kochia (*Kochia scoparia*). Weed Sci Soc Am Abstr 41:124
Mickelson J, Bussan A, Davis E, Hulting A, Dyer W (2004) Postharvest kochia (*Kochia scoparia*) management with herbicides in small grains. Weed Technol 18:426–431
Mohler CL (2001) Weed life history: identifying vulnerabilities. In: Liebman M, Mohler CL, Staver CP (eds) Ecological management of agricultural weeds. Cambridge University Press, UK
Morrison IN, Devine MD (1994) Herbicide resistance in the Canadian prairie provinces: five years after the fact. Phytoprotection (Suppl) 75:5–16
Nussbaum ES, Wiese AF, Crutchfield DE, Chenault EW, Lavake D (1985) The effects of temperature and rainfall on emergence and growth of eight weeds. Weed Sci 33:165–170
O'Donovan JT, McAndrew DW (2000) Effect of tillage on weed populations in continuous barley (*Hordeum vulgare*). Weed Technol 14:726–733
O'Donovan JT, McAndrew DW, Thomas AG (1997) Tillage and nitrogen influence weed population dynamics in barley (*Hordeum vulgare*). Weed Technol 11:502–509
O'Donovan JT, Blackshaw RE, Harker KN, McAndrew DW, Clayton GW (2001) Canada thistle (*Cirsium arvense*) management in canola (*Brassica rapa*) and barley (*Hordeum vulgare*) rotations under zero tillage. Can J Plant Sci 81:183–190
O'Donovan JT, Blackshaw RE, Harker KN, Clayton GW, Maurice DC (2005) Field evaluation of regression equations to estimate crop yield losses due to weeds. Can J Plant Sci 85:955–962

O'Donovan JT, Blackshaw RE, Harker KN, Clayton GW, Moyer JR, Dosdall LM, Maurice DC, Turkington TK (2007) Integrated approaches to managing weeds in spring-sown crops in western Canada. Crop Prot 26:390–398

Olfert O, Weiss RM (2006) Impact of climate change on potential distributions and relative abundances of *Oulema melanopus*, *Meligethes viridescens* and *Ceutorhynchus obstrictus* in Canada. Agric Ecosyst Environ 113:295–301

Olfert O, Hinks CF, Craig W, Westcott ND (1990) Resistance of *Kochia scoparia* to feeding damage by grasshoppers (Orthoptera: Acrididae). J Econ Entomol 83:2421–2424

Peterson DE (1999) The impact of herbicide-resistant weeds on Kansas agriculture. Weed Technol 13:632–635

Pimentel D, Lach L, Zuniga R, Morrison D (2000) Environmental and economic costs of nonindigenous species in the United States. BioScience 50:53–65

Pyšek P, Richardson DM, Rejmánek M, Webster GL, Williamson M, Kirschner J (2004) Alien plants in checklists and floras: towards better communication between taxonomists and ecologists. Taxon 53:131–143

Roberts HA, Feast PM (1972) Fate of seeds of some annual weeds in different depths of cultivated and undisturbed soil. Weed Res 12:316–324

Roberts HA, Feast PM (1973) Emergence and longevity of seeds of annual weeds in cultivated and undisturbed soil. J Appl Ecol 10:133–143

Rogers DJ, Reid DE, Rogers JJ, Addison SJ (2007) Prediction of the naturalisation potential and weediness risk of transgenic cotton in Australia. Agric Ecosyst Environ 119:177–189

Sakai AK, Allendorf FW, Holt JS, Lodge DM, Molofsky J, With KA, Baughman S, Cabin RJ, Cohen JE, Ellstrand NC, McCauley DE, O'Neil P, Parker IM, Thompson JN, Weller SG (2001) The population biology of invasive species. Annu Rev Ecol Syst 32:305–332

Silvertown J (1980) Leaf-canopy-induced seed dormancy in a grassland flora. New Phytol 85:109–118

Soon YK, Clayton GW (2002) Eight years of crop rotation and tillage effects on crop production and N fertilizer use. Can J Soil Sci 82:165–172

Stallings GP, Thill DC, Mallory-Smith CA, Shafii B (1995) Pollen-mediated gene flow of sulfonylurea-resistant kochia (*Kochia scoparia*). Weed Sci 43:95–102

Swanton CJ, Harker KN, Anderson RL (1993) Crop losses due to weeds in Canada. Weed Technol 7:537–542

Thomas AG, Leeson JY (2007) Tracking long-term changes in the arable weed flora of Canada. In: Clements DR, Darbyshire SJ (eds) Invasive plants: inventories, strategies, and action. Topics in Canadian Weed Science, Vol. 5. Canadian Weed Science Society – Société canadienne de malherbologie, Sainte-Anne-de Bellevue, QC, pp 43–69.

Thompson CR, Thill DC, Shafii B (1994) Germination characteristics of sulfonylurea-resistant and -susceptible kochia (*Kochia scoparia*). Weed Sci 42:50–56

Via S, Gomulkiewicz R, De Jong G, Scheiner SM, Schlichting CD, Van Tienderen PH (1995) Adaptive phenotypic plasticity: consensus and controversy. Trends Ecol Evol 10:212–217

Warwick SI, Xu R, Sauder C, Beckie HJ (2008) Target-site mutations and single nucleotide polymorphism genotyping in ALS-resistant kochia (*Kochia scoparia*). Weed Sci. 56:797–806

White DJ, Haber E, Keddy C (1993) Invasive plants of natural habitats in Canada. Canadian Wildlife Service, Environment Canada/Canadian Museum of Nature, Ottawa, ON. 121 p

Chapter 9
Invasive Plant Species and the Ornamental Horticulture Industry

Alex X. Niemiera and Betsy Von Holle

Abstract The ornamental horticulture industry is responsible for the introduction, propagation, and transport of thousands of nonnative plant species, most of which stay in their intended locations or spread without significant environmental impacts. However, some nonindigenous plant species have proved to be particularly invasive and quite environmentally deleterious. The economically and politically powerful horticulture industry is faced with the dichotomous dilemma of the freedom to import and propagate plant species juxtaposed with the responsibility to be a diligent land steward. We discuss the various fundamental biological factors of plant invasion, as well as the environmental impacts, probability, prediction, and ranking of invasive species. We also review the role of the nursery industry in importing nonnative species, the perception of the problem by nursery personnel, and the impact of governmental and self-regulation. We conclude with recommendations for the ornamental plant industry to mitigate its role in dispersing invasive, nonnative plant species.

Keywords Ornamental horticulture industry • nonindigenous species • invasive plants • voluntary regulation

Abbreviations ANLA: American Nursery and Landscape Asssociation; ANLA – HRI: American Nursery and Landscape Association – Horticultural Research Institute; APHIS: Animal and Plant Health Inspection Service; AQIS: Australian Quarantine and Inspection Service; BANR: Board on Agriculture and Natural Resources; Cal-IPC: California Invasive Plant Council; EPPC: Exotic Pest Plant Council; EPPO: European and Mediterranean Plant Protection Organization; EMG: Extension Master Gardener; FL-EPPC: Florida Exotic Pest Plant Council; GISP:

A.X. Niemiera
Department of Horticulture, Virginia Polytechnic Institute and State University, Blacksburg, VA 24061
niemiera@vt.edu

B.Von Holle (✉)
Department of Biology, University of Central Florida, Orlando, FL 32816, USA
vonholle@mail.ucf.edu

Inderjit (ed.), *Management of Invasive Weeds*,

Global Invasive Species Programme; IRA: Import Risk Analysis; NGIA: Nursery and Garden Industry Australia; NIS: Nonindigenous species; SE-EPPC: Southeast Exotic Pest Plant Council; US: United States; USDA: United State Department of Agriculture; WRA: Weed Risk Assessment

9.1 Introduction

Nonindigenous species (NIS), also termed nonnative, exotic, and alien, are the subject of a considerable amount of interest, research, and debate. Many nonnative plant species are incontrovertibly a great benefit to society by serving as food, timber, and ornamental plants (Ewel et al. 1999). However, other nonnative plant species are particularly invasive, and therefore a bane to society when they negatively impact native biodiversity and cause huge economic expenditures (Parker et al. 1999; Pimentel et al. 2000).

The topic on invasive, nonindigenous plants encompasses a great breadth of issues and stakeholders. There is a vast literature dealing with the ecological, economic, regulatory, control, management, and social aspects of invasive, nonnative plants. Vested stakeholders include scientists, environmental groups, land managers, regulatory officials, businesses in ornamental horticulture, seed, forest products, and the gardening public. The scale of stakeholder interests ranges from international to local arenas. Stakeholders' interest greatly affects their perspective of the topic on nonnative plants; even the definitions of terms within the invasive NIS lexicon are greatly affected by a stakeholder's interest. Differences in stakeholders' perception can lead to adversarial interactions (Drake 2005). These interactions are precipitated by the intersection of science, conflicting value systems, environmental ethics, and public policy (Lodge and Shrader-Frechette 2003). A particularly visible and pertinent example of a vested interest that precipitates differences in opinion is the ornamental horticulture industry, whose businesses import, propagate, sell, and plant mostly nonnative flora. This industry is especially vested in this issue since it is responsible for the introduction and spread of thousands of nonnative plant species, most of which stay in their intended locations or spread without significant environmental negative impacts, while other nonindigenous plant species have proved to be particularly invasive and quite environmentally deleterious. Thus, the juxtaposition of the industry's powerful and fruitful economic impacts and the environmental and regulatory agencies' desire to protect natural areas from invasive NIS sets the stage for a conflict with no clear compromise or resolution. The ornamental horticulture industry is by no means the only stakeholder in the fray of the invasive plant debate. Other parties that are in the midst of the contentious invasive plant issue include botanical gardens, gardeners and garden clubs, public agencies that plant and manage landscapes, and horticultural educational programs.

There is a relatively weak link between the scientific realm of invasive plant biology and the ornamental horticulture industry. This is primarily because the science of plant invasion biology is not often effectively communicated to individuals outside of the research realm (Allendorf and Lundquist 2003; Jordan et al. 2003). The intersection of science, conflicting value systems, environmental ethics, public policy, and articles in the popular press have given mixed messages, confusion, and tension in the NIS theater (Lodge and Shrader-Frechette 2003). Communication shortcomings are especially evident for audiences in the ornamental horticulture industry and the gardening public. Ornamental horticulture personnel are generally aware of the issue of invasive NIS; however, there is a need for the industry to grasp the fundamental concepts of plant invasion biology and to address and formulate strategies concerning the sale and planting of potentially invasive NIS. The nursery-mediated spread of invasive plants is a major challenge and concern for the future of the nursery industry (Green and Green 2003). The objective of our paper is to convey the salient aspects of plant invasion biology that are relevant to the ornamental horticultural audience. We will cover the fundamental aspects of invasive nonnative plants, the role of the nursery and landscape industry in invasive nonindigenous plant species, the reasons why the ornamental horticulture industry should care about NIS, and recommendations for the ornamental plant industry to mitigate its role in dispersing invasive, nonnative plant species.

9.2 Fundamental Aspects of Invasive NIS

9.2.1 What is Plant Invasion?

Biological invasion is a phenomenon that is old as life itself (Drake 2005). However, the rate of invasion has greatly accelerated in the past century due to anthropogenic factors such as removal of dispersal barriers, international travel, and enterprise (Drake 2005). The process of invasion (area occupied vs. time) exhibits the pattern of a sigmoidal curve in which the initial slow growth, exponential growth, and the flattening of the curve represent introduction, colonization, and naturalization, respectively (Radosevich et al. 2003). Radosevich et al. (2003) state that a species is naturalized "when it successfully establishes new self-perpetuating populations, is dispersed widely throughout a region, and is incorporated into the resident flora." The time from first arrival to the rapid occupation by the naturalizing species is termed *lag time* and may occur in years or decades (Kowarik 1995). Lag time duration can be influenced by environmental factors of the recipient habitat that improve conditions for the invading organism, detection, invasion pressure, dispersal pathways, introduction of new pollinators, or genetic changes that improve fitness of the organism (Bryson and Carter 2004; Crooks and Soule 1999; Mack et al. 2000). Lag times for woody plants can exceed 100 years (Kowarik 1995); however, lag times for herbaceous perennials are believed to be much shorter than for woody plants (Reichard and White 2001). Detection of invaders and quantification of their invasiveness is a function of human perception. Species may be cryptic and widespread and therefore go unnoticed for years until people start to look for them.

9.2.2 Terminology

The interpretation of *invasive species* by horticulturalists, policy makers, and scientists often varies depending on stakeholder interests. The American Nursery and Landscape Association (ANLA, http://www.anla.org/industry/index.htm. Accessed 23 May 2008) and the Weed Science Society of America define invasive plants as "plants that have or are likely to (1) spread into native plant communities and cause environmental harm by developing self-sustaining populations and disrupting those systems; or, (2) spread into managed plant systems and cause economic harm" (Hall 2000). The legal definition of an invasive species, and the official position of the U.S. Government (Federal Register – Presidential Documents 1999), is "an alien species whose introduction does or is likely to cause economic or environmental harm or harm to human health." In a strict interpretation, any plant outside its native ecosystem is considered nonindigenous. For example, black locust (*Robinia pseudoacacia*) is a common tree native to the central US Appalachian and Ozark Mountains but is considered an invasive species in California. Thus, plants native to one state can be invasive in another state.

Richardson et al. (2000) define a minimum set of terms that describes the invasion/naturalization process of plant species: "*Introduction* means that the plant (or its propagule) has been transported by humans across a major geographical barrier. *Naturalization* starts when abiotic and biotic barriers to survival are surmounted and when various barriers to regular reproduction are overcome. *Invasion* further requires that introduced plants produce reproductive offspring in areas distant from sites of introduction (approximate scales: >100 m over <50 years for taxa spreading by seeds and other propagules; >6 m per 3 years for taxa spreading by roots, rhizomes, stolons, or creeping stems)." Colautti and MacIsaac (2004) further refine the definition of "invasive species" with biogeographically based terminology, where an invasive species can be placed in three stages of invasion: widespread but rare (stage IVa), localized but dominant (stage IVb), or widespread and dominant (stage V). Thus, invasion biologists are moving toward more explicit terms to accurately define an invasive species. The dynamic nature of the lexicon is characterized by certain terms being abandoned due to their potential xenophobic link such as *alien* being replaced by the more objective *nonindigenous* (Simberloff 2003).

9.2.3 How Do Invasive Species Harm the Environment?

The definition of the impact of an invasive species has evolved quickly within the last few years yet policy makers and the gardening public have not been apprised of these scientific advances in terminology. Parker et al. (1999) characterize impact on the basis of range, abundance, and the per-capita or per-biomass effect of the invader. Daehler (2001) contends that the notion of impact depends on the variable being studied and the scale of study. He concludes that, "All species that meet the spread criterion probably have ecological and environmental impacts, although for most non-indigenous species, these impacts have not been adequately quantified." Davis and Thompson (2001) counter Daehler's contentions by stating that, "outside of the discipline of ecology, 'invasive species' are usually explicitly defined on the basis of their impact." Davis and Thompson also contend that consolidating all NIS into the "invader" category contributes "to a belief that invasions are a unique ecological phenomenon, which we believe has hindered ecologists' efforts to understand the invasion process." They see value in segregating invasive plants on the basis of impact which may then lead to discovering the traits that are unique to "high impact" invaders. Lodge et al. (2006) aptly note that a best attempt to quantify net "harm" by an invasion to the environment, industry, or to human health requires the collective input of economists, public health experts, and ecologists.

To lump all invasive NIS into one group is, from a practical perspective, too inclusive since the impact of invasive NIS can range from relatively innocuous to very environmentally disruptive (Fox et al. 2003). This is an especially relevant point considering that the ornamental horticulture industry sells many plants that spread outside of planted sites and that the level of environmental impact of these widespread species has rarely been determined. Coulatti (2005), discussing the inclusion of impact in the invasive definition, concludes "there is a large intellectual rift between ecologists on one side, and resource managers and politicians on the other. This creates confusion for newcomers to the discipline, and impedes the rapid and unambiguous dissemination of knowledge from ecological experiments to the formation of strategies designed to protect natural habitats from problematic invaders."

Although NIS are a major environmental concern, the proportion of NIS plants that become invasive is quite small. Rejmanek et al. (2005) make a pertinent case that "not all naturalized plant taxa, and not even all invaders, are harmful…" Williamson and Fitter (1996) developed the "tens rule," which states that one in ten imported plant or animal species (brought into the country) appear in the wild (introduced, feral), and one in ten of those become established (self-sustaining population), and one in ten of established plants become a pest (negative economic effect). Thus, if 1,000 species were imported, then 100 species would escape into the wild, 10 species would establish in the wild, and only one species would be become a pest. These authors acknowledge that this is a relatively gross prediction and qualified that 1 in 10 actually represents the range of 1 in 5 to 1 in 20. They noted that crop plants did not follow this rule and had a higher incidence of becoming a pest than predicted by the tens rule. Lockwood et al. (2001) determined the proportion of naturalized (self-sustaining populations) NIS that were classified as "the most harmful exotics" or "natural area invaders" in three US states. These authors found that 5.8, 9.7, and 13.4% of nonnative plants in California, Florida, and Tennessee, respectively, were natural area invaders. Thus, their findings are in general agreement with the tens rule of Williamson and Fitter (1996). Despite the relatively low percent of plants that ultimately become serious invaders, the large number of garden plants for sale makes the potential invasive, nonindigenous plant list quite sizable. *Dave's Garden – Plant Files* (*Dave's Garden – Guides and Information*, http://davesgarden.com/guides/. Accessed 23 May 2008), an Internet (worldwide) database of garden plants, lists 38,779 species and 100,685 cultivars. Isaacson (1996), in an inventory of North American seed and nursery catalogs (1988–1989), records almost 60,000 plant taxa sold. Applying the tens rule to the 38,779 species number, approximately 3,800 plants would escape, 380 species would establish in the wild, and 38 would become pests. Thirty-eight species can be construed as a relatively small number; however, this apparently low count belies the negative ecological effects of even a single species. One only has to consider the serious environmental effects caused by nonindigenous US landscape species such as Japanese honeysuckle (*Lonicera japonica*), English ivy (*Hedera helix*), cape ivy (*Delaireia odorata*), and Chinese tallow (*Triadica sebifera*) to realize that a single species can cause ecological havoc. The tens rule also does not take into account the unique situation of garden plants in which plants are sold year after year and planted in all parts of the country. Such repeated introductions (invasion pressure) will be discussed in the next section. Thus, the many NIS that are queued in lists of nongovernmental organizations and states vary considerably in their reproduction, rate and ecological region of spread, and impact. Hence, management decisions and regulatory actions should be species and region specific.

There are numerous governmental and nongovernmental lists, which queue invasive plant species by locality (i.e., county, state, region). These lists vary considerably in the criteria used to list a species and to rank a species' invasiveness. Thus, the usefulness of some of these lists is questionable. In an attempt to assess the criteria for published invasive plant lists, Fox and Gordon (2004) conducted a meta-analysis of 113 invasive plant lists from states, regions, and countries. They found that there was a gross lack of consistency of invasive criteria used to classify species as invasive or to rank invasiveness. Only 10% of the lists used invasive scoring systems that provided consistent application of criteria whereas two-thirds of the lists incorporated vague terms to describe environmental impact. In essence, their analysis showed that most invasive plant lists lack verifiable criteria that offer consistent interpretation and application. These authors, as part of a workshop effort, endorsed a standard system for invasive plant lists that (1) has a robust, scientific basis, (2) only lists species already present in an area, and (3) is flexible enough to be useful relative to the purpose of the list (e.g., regulatory or advisory). Fox and Gordon (2004) acknowledged the formidable challenges (i.e., complexity, collaboration between agricultural and natural system experts, continued data acquisition) to develop a standardized system with flexible options. Other than plants banned by the federal or state governments, most invasive plants lists in the USA carry no regulatory weight and serve to advise against the use of listed species. In contrast, New Zealand has a three-part invasive plant list system, i.e., banned plants, plants that require monitoring, and species-specific and site-specific weed control, which are clearly delineated and defined (Timmins 2004). New Zealand has 2,350 indigenous land plant species and 2,020 nonindigenous naturalized plant species; over 70% of invasive weeds were imported as ornamental plants (Department of Conservation 2002). While the USA has not adopted such an approach, some state governments, such as Montana, have tried stricter regulations (Simberloff et al. 2005). Additionally, New Hampshire (as of January 2007; New

Hampshire Administrative Code, Chapter Agr 3802.1., http://www.gencourt.state.nh.us/Rules/agr3800.html. Accessed 9 June 2008) and Massachusetts (as of January 2009; Massachusetts Prohibited Plant List, http://www.mass.gov/agr/farmproducts/proposed_prohibited_plant_list_v12–12–05.htm. Accessed 9 June 2008) ban the propagation, sale, purchase, or distribution of three common invasive nonindigenous landscape species (*Acer platanoides*, *Berberis thunbergii*, and *Euonymus alatus*).

9.2.4 Predicting Invasive Potential

Predicting which species will be invasive in a particular area is a very difficult task due to the complexity of nature (Drake 2005). There has been an abundance of work to determine which plant characteristics and what ecological factors lead to plant invasion (Dekker 2005; Kolar and Lodge 2001; Rejmanek and Richardson 2005; Myers and Bazely 2003). The interest in this subfield of invasion biology is substantiated by the fact that the number of scientific papers addressing invasion prediction increased fivefold from 1986 to 1999 (Kolar and Lodge 2001). At present, the most reliable and powerful predictor of a species' invasiveness is its record of invasiveness in other nonnative sites (Wittenberg and Cock 2001). Many prediction schemes have been developed to assess the potential of plant taxa to be invasive. These approaches to understanding the invasive potential significantly increase our ability to predict which taxa will be invasive. Prediction models have correctly identified (postpriori) 80–90% of invasive NIS (Reichard and Hamilton 1997; Widrlechner et al. 2004; Pheloung et al. 1999; Daehler and Carino 2000). The shortcoming of these models is that they have a relatively high rate (≥10%) of false positives (identifying a noninvasive species as invasive). Perhaps this high rate of false positives is the price we should pay to exclude invaders from our natural areas. Another shortcoming of invasive potential prediction is the knowledge needed for most of these schemes (plant and region specific), and scheme methodology has not been integrated so it can easily be used by those who are not well versed in ecology (Rejmanek et al. 2005). Mack (2005) emphasizes the need for prediction schemes to include, among other variables, the role humans play in overcoming the effect of environmental stochasticity on immigrant plant populations.

Prediction based on biological characteristics can reliably foretell if a plant will be invasive (i.e., establishment and spread); however, prediction is less reliable in forecasting the impact a taxon will have on an environment (Rejmanek et al. 2005). Rejmanek et al. (2005) note that "it is important to realize that invasiveness and impact are not necessarily positively correlated." These authors are in favor of categorizing invasive NIS that have had a profound effect on biodiversity, about 10% of invasive plants, with the term "transformer species," a term first proposed by Wells et al. (1986). Transformer species, because of their impact, would receive the majority of resources for containment, eradication, and control.

Prediction success can be hindered by the phenomenon of "invasional meltdown," a term coined by Simberloff and Von Holle (1999). Invasional meltdown is "the process by which a group of NIS facilitate one another's invasion in various ways, increasing the likelihood of survival and/or of ecological impact, and possibly the magnitude of impact." This phenomenon is especially relevant to the ornamental horticulture industry, which introduces hundreds of NIS as well as pest and pathogen "hitchhikers" on these ornamental plants into our landscapes that may synergistically interact in the future. Once a NIS is introduced, the unforeseen suite of future complex interactions greatly increases the difficulty of predicting the consequences of invasion (Mooney 2005).

NatureServe, in cooperation with The Nature Conservancy and the US National Park Service, developed a systematic assessment protocol that uses a set of questions and scientific documentation to rank invasive nonnative plant species (Morse et al. 2004). The Invasive Species Assessment Protocol is a tool for assessing, categorizing, and listing nonindigenous plant species on the basis of their impact on biological diversity. Each species has an overall ranking, which is composed of subrankings from four areas: (1) ecological impact, (2) current distribution and abundance, (3) trend in distribution and abundance, and (4) management difficulty. To date, 452 nonnative plant species occurring outside cultivation in the USA are ranked; the goal of the program is to rank 3,500 of these nonindigenous plant species. This objective ranking system and the documented list of invasive nonnative plant species serve as an effective decision support system for the ornamental horticulture industry to adopt for deciding which nonindigenous plant species to stop selling. The Invasive Species Assessment Protocol determines the ranking on a national distribution, and therefore the rank must be interpreted in the context of a specific region since a plant may not be problematic in all or most areas. However, in many cases a text description of the invasive nature of each species mentions the regional nature of invasiveness. In an attempt to consistently describe and categorize invasive nonnative plant species, some states, e.g., Florida (Fox et al. 2005), and Virginia (Heffernan et al. 2001), have also developed relatively rigorous assessment systems. Additionally, the Exotic Pest Plant Councils (EPPCs) rank nonnative plant species by their impact within specific regions or states [e.g., SE-EPPC (http://www.se-eppc.org/), FLEPPC (http://www.fleppc.org/list/list.htm), Cal-IPC (http://www.cal-ipc.org/)].

The call for improved and widespread prediction tools has been mandated by the US government. In an attempt to gain control of the importation of potentially NIS, the Animal and Plant Health Inspection service (APHIS, a branch of the USDA), the entity responsible for preventing the introduction of plant and animal pests, commissioned the National Research Council's Board on Agriculture and Natural Resources (BANR) to comprehensively review scientific knowledge regarding invasive NIS. BANR was charged to develop "risk assessments, identify potential invaders, and guide the strategic allocation of its resources to safeguarding plant life in the United States." In response to this charge, BANR established the Committee on the Scientific Basis for Predicting the Invasive Potential of Nonindigenous Plants and Pests in the USA, which was composed of experts in disciplines related to the invasive

NIS problem. This committee put forth 12 recommendations that strengthen the scientific basis of predicting the invasive potential of plants. The Global Invasive Species Programme (GISP), an international consortium of scientific, government, and foundation groups, has also recommended NIS prediction and screening schemes (Wittenberg and Cock 2001). Following an evaluation of US national policies and practices on biological invasions, and in light of current scientific and technical advances, the Ecological Society of America put forth six recommendations that require government action to "help prevent invasions, respond rapidly to new invasions and control and limit damage from existing invasions" (Lodge et al. 2006).

In an attempt to reduce the risks of the introduction of noxious weeds and host pests associated with the importation of plants for planting, APHIS is undertaking a revision of the regulations (Code of Federal Regulations, 7 CFR Part 319.37) governing the import of plants for planting, commonly referred to as Q-37 (USDA, APSHIS Import and Export – Plant Import Information –Importation of Plants for Planting – Revision of the Nursery Stock Quarantine, http://www.aphis.usda.gov/import_export/plants/plant_imports/Q37_revision.shtml. Accessed 9 June 2008). Revision of Q-37 was necessary because trade in plants has now expanded to greater global coverage of imports, with increasing numbers and magnitude of plant taxa imported, making monitoring of plant material less likely. Following the basic tenet behind propagule pressure, the increasing magnitude of imported plant material increases the likelihood of new successful invasive species to the USA (Reaser et al. 2008).

9.3 The Role of the Ornamental Horticulture Industry in Invasive NIS

The USA is the world's foremost producer of and market for nursery and floriculture crops (ANLA). These industries provide entrepreneurial opportunities, supply jobs for tens of thousands of employees, and generate large tax revenues for the government. Because of their substantial economic effect, these industries are a politically potent force. The horticulture industries sell hundreds of NIS. Most of these taxa have graced our landscapes with untold aesthetic and environmental value. However, there is no ambiguity that these industries are responsible for introducing a relatively high percentage of the invasive NIS that have negative economic and environmental impacts ranging from minor to major. Reichard (1997) calculated that 85% of the 235 known invasive woody NIS in the USA were introduced by the nursery industry as landscape material. Randall and Marinelli (1996), using invasive NIS lists from the Nature Conservancy and the National Association of EPPCs, determined that half of the 300 invasive NIS in the USA (excluding Hawaii) were imported for horticultural purposes. Bell et al. (2003), using data of six nongovernmental organizations that listed invasive NIS, determined that 34–83% of the total number of invasive taxa in the USA had a horticultural origin. In Florida, at least 47% of plants that are negatively affecting the environment were introduced for ornamental purposes; an additional 27% are suspected of such origins (Fox et al. 2003). This phenomenon is not unique to the USA. Mack and Enberg (2002) found that 65% of invasive plants introduced into Australia between 1971 and 1995 were introduced as ornamentals. Thus, the ornamental horticulture industry unwittingly introduced invasive NIS to market. This is not a recent phenomenon. While some plant species were imported into North America for ornamental purposes prior to the seventeenth century, aesthetic plant importation was most evident in and after the eighteenth century (Mack 2003): By 1860, imported ornamental plant species significantly outnumbered imported utilitarian species (Mack and Enberg 2002).

One of the most lucrative areas of ornamental plant sales is new plant introductions. This is evidenced by the emergence of many nursery businesses whose major marketing focus is novel plant introductions. Additionally, the American Nurseryman (1999), a leading trade journal for nursery and landscape professionals, devotes an entire bimonthly journal issue each year to new plant introductions. While no quantification of the nativity or residence time in the USA of these introductions has been made, the new-to-the-trade plant phenomenon stimulates efforts to seek out new introductions, many from outside the USA, and yield substantial profits. A quick glance at most garden plant catalogs and mail order web sites verifies this trend. Internet-based sales of ornamental plants are now a sizeable provider of garden plants, which facilitates the importation of NIS. Dave's Garden – Garden Watchdog (Dave's Garden – Garden Watchdog, http://davesgarden.com/products/gwd/. Accessed 23 May 2008), an internet site directory of gardening (plants and plant-related items) mail order vendors lists 6,257 businesses, most of which have a web site. Internet sales of NIS are sources of invasive, potentially invasive, and even some illegal plant taxa (Clayton 2004). To halt the sale of prohibited invasive NIS on the internet, researchers from North Carolina State University and the USDA have developed Web application software that searches the Internet for vendors who sell illegal invasive NIS (NC State University News Release 2003). One would assume that imposing strict limitations on new plant introductions, especially those that are likely to become invasive (determined via screening procedures) or carry nonnative pests and pathogens, would increase our success rate in preventing the release of invasive NIS. However, Simberloff et al. (2005) point out two limitations to this assumption. First, there is a great divergence in opinion on the impact of invasive NIS between stakeholders. Embedded in these differences of opinion are the conflicts of interest of the regulatory agency tasked to govern the flow of NIS (USDA), which ironically is funded, in part, by the tax money generated by the sale of NIS. Second, the central aspects of risk assessment, predicting specific negative consequences and estimating their probability, are greatly affected by the unpredictability of impacts of introduced species.

Members of the ornamental horticulture industry recognize the need for action regarding invasive NIS. Hall (2000) surveyed ANLA members and found the following results: (1) Sixty-eight percent of respondents were in favor of the government screening new NIS introductions. However, respondents wanted policies to be more regionally directed than political boundary directed, to have a more effective enforcement and implementation of policies, and have more scientific proof justifying decisions and species placed on invasive lists. (2) Over half of the respondents introduction, widely propagated and distributed taxa will have an unknown potential for harming the environment.

The probability of invasion increases with time (i.e., residence time of NIS) because more propagules are spread and the probability of founding new populations increases (Rejmanek et al. 2005). Thus, the mass propagation, distribution, and planting of a species serve to greatly increase invasion pressure and is most apt to shorten the lag time for a potentially invasive NIS to become invasive. Additionally, "invasional meltdown" may decrease this lag time for invasion and is another reason why the ornamental horticulture industry should take a more active role in invasive NIS issues. For example, figs cannot reproduce without the presence of a coevolved pollinator. Of the 60 species of fig (*Ficus*) occurring in south Florida, 20 of them are widely planted, and the specific pollinating wasps for three of these widely planted species were recently introduced into the area. The introduction of these three nonnative species of pollinating wasps has resulted in the reproduction of all three associated, introduced fig species and the rapid spread of one of these species (*Ficus microcarpa*) (Kaufman et al. 1991).

9.5 What Should the Ornamental Horticulture Industry Do?

9.5.1 *Voluntary Regulation*

The ornamental horticulture industry has taken steps to address and mitigate its role in being purveyors of invasive NIS. In 2001, a coalition of horticulture entities met at the Missouri Botanical Garden for a meeting entitled "Linking Ecology and Horticulture to Prevent Plant Invasions" and formulated the Saint Louis declaration, which consisted of a two-part treaty, Findings and Proceedings and Voluntary Codes of Conduct (Baskin 2002). The latter was a list of measures for various sectors of the ornamental horticulture industry (government, nursery professionals, gardening public, landscape architects, and botanic gardens and arboreta) "to curb the use and distribution of invasive plant species through self-governance and self regulation." A follow-up meeting "Linking Ecology and Horticulture to Plant Invasions II" was held in Chicago in 2002 (Fay 2003). Many of the major ornamental horticulture organizations endorsed the Voluntary Codes of Conduct. These codes have helped to develop measures to reduce the sale of invasive NIS and form partnerships such as the California Horticultural Invasives Prevention (Cal-HIP, http://www.cal-ipc.org/landscaping/calhip.php. Accessed 23 May 2008) (California Invasive Plant Council) and Washington State Nursery and Landscape Association ask Force (Washington Invasive Species Coalition, http://www.invasivespeciesco-ition.org/GardenPlants/TaskForce. Accessed 9 June 2008). In 2005, the orticultural Research Institute, the research arm of the ANLA, granted 15% of ir $220,000 research budget to invasive plant research (ANLA – HRI, http:// w.anla.org/pdffiles/Projlistingwcharts.pdf. Accessed 23 May 2008). An exam- of proactive behavior regarding invasive NIS occurred in Florida in which were willing to participate in programs to educate themselves, and almost one-third of the respondents were willing to actively educate (e.g., hold workshops) their customers. (3) Sixty-four and 29% of the respondents said they would definitely or maybe, respectively, remove invasive NIS from their stock. (4) Over half of the respondents were willing to have new, noninvasive cultivars to replace invasive NIS. (5) About a third of the respondents said that they would stop selling invasive plants only if they knew that other businesses had stopped selling invasive NIS. This last finding is suggestive of limited nursery participation in the event of a voluntary system for banning the sale of invasive NIS.

The public is generally unaware of the negative ecological and economic impacts of invasive NIS (Colton and Alpert 1998). However, surveys have found mixed evidence for awareness of the gardening public about invasive garden NIS (Wolfe and Dozier 2000; Kelly et al. 2006). In an Internet survey of gardeners, Reichard and White (2001) came to the conclusion that, "Because the preference to buy noninvasive species is correlated with familiarity, as the general plant-buying public becomes more aware of invasions, nurseries and the seed trade industry will have to alter their practices to ensure that invasive species are not sold."

9.4 Why Should the Horticultural Industry Care about Nonnative Species?

The very nature of the ornamental horticulture industry (selling, transporting, and cultivating NIS) has the potential to foster the invasion process. One of the most important factors that contribute to an area being invaded is invasion pressure, i.e., th
large numbers or frequency of introduction of NIS (Lockwood et al. 2005; Von Ho
and Simberloff 2005). Supporting the invasion pressure contention and emphasiz
the role of the ornamental horticulture industry in invasion pressure, Pembe
(2000) investigated the naturalization rate of NIS related to the number of years
was sold in the nursery trade; he found that the rate of naturalization increased
period of sale increased. For example, only 1.9% of plants naturalized that w
for 1 year, whereas 30.9% of plants naturalized that were sold for 10 years
Once sold, garden plants are cultivated. This cultivation is an important
overcoming the environmental resistance to invasion and favors the invasi
(Kowarik 2003; Mack 2000, 2005). Additionally, desirable garden plant
tics, such as a fast growth rate, abundance of fruit, and tolerance of
conditions (e.g., drought, poor soil), are also characteristics of succe
Mack (2005) contends that the horticulture industry is in a favora
devise a flexible, rapid, science-based system to screen NIS for inv
could help mitigate its role in supporting the process of invasion.

Another reason why the NIS issue requires attention by the or
ture industry is the previously mentioned phenomenon of "l
period of a slow spread rate prior to exponential rate of sp
a predictive analysis performed for invasive potential of

growers agreed to voluntarily stop growing 45 potentially invasive NIS (Wirth et al. 2004). However, growers continue to grow and sell 14 invasive NIS, which are considered highly ornamental, widely used in landscapes, and of significant value to Florida growers. The economic impact of discontinuing the sale of these 14 species was an estimated $59 million and 800 jobs (Wirth et al. 2004). However, a full economic and public policy analysis should include a cost–benefit analysis of control costs of the 14 species in natural environments and private properties as well as costs of implementation and enforcement of any regulatory actions (Wirth et al. 2004). The Florida situation poses a typical balance between economic benefits and costs, as well as environmental costs, and regulatory action.

The ornamental horticultural industry can greatly enhance its image by taking a noticeable and active role in addressing and providing solutions to the problems posed by the invasive NIS it currently sells. Bell et al. (2003) proposed that the nursery industry address four issues to effectively respond to the problem of invasive plants. These were (1) recognize the importance of the problem to natural landscapes, (2) recognize that ornamental plant nurseries are involved, (3) establish a dialog with public agencies and private groups concerned about invasive plants, and (4) be willing to participate in programs to eliminate or reduce sales of problem species. One potential complication to the wholesale adoption of the nursery industry to stop selling NIS is that the industry is relatively fragmented, as it comprises many small businesses. Many of these do not belong to national, state, or regional trade associations (based on personal observations and communications with industry personnel). The main obstacles of these businesses to participating in preventative measures proposed by the St. Louis Voluntary Codes of Conduct were listed as "the lack of information," "limited personnel," and "too time consuming" (Burt et al. 2007). We (authors) feel that a proactive stance on the invasive NIS topic would be an act of responsible land stewardship and will not result in a loss of profit if alternative noninvasive taxa are properly marketed. Clearly, there is a need for effectively communicating the fundamentals of invasive plants to industry personnel.

9.5.2 Nonvoluntary Regulation

Australia regulates exotic species import via an Import Risk Analysis (IRA) system (*AQIS Import Risk Analysis Handbook*, http://www.daff.gov.au/__data/assets/pdf_file/0011/399341/IRA_handbook_2007_WEB.pdf. Accessed 23 May 2008). The IRA system operates on a politically independent and scientifically based process. The Australian Weed Risk Assessment system (WRA), a portion of the IRA, (Biosecurity Austrailia – The Weed Risk Assessment System, http://www.daff.gov.au/ba/reviews/weeds/system. Accessed 23 May 2008) is a methodology to determine whether a NIS should be imported into Australia. Answers to questions on various plant aspects are given numerical scores, which are used to determine an outcome: to *accept*, *reject*, or *further evaluate* the species.

The Nursery and Garden Industry Australia (NGIA) as well as many state nursery and garden assocations (in Australia) have proactively formed working alliances with state governments on restricting the distribution and sale of invasive species (EPPO Reporting Service – Invasive Plants 2007/061). The main foci of the NGIA's "Invasive Policy Position" (Nursery and Garden Industry – Invasive Plants Policy Position, http://www.ngia.com.au/docs/pdf/your_associations/NGIA_invasiveweedspolicy.pdf. Accessed 9 June 2008) are (1) that government takes a fair approach to ascribing blame for invasive plants to nursery and garden groups, (2) the development of mutually agreed upon national and state prohibited plant species lists, (3) reliable and independent methods for assessing invasiveness, (4) government recognition of the industry's invasive plant regulation efforts, (5) government approval and support of industry-based communciation and awareness programs that target industry and consumer groups, and (6) government support for a secure and sustainable nursery and garden industries.

An example of nursery industry and government collaboration is the Australian "Grow Me Instead" Program (Nursery & Garden Industry – Grow Me Instead! http://www.ngia.com.au/home_gardeners/growme_instead.asp. Accessed 9 June 2008). The purposes of this program are to (1) identify garden species that are invasive, (2) identify suitable native and nonnative alternative species, and (3) educate the public via nursery industry programs with the ultimate goal of ceasing the sale of invasive plant species. This best management practices approach to the invasive plant problem exemplifies an advocative relationship between government and industry. Such a progressive relationship is apt to reduce the sale of invasive species and avoids the more typical adversarial relationship between the nursery and governments because the nursery industry is taking an *active* role in educating the public and managing the sale of invasive species.

Despite the proactive and proenvironment measures taken by the NGIA described earlier, there are some areas of cooperation that are not evident. A recent report, "Poisonous and Invasive Plants in Australia: Enabling Consumers to Buy Safe Plants" (Thomson 2007), calls for the NGIA to, in part, develop a plant labeling code of practice, which will give consumers concise information on a species' poisionous and invasive properties. The NGIA issued labeling guidelines in 2007 (Nursery & Garden Industry Australia – National Plant Labelling Guidelines 2007) but did not issue a code of practice. The lack of such a code makes implementation of labeling recommendations unlikely.

9.5.3 Biological Measures to Control Invasive NIS

There are some strategies and efforts to induce sterility into popular ornamental invasive NIS (Egolf 1981, 1986, 1988; Li et al. 2004; Olsen and Ranney 2005). As an example, the triploid *Hibiscus syriacus* L. "Diana" sets very little fruit (Dirr 1998), which is in contrast to the diploid species that sets a large amount of fruit and prolifically reproduces itself (AXN, personal observation). Inducing sterility, either by breeding or molecular tools, could diminish invasion risk of a seed-dispersed species that, due to its popularity and economic impact, would not be removed from the ornamental horticulture trade. However, extensive research on the efficacy and stability of sterility systems as well as the realized prevention of invasiveness should be conducted before sterile, noninvasive cultivars are released (Anderson et al. 2006).

9.5.4 Volunteers to Assist in Controlling Invasive NIS

Volunteers are a valuable resource for early detection of invasive NIS (Simberloff 2003; Wittenberg and Cock 2001). Personnel associated with the horticulture and landscaping industries are well qualified to detect emergent invasive species. In addition to those employed to survey and scout for spreading NIS, Wittenberg and Cock (2001) suggest that gardeners, landscape managers, fisherman, land surveyors, hikers, and others who venture into natural habitats be trained to identify or seek identification for known or new invasive NIS. The Federal Interagency Committee for the Management of Noxious and Exotic Weeds in conjunction with other governmental, state, and local partners has proposed a National Early Detection and Rapid Response System of invasive plants in the USA (Westbrooks 2004). Early detection and reporting of suspected new invasive species by volunteers is the foundation of the system. Hegamyer et al. (2003) present several successful case studies on the use of volunteers for early detection of invasive species, some of which were performed by Extension Master Gardeners (EMGs). In the USA, there are approximately 90,000 active EMGs trained in aspects of plant science and land stewardship. Thus, EMGs potentially represent a sizable, effective volunteer force, especially in view of their knowledge of plant science and garden species.

9.5.5 Controlling Invasive NIS – Prevention and Eradication

The horticulture industry is uniquely situated to work with the scientific community to more accurately predict which NIS will be invasive. Mack (2005) recommends that we must go beyond the traditional use of invasive plant traits and the invasive history of species (criteria promoted in the *Voluntary Code of Conduct for Nursery Professionals*) and undertake field trials that (1) identify those species that easily self-propagate (sexually or asexually) with minimal or no cultivation, and (2) identify and report species that routinely escape cultivation. These measures will yield valuable data on those species that are apt to establish populations outside their planted range. Mack (2005) encourages nurseries to serve as test sites, a capacity that nurseries already serve in evaluating plant traits, to determine those species that require minimal or no cultivation and have the capacity to emigrate from their planted sites and generate new populations. Simberloff et al. (2005) proposed that the decision of whether a species should be introduced should be based on "a solid understanding of what regulates populations in their native range." Both of these authors readily admit that these approaches are unfeasible unless society is willing to cease introductions while species-specific information is collected. The horticulture and landscape industry would certainly view the call to cease plant introductions as radical. However, a first step toward accurate prevention of the introduction of invasive nonnative species would be an "International Invasive Plant Data Center" that would create and update a global database of invasive nonnative plant species (Rejmanck et al. 2005). Because the history of invasiveness in one region is the best predictor of invasive potential in another region, a comprehensive and up-to-date invasive plant database will be useful in determining which taxa might be safely introduced into new areas.

While accurate prediction of invasive potential and prohibiting the importation of invasive species is the best case scenario, early detection and eradication of escaped species is the next best strategy. Eradication is possible if the invasive species is detected early enough and enough resources are dedicated to its removal (Simberloff 1997). In terms of early eradication of invasive NIS, success is optimized by meeting the following criteria: (1) limited distribution of the target species or organism, (2) adequate eradication resources, (3) clear legal grounds for action and unambiguous lines of authority, (4) the biology of the organism must be understood to develop an effective extirpative strategy, and (5) eradication should not do more harm than good (Simberloff 2003).

9.5.6 Information Sources

Where do horticulture industry personnel and the gardening public get science-based information regarding a NIS? The USDA's *National Invasive Species Information Center: Gateway to invasive species information, covering Federal, State, local, and international sources* web site is a comprehensive site covering most aspects of invasive plants and animals (USDA National Invasive Species Information Center, http://www.invasivespeciesinfo.gov/. Accessed 9 June 2008). However, there is no single information source in which nursery/landscape industry personnel and the gardening public can obtain information that focuses on the issues of landscape plants and nursery industry-related invasive issues. A search of the Internet (Google™) for "invasive landscape plants" or "invasive garden plants" yields a listing of 1.1 million and 152,000 web sites, respectively. This enormous amount of information to consider will likely overwhelm those seeking specific information. Thus, a well-advertised web site targeted at nursery/landscape industry personnel and the gardening public is vital to public education regarding NIS. These groups need to know (1) the fundamentals of invasive NIS biology, (2) landscape species that have been documented to be invasive and their relative impact, (3) the region(s) in which these invasive NIS are a problem, (4) alternative, noninvasive species for each region to be used in place of invasive NIS, and (5) because NIS establishment and impacts will vary significantly in response to climate and physiographic region, regional and state resources are especially important (Fox et al. 2003).

Another effective strategy for the education of industry regarding nonnative species would be for the US Cooperative Extension Service to develop and conduct an education program for nursery businesses. The program would be targeted at industry members regarded as innovators. This innovative group would serve as the first adopters of the educational objectives, and then serve as a model for other businesses, thereby encouraging widespread adoption. Harrington et al. (2003) concluded that educating ornamental horticulture personnel and the public should be a major focus in mitigating the invasive plant problem. McKinney (2004) found a high correlation ($r^2 = 0.69$) between the number of introduced species (plants and animals) of an area and the human population in that area. He contends that educating the general public about the dangers of exotic species importation "may be the only way to reduce rates of introduction." The ornamental horticulture industry should move to make their efforts in addressing invasive NIS more visible and public than in the past. Educating personnel on the fundamental aspects of invasive NIS and referencing NatureServe's list of documented invasive taxa (NatureServe 2005) as plants not to be sold (based on regional observations) would be a significant first step in developing a best management practices strategy. Other helpful resources are California's "Don't Plant a Pest" (California Invasive Plant Council – Don't Plant a Pest, http://www.cal-ipc.org/landscaping/dpp/. Accessed 23 May 2008) and Washington's "Garden Wise" (Washington Invasive Species Coalition – Garden Wise, http://www.invasivespeciescoalition.org/GardenPlants/index_html/view?searchterm = water%20wise. Accessed 9 June 2008) educational programs that target nursery professionals and gardeners who wish to plant noninvasive species in their landscapes. Both programs offer noninvasive alternatives to popular invasive garden species. Adopting these types of educational resources by the ornamental horticulture industry would help conserve native biodiversity and be evidence of responsible land stewardship. Thus, the ornamental horticultural industry is uniquely situated to work with the scientists and policy makers to increase public understanding of invasive species as well as decrease the introduction and spread of high-impact invasive plant species.

References

Allendorf FW, Lundquist LL (2003) Introduction: population biology, evolution, and control of invasive species. Conserv Biol 17:24–30

American Nurseryman (1999) Invasives roundtable. Am Nurseryman 190:54–77

Anderson NO, Gomez N et al. (2006) A non-invasive crop ideotype to reduce invasive potential. Euphytica 148:185–202

Baskin Y (2002) The greening of horticulture: new codes of conduct aim to curb plant invasions. BioScience 52:464–471

Bell CE, Wilen CA et al. (2003) Invasive plants of horticultural origin. HortScience 38:14–16

Bryson CT, Carter R (2004) Biology of pathways for invasive weeds. Weed Technol 18:1216–1220

Burt JW, Muir AA et al. (2007) Preventing horticultural introductions of invasive plants: potential efficacy of voluntary initiatives. Biol Invasions 9:909–923

Canton BP (2005) Availability in Florida nurseries of invasive plants on a voluntary "do not sell" list. United States Department of Agriculture, Animal and Plant Health Inspection Service, Raleigh, NC
Clayton M (2004) Now, invasive species stream in online. Christian Science Monitor Oct. 28 (Features: Planet). The Christian Science Monitor, p. 13
Colautti RI (2005) In search of an operational lexicon for biological invasions. In: Inderjit (ed) Invasive plants: ecological and agricultural aspects. Birkhäuser, Germany
Colautti RI, MacIsaac HJ (2004) A neutral terminology to define 'invasive' species. Divers Distrib 10:135–141
Colton TF, Alpert P (1998) Lack of public awareness of biological invasions by plants. Nat Area J 18:262–266
Crooks JA, Soulé ME (1999) Lag times in population explosions of exotic species: causes and implications. In: Sandlund OT, Schei PJ et al. (eds) Invasive species and biodiversity management. Kluwer, Dordrecht, Netherlands, pp. 103–125
Daehler CC (2001) Two ways to be an invader, but one is more suitable for ecology. Bull Ecol Soc Am 82:101–102
Daehler CC, Carino DA (2000) Predicting invasive plants: prospects for a general screening system based on current regional models. Biol Invasions 2:93–102
Davis MA, Thompson K (2001) Invasion terminology: should ecologists define their terms differently than other? No, not if we want to be of any help! Bull Ecol Soc Am 82:206
Dekker J (2005) Biology and anthropology of plant invasions. In: Inderjit (ed) Invasive plants: ecological and agricultural aspects. Birkhäuser, Germany, pp. 235–250
Department of Conservation (2002) Space invaders – a summary of the Department of Conservation's strategic plan for managing invasive weeds. http://www.doc.govt.nz/upload/documents/conservation/threats-and-impacts/weeds/space-invaders.pdf. Accessed 23 May 2008
Dirr MA (1998) Manual of woody landscape plants – their identification ornamental characteristics, culture, propagation, and uses. Stipes, Champaign, IL
Drake JA (2005) Plant invasions: ecological and agricultural aspects. In: Inderjit (ed) Invasive plants: ecological and agricultural aspects. Birkhäuser, Germany, pp. XIII–XIV
Egolf DR (1981) 'Helene' Rose of Sharon (Althea). HortScience 16:226–227
Egolf DR (1986) 'Minerva' Rose of Sharon (Althea). HortScience 21:1463–1464
Egolf DR (1988) 'Aphrodite' Rose of Sharon (Althea). HortScience 23:223
EPPO Reporting Service – Invasive Plants (2007). http://archives.eppo.org/EPPOReporting/2007/Rse-0703.pdf. Accessed 23 May 2008
Ewel JJ, O'Dowd DJ et al. (1999) Deliberate introductions of species: research needs. BioScience 49:619–630
Fay K (2003) Linking ecology and horticulture to prevent plant invasions. II. Workshop at the Chicago Botanic Garden. http://www.centerforplantconservation.org/invasives/Download%20PDF/CBG_Proceedings.pdf. Accessed 23 May 2008
Federal Register – Presidential Documents (1999) Vol. 64, no. 25. http://www.invasivespecies.gov/laws/eo13112.pdf. Accessed 23 May 2008
Fox AM, Gordon DR (2004) Criteria for listing invasive plants. Weed Technol 18:1309–1313
Fox AM, Gordon DR et al. (2003) Challenges of reaching consensus on assessing which non-native plants are invasive in natural areas. HortScience 38:11–13
Fox AM, Gordon DR et al. (2005) IFAS assessment of the status of non-native plants in Florida's natural areas. http://aquat1.ifas.ufl.edu/finalassessjun05.pdf. Accessed 9 June 2008
Green JL, Green AK (2003) The nursery industry in the United States – always new frontiers. HortScience 38:994–998
Hall M (2000) IP plants: invasive plants and the nursery industry. Undergraduate senior thesis in environmental studies. http://www.brown.edu/Research/EnvStudies_Theses/full9900/mhall/IPlants/IPlants_Frames.html. Accessed 9 June 2008
Harrington RA, Kujawski R et al. (2003) Invasive plants and the green industry. J Arboric 29:42–48

Heffernan KE, Coulling PP et al. (2001) Ranking invasive exotic plant species in Virginia. Natural Heritage Technical Rep. 01-13. Virginia Dept of Conservation and Recreation, Division of Natural Heritage, Richmond. http://www.dcr.virginia.gov/dnh/rankinv.pdf. Accessed 9 June 2008

Hegamyer K, Nash SP et al. (eds) (2003) The early detectives: how to use volunteers against invasive species, case studies of volunteer early detection programs in the U.S. http://www.invasivespeciesinfo.gov/toolkit/detect.shtml. Accessed 9 June 2008

Isaacson RT (ed) (1996) Andersen horticultural library's source list of plants and seeds: a completely revised listing of 1988–89 catalog compiled by Richard T. Isaacson and the staff of the Andersen Horticultural Library, 4th edn. Andersen Horticultural Library, University of Minneapolis, Minneapolis, MN

Jordan N, Becker R et al. (2003) Knowledge networks: an avenue to ecological management of invasive weeds. Weed Sci 51:271–277

Kauffman S, McKey DB et al. (1991) Adaptations for a two-phase seed dispersal system involving vertebrates and ants in a hemiepiphytic fig (*Ficus microcarpa*: Moraceae). Am J Bot 78:971–977

Kelly KM, Conklin JR et al. (2006) Invasive plant species: results of a consumer awareness, knowledge, and expectations survey conducted in Pennsylvania. J Environ Hort 24:53–58

Kolar CS, Lodge DM (2001) Progress in invasion biology: predicting invaders. Trends Ecol Evol 16:199–204

Kowarik I (1995) Time lags in biological invasions with regard to the success and failure of alien species. In: Pysek P, Prach K et al. (eds) Plant invasions: general aspects and special problems. SPB Academic, Amsterdam

Kowarik I (2003) Human agency in biological invasions: secondary releases foster naturalisation and population expansion of alien plant species. Biol Invasions 5:293–312

Li Y, Cheng Z, et al. (2004) Invasive ornamental plants: problems, challenges, and molecular tools to neutralize their invasiveness. Crit Rev Plant Sci 23:381–389

Lockwood JL, Simberloff D (2001) How many, and which, plants will invade natural areas. Biol Invasions 3:1–8

Lockwood JL, Cassey P et al. (2005) The role of propagule pressure in explaining species invasions. Trends Ecol Evol 20:223–228

Lodge DM, Shrader-Frechette K (2003) Nonindigenous species: ecological explanation, environmental ethics, and public policy. Conserv Biol 17:31–37

Lodge DM, Williams S et al. (2006) Biological invasions: recommendations for U.S. policy and management. Ecol Appl 16:2035–2054

Macdonald IA, Loope LL et al. (1989) Wildlife conservation and invasion of nature reserves by introduced species: a global perspective. In: Drake JA, Mooney HA et al. (eds) Biological invasions. A global perspective. SCOPE 37. Wiley, Chichester, UK

Mack RN (2000) Cultivation fosters plant naturalization by reducing environmental stochasticity. Biol Invasions 2:111–122

Mack RN (2003) Plant naturalizations and invasions in the eastern United States: 1634–1860. Ann Missouri Bot Gard 90:77–90

Mack RN (2005) Predicting the identity of plant invaders: future contributions from horticulture. HortScience 40:1168–1174

Mack RN, Erneberg M (2002) The United States naturalized flora: largely the product of deliberate introductions. Ann Missouri Bot Gard 89:176–189

Mack RN, Simberloff D, et al. (2000) Biotic invasions: causes, epidemiology, global consequences, and control. Ecol Appl 10:689–710

McKinney ML (2004) Citizens as propagules for exotic plants: measurement and management implications. Weed Technol 18:1480–1483

Mooney HA (2005) Invasive alien species: the nature of the problem. In: Mooney HA, Mack RN et al. (eds) Invasive alien species: a new synthesis. Island Press, Washington

Morse LE, Randall JM et al. (2004) An invasive species assessment protocol: evaluating non-native plants for their impact on biodiversity, Version 1. NatureServe, Arlington, VA. http://www.natureserve.org/library/invasive_species_assessments.pdf. Accessed on June 9 2008

Myers JH, Bazely D (2003) Ecology and control of introduced plants. Cambridge University Press, Cambridge

NC State University News Release (2003) http://www.ncsu.edu/news/press_releases/03_07/196.htm Accessed 9 June 2008

Nursery & Garden Industry Australia – National Plant Labelling Guidelines (2007) http://www.ngia.com.au/docs/pdf/your_associations/NGIA_Labelling_Guidelines_AUG07v1.1.pdf. Accessed June 9 2008

Olsen RT, Ranney TG (2005) Breeding for non-invasive landscape plants. http://www.ces.ncsu.edu/depts/hort/nursery/short/2005_short_cs/breeding_noninvasive.htm. Accessed 9 June 2008

Parker IM, Simberloff D et al. (1999) Impact: toward a framework for understanding the ecological effects of invaders. Biol Invasions 1:3–19

Pemberton RW (2000) Naturalization patterns of horticultural plants in Florida. In: Spencer NR (ed) Proceedings of the X International Symposium on Biological Control of Weeds. Montana State University Bozeman, Montana

Pheloung PC, Williams PA (1999) A weed risk assessment model for use as a biosecurity tool evaluating plant introductions. J Environ Manage 57:239–251

Pimental D, Lach L et al. (2000) Environmental and economic costs of nonindigenous species in the United States. BioScience 50:53–64

Radosevich SR, Stubbs MM et al. (2003) Plant invasions – process and patterns. Weed Sci 51:254–259

Randall JM, Marinelli J (eds) (1996) Invasive plants: weeds of the global garden. Brooklyn Botanic Garden, New York. http://www.bbg.org/gar2/topics/sustainable/handbooks/invasiveplants/. Accessed 9 June 2008

Reaser J, Meyerson L et al. (2008) Saving camels from straws: how propagule pressure-based prevention policies can reduce the risk of unintentional introductions. Biol Invasions 10:1085–1098

Reichard SH (1997) Prevention of invader plant introductions on national and local levels. In: Luken JO, Thiere JW et al. (eds) Assessment and management of plant invasions. Springer, New York

Reichard SH, Hamilton CW (1997) Predicting invasions of woody plants introduced into North America. Conserv Biol 11:193–203

Reichard SH, White P (2001) Horticulture as a pathway of invasive plant introductions in the United States. BioScience 51:103–113

Rejmanek M, Richardson DM (2005). Plant invasions and invisibility of plant communities. In: van der Maarel E (ed) Vegetation ecology. Blackwell Science, Oxford, UK

Rejmanek M, Richardson DM et al. (2005) Ecology of invasive plants: state of the art. In: Mooney HA, Mack RN et al. (eds) Invasive alien species: a new synthesis. Island Press, Washington

Richardson DM, Allsopp N et al. (2000) Plant invasions – the role of mutualisms. Biol Rev 75:65–93

Simberloff D (1997) Eradication. In: Simberloff D, Schmitz DC (eds) Strangers in paradise: impact and management of nonindigenous species in Florida. Island Press, Washington, DC

Simberloff D (2003) Confronting introduced species: a form of xenophobia. Biol Invasions 5:179–192

Simberloff D, Von Holle B (1999) Positive interactions of nonindigenous species: invasional meltdown? Biol Invasions 1:21–32

Simberloff D, Parker IM et al. (2005) Introduced species policy, management, and future research needs. Front Ecol Environ 3:12–20

Thomson N (2007) Poisonous and invasive plants in Australia: enabling consumers to buy safe plants. WWF-Australia Issues Paper. WWF-Australia, Sydney

Timmins SM (2004) How weed lists help protect native biodiversity in New Zealand. Weed Technol 18:1292–1295

Von Holle B, Simberloff D (2005) Ecological resistance to biological invasion overwhelmed by propagule pressure. Ecology 86:3212–3218

Wells MJ, Poynton RJ, et al. (1986) The history of introduction of invasive alien plants to southern Africa. In: Macdonald IAW, Kruger FJ et al. (eds) The ecology and management of biological invasions in southern Africa. Oxford University Press, Cape Town, South Africa
Westbrooks RG (2004) New approaches for early detection and rapid response to invasive plants in the United States. Weed Technol 18:1468–1471
Widrlechner MP, Thompson JR et al. (2004) Models for predicting the risk of naturalization of non-native woody plants in Iowa. J Environ Hort 22:23–31
Wilson EO (1992) The diversity of life. Harvard University Press, Cambridge, Massachusetts
Williamson M, Fitter A (1996) The varying success of invaders. Ecology 77:1661–1666
Wirth FF, Davis KJ et al. (2004) Florida nursery sales and economic impacts of 14 potentially invasive landscape plant species. J Environ Hort 22:12–16
Wittenberg R, Cock MJW (eds) (2001) A toolkit of best prevention and management practices. CAB International, Wallingford, Oxon, UK
Wolfe EW, Dozier H (2000) Development of a scale for monitoring invasive plant environmentalism. J Appl Measure 1:219–237

Chapter 10
Biological Control of Invasive Weeds in Forests and Natural Areas by Using Microbial Agents

Alana Den Breeÿen and Raghavan Charudattan

Abstract Biological control of forest weeds by using microbial plant pathogens has been tried in a few cases with some notable success. Diverse weed targets such as broad-leaved exotic invasive tree species, native tree and shrub species that recolonize following clearcutting, and invasive shrubs, annual and perennial herbs, and vines have been targeted. Examples of several programs, some highly successful and others with outcomes still uncertain, are described including the control of *Acacia saligna* by the introduced rust fungus *Uromycladium tepperianum* in South Africa, broad-leaved tree species by *Chondrostereum purpureum* in the Netherlands and Canada, *Clidemia hirta* in Hawaii by *Colletotrichum gloeosporioides* f.sp. *clidemiae, Ageratina riparia* by the foliar smut fungus *Entyloma ageratinae* in Hawaii and New Zealand, *Hedychium gardnerianum* by the bacterium *Ralstonia solanacearum* in Hawaii, *Passiflora tarminiana* by *Septoria passiflorae* in Hawaii, *Imperata cylindrica,* an exotic invasive grass in the southeastern USA, by *Bipolaris sacchari* and *Drechslera gigantea, Eichhornia crassipes* by fungal pathogens in integration with arthropods, and *Solanum viarum* by *Tobacco mild green mosaic tobamovirus* (TMGMV). This list of examples is not complete but it is meant to illustrate the classical vs. bioherbicide strategies, integrated control using pathogens and insects, and different types of pathogens (biotrophic vs. necrotrophic fungi, a bacterium, and a virus). Pathogens that are easily disseminated from initial release sites through rapid buildup of secondary disease cycles have produced some of the highly successful programs compared with pathogens that require postrelease augmentation in the form of multiple releases, inundative applications, or technological aids.

Keywords Bioherbicide • Classical biocontrol • Forest weeds • Weeds in natural areas • Plant pathogens

---

A.D. Breeÿen and R. Charudattan(✉)
Plant Pathology Department, University of Florida, Gainesville, FL 32611-0680, USA
rcharu@ufl.edu

Inderjit (ed.), *Management of Invasive Weeds,*

10.1 Introduction

Invasive weeds cause significant economic losses and ecological disruptions in both commercial and native forest ecosystems. They also cause serious problems in natural areas, some of which are wooded, contain scrublands, wetlands, grasslands, and biologically diverse ecosystems. Invasive weeds by their very nature are highly successful colonizers and compete for resources such as light, nutrients, and water, which results in the suppression of young or naturally regenerating trees (Green 2003). The replacement of a relatively diverse native ecosystem by monotypic stands of alien species is a serious disruption of the ecosystem (Randall 1996). In Hawaii, 946 out of a total of 2,690 plant species are alien, with about 800 native species endangered and over 200 endemic species believed to be extinct because of alien species (Pimentel et al. 2005; Vitousek 1988). In Florida, about 25% of some 3,500 plant species are nonindigenous and about 60 species are said to be highly invasive (Simberloff 1997). According to Campbell (1998), an estimated 138 alien tree and shrub species including salt cedar (*Tamarix pendantra* Pall., *Tamaricaceae*), eucalyptus (*Eucalyptus* spp., *Myrtaceae*), Brazilian pepper (*Schinus terebinthifolius* Raddi, *Anacardiaceae*), and Australian melaleuca [*Melaleuca quinquenervia* (Cav.) S.T. Blake, *Myrtaceae*], have invaded native US forest and shrub ecosystems (Pimentel et al. 2005). *Melaleuca quinquenervia* was spreading at a rate of 11,000 ha/year throughout the forest and grassland ecosystems of the Florida Everglades, causing damage to the natural vegetation and wildlife (Pimentel et al. 2005). This rate of spread appears to be abating with the introduction of two insect biocontrol agents, the melaleuca weevil, *Oxyops vitiosa* (Center et al. 2000), and the melaleuca psyllid, *Boreioglycapsis melaleucae* (Pratt et al. 2004).

Successful biological control programs ultimately reduce, and sometimes eliminate, the need for conventional methods of control for invasive weed species that have escaped from their natural pests and pathogens. The benefit-to-cost ratio of successful biological control can be very high, especially for countries repeating earlier successful introductions from another country (Waterhouse 1998). In this chapter, we describe several successful and early-stage biological control programs using microbial agents to illustrate the application strategies (i.e., inoculative [classical biocontrol] vs. inundative [bioherbicide] approaches) and the weed–pathogen systems involved. For overviews of biological control of weeds by using plant pathogens, the readers are referred to Charudattan (2001a), Gardner (1992), and Yandoc-Ables et al. (2006a, b).

10.2 Classical Biological Control

Classical biological control of weeds involves the introduction, establishment, and self sustenance of pathogens from the native range of the target weed into an area where the weed has naturalized and become a problem (Briese 2000). The aim is to achieve a long-term equilibrium between the population of natural enemies and the weed, and in the process trigger a reduction of the weed population below an economic or ecological threshold (see Norris 1999). In a rangeland, pasture, or natural situation, a reduction in the density of an invasive weed will tend to open niches for other desirable native or introduced plant species to reestablish, thereby restoring the ecosystem. In some cases, such as managed forest systems, this natural restoration may need to be supervised to maintain desirable silvicultural and forestry objectives. In certain situations, native species need to be introduced proactively, aiding reestablishment of desirable over less desirable species. It is also important to be vigilant to prevent colonization of open niches by other invasive species. Barton (2004), while examining the safety record of pathogens used in classical biological control of weeds, has reviewed this topic and provided an in-depth analysis of the overall success of this approach.

10.2.1 Acacia saligna–Uromycladium tepperianum

One of the most successful classical biocontrol programs documented is the control of *Acacia saligna* (Labill.) H.L. Wendl; *Fabaceae* in South Africa by the rust fungus *Uromycladium tepperianum* (Sacc.) McAlpine, which is indigenous to Australia. As one of the most serious invasive weeds in South Africa, *A. saligna* (Port Jackson willow) forms dense stands that replace indigenous vegetation and interfere with agricultural practices and management of natural areas (Morris 1997). As the weed is difficult and costly to control mechanically and chemically, it was an excellent candidate for biological control. In 1987, a gall-forming rust fungus, *U. tepperianum*, was imported into South Africa after extensive host-range testing undertaken in Australia confirmed its specificity (Morris 1987). Morris (1987) studied the teliospore germination, early stages of host infection, host specialization, and the reactions of some African *Acacia* and *Albizia* species to inoculation with *U. tepperianum* in Australia. Cross-inoculation of teliospores isolated from *A. implexa* Benth., *A. saligna*, and *P. lophantha* Willd. spp. *lophantha* with the same three species suggested that distinct genotypes of the rust occur on these species. As the reactions were distinguishable at the host species level, they should, according to Anikster (1984), be termed *formae speciales*. Morris (1987) showed that normal galls only developed on the species from which the teliospores were collected even though several known *U. tepperianum* host species were included in the study. Although the results indicated that these rust genotypes may be specific to a single host species, Morris (1987) recommended that a larger range of recorded host species be tested prior to formally designating the *formae speciales*.

Since the initial releases in the late 1980s, *U. tepperianum*, which causes extensive gall formation and subsequently spreads, has established throughout the range of *A. saligna* (Morris 1997). Fifteen years of monitoring (1991–2005) showed that tree density declined by between 87% and 98% with a reduction in canopy mass compared with data recorded prior to the *U. tepperianum* release (Wood and Morris 2007). The introduction of the seed-feeding weevil, *Melanterius compactus* Lea (Coleoptera: Cuculionidae), should accelerate a continuous decline in stand density over time. According to Wood and Morris (2007), *U. tepperianum* remains a highly effective biological control agent against the alien invasive tree *A. saligna* in natural areas and forests in South Africa (Fig. 10.1).

Fig. 10.1 Biological control of *Acacia saligna* (Port Jackson willow) by the gall-forming rust fungus *Uromyces tepperianum* in Western Cape Province of South Africa. *Top left*: a heavily galled, dead branch; *top right*: a heavily galled tree; and *bottom left*: a dead tree killed primarily by the rust disease in combination with other biotic and abiotic stresses (pictures credit: R. Charudattan)

10.2.2 *Passiflora tarminiana–Septoria passiflorae*

Passiflora tarminiana Coppens & V. E. Barney (banana passion flower, banana poka vine; *Passifloraceae*), native to the Andes, is an aggressive invasive tropical vine in Hawaii. It is invasive in disturbed areas and its effects include loss of biodiversity, smothering of trees, and the encouragement of other invasive species such as feral pigs that feed on the fruit (ISSG Database 2005). By 1983, more than 50,000ha of wet and mesic forests in Hawaii had become infested with *P. tarminiana*, costing the Division of Forestry and Wildlife $90,000 annually to control this weed in just one area, the Hilo Forest Reserve (Trujillo 2005). In 1991, *Septoria passiflorae* Syd. was isolated from infected Hawaiian banana poka seedlings grown in Colombia. Subsequent to the pathogen's importation to Hawaii, host-range tests, completed in 1994, confirmed its specificity to banana poka vine (Trujillo et al. 1994). *Septoria passiflorae* was approved as a biocontrol agent for banana poka vine in 1995. Field inoculations done in 1996–1997 using spore suspensions resulted in significant control of banana poka vines in the islands of Hawaii, Maui, and Kauai (Trujillo 2005). In the Hilo Forest Reserve, Hawaii, biomass was reduced between 40 and 60% annually after the first inoculations, and 4 years later, a 95% biomass reduction was observed (Trujillo 2005) The introduction of *S. passiflorae* has resulted in the preservation of some endangered species and regeneration of the indigenous *Acacia koa* forest (the source of the most valuable timber species in Hawaii) while saving millions of dollars in weed-control cost and forest revitalization programs in Kauai, Maui, and Hawaii (Trujillo 2005).

10.2.3 *Ageratina riparia–Entyloma ageratinae*

Ageratina riparia (Regel) R. M. King & H. Rob. (mist flower, Hamakua "Pa-makani," *Asteraceae*), a low-growing perennial with clusters of tiny white flowers, was accidentally introduced into Hawaii in 1925. By 1972, mist flower infestations had spread over 100,000ha of range- and forestlands on the Hawaiian Islands (Morin et al. 1997; Trujillo 1985). The foliar smut fungus, *Entyloma ageratinae* Barreto and Evans was introduced to Hawaii from Jamaica in 1975 to control this aggressive weed. Initially misnamed as a *Cercosporella* sp., the pathogen was renamed *E. ageratinae* sp. nov. Barreto and Evans (Barreto and Evans 1988), based on physiological host reactions rather than on morphological differences [Trujillo (2005) calls the fungus *E. compositarum* f. sp. *ageratinae*]. After extensive host-range studies confirmed that the pathogen was specific to mist flower, field inoculations were made at infested sites throughout the Hawaiian Islands (Trujillo 2005). Complete control of the plant was achieved in less than 7 months in wet areas and within 3–8 years in dry areas. Biological control of mist flower has been an outstanding success in Hawaii with extensive rehabilitation of the Hawaiian rangelands (Trujillo 2005).

Entyloma ageratinae was released in New Zealand in 1998 to suppress mist flower (Barton et al. 2007). Within 2 years, it had established at all the initial release sites and was found up to 80 km from the nearest release site (Barton et al. 2007). Within 5.5 years, it had spread throughout the mist flower sites in the North

Fig. 10.2 Biological control of *Ageratina riparia* (known as mist flower or by its Hawaiian name, Hamakua pamakani) by the foliar smut fungus *Entyloma ageratinae* in New Zealand. *Top row*: The effect of the smut pathogen before (*left*) and after (*right*) its release and establishment in 1998. The mean percentage of mist flower cover (the predominant plant in the dark green understory in the picture on the left) determined from 51 sites decreased from 81 to 1.5% in 5 years. *Bottom*: a smut-infected mist flower leaf showing necrotic lesions and white smut growth and sporulation. Picture credit: Martin Heffer (foliar disease symptoms) and Alison Gianotti (before and after pictures). The images were provided by Jane Barton, Landcare Research, Auckland, New Zealand and reprinted here with permission from Landcare Research, New Zealand

Island, the most distant site being found 440 km from the nearest release. The disease developed rapidly, causing severe defoliation and significant decline in plant height. As the mist flower cover significantly decreased as a result of the disease, the species richness and mean percentage cover of native plants increased. It is interesting to note that no significant change was observed in the species richness and mean percentage cover of exotic plants (excluding mist flower). According to Barton et al. (2007) the introduction of *E. ageratinae* as a biological control agent of mist flower in New Zealand has been very successful (Fig. 10.2).

10.3 Bioherbicides

The bioherbicide approach utilizes indigenous plant pathogens isolated from weeds and mass-produced in culture (Charudattan 1988). Pathogenicity tests are conducted on the weed under a range of environmental conditions whereby field efficacy and host-range tests are completed (Ayres and Paul 1990). Weed suppression is obtained by applying the pathogen at rates sufficient to cause high infection levels (Templeton 1982). Pathogen survival between growing seasons is often insufficient to maintain high infection rates and as a result new epidemics are not initiated in the subsequent seasons which makes it necessary to repeat annual bioherbicide applications (TeBeest et al. 1992).

10.3.1 Broad-Leaved Tree Species: Chondrostereum purpureum, Cylindrobasdium laeve

The use of endemic wood-rotting fungi as bioherbicides to control invasive woody weed species was successfully implemented in the Netherlands and South Africa (Green 2003). Applied directly on freshly cut stump surfaces, these fungi prevent resprouting of inoculated tree stumps. Both fungi are naturally present in native forests and normally infect through wounds and natural openings in weakened trees. They infect the cambial tissue and kill the tree. They are nearly 100% effective when applied as a bioherbicidal preparation to cut stumps, although the stumps may resprout before being killed.

Chondrostereum purpureum (Pers. ex Fr) Pouzar was developed as a control for *Prunus serotina* Ehrh. (*Rosaceae*), a North American tree that is invasive in parts of Northern Europe. A mycoherbicide preparation under the name BIOCHON was developed in the Netherlands, but the cost of registering it for commercial use precluded further development. An attempt was underway to offer it for use as a natural wood-decay promoter, but the Dutch authorities did not permit such unregistered use. There might be some recent interest in using this fungus in German forests (De Jong 2000; De Jong et al. 1990, 2007).

In Canada, two stump treatment products, Myco-tech™ and Chontrol™ (Yandoc-Ables et al. 2006a), containing native isolates of *C. purpureum* have been developed and registered. Through extensive epidemiological studies, it has been established that the this fungus posed no threat to susceptible crops grown further than 500 m from the application site (De Jong et al. 1990). The use of this control method will only have a limited impact on nontarget trees since the fungus spores are ubiquitous and healthy trees are resistant to attack. However, Setliff (2002) showed that *C. purpureum* is an important pathogen with epidemic potential in forest tree species especially in the Betulaceae (*Betula* and *Alnus*) and the Salicaceae (*Populus* and *Salix*). Using this fungus as an inundative biological control agent should be considered within this perspective rather than as a sporadic pathogen of fruit trees. A thorough investigation into the ecological impacts of *C. purpureum* for each geographical region should be undertaken with the purpose of identifying low disease risk areas where *C. purpureum* can be applied with the minimum environmental impact (Setliff 2002).

Exploratory studies in New Zealand have shown that *C. purpureum* could be used to prevent resprouting of cut gorse (*Ulex europaeus* L., *Fabaceae*) bushes (Bourdôt 2007). One year after treatment of stumps with mycelium on agar or mycelium formulation, basidiocarps of the fungus were present on 15 of the 16 treated stumps. Seventeen months after treatment, half of the *C. purpureum*-treated stumps were dead as compared with the untreated stumps (Fig. 10.3). These results confirm that *C. purpureum* has potential as a biological control agent for gorse (Bourdôt 2007; Bourdôt et al. 2006).

In South Africa, a basidiomycete, *Cylindrobasidium laeve* is registered as the bioherbicide, Stumpout. It is applied directly to freshly cut stumps of several invasive *Acacia* spp. According to Morris et al. (1999), Stumpout kills 95–100% of resprouting stumps.

10.3.2 *Clidemia hirta–Colletotrichum gloeosporioides f.sp. clidemiae*

Clidemia hirta (L.) D. Don (Koster's curse, *Melastomaceae*) is a perennial shrub native to Central and South America, and the islands of the West Indies (Wester and Wood 1977, Peters 2001). It is invasive in tropical forest understories particularly in the Hawaiian Islands, and its negative effect on the native ecosystem (Gerlach 2006) has led to the fear that similar effects will be felt in the other introduced regions including the Seychelles, the Comoros Island, Réunion, Mauritius, the Malaysian Peninsula, and parts of Micronesia (Palau) (Gerlach 2006). First reported in Oahu in 1941 (Trujillo 2005), *Clidemia* populations reached significant proportions by 1984 with infestation extending over 60,000 ha of forested areas of Kauai, Molokai, Maui, and Hawaii (Trujillo 2005). All management strategies including physical removal and herbicide treatments were not practical and generally failed especially when initiated after first fruit set.

A fungus, *Colletotrichum gloeosporioides* (Penz.) Penz. f. sp. *clidemiae* was isolated from diseased clidemia leaves collected in Panama in 1985 (Trujillo 2005).

Fig. 10.3 Biological control of gorse (*Ulex europaeus*) by using cut-stump treatment with *Chondrostereum purpureum* in New Zealand. Top: Two adjacent gorse plants 12 months after treatment. The plant on the left (*green* marker) was only cut but not treated with the fungus (control treatment) and the stump on the right (*yellow* marker) was cut and immediately treated with *C. purpureum*. The arrows point to gorse regrowing in the cut-only treatment (*left*) and dead stumps with no sign of regrowth in the fungus treatment (*right*). *Bottom*: A closeup of one of the cut and treated gorse plants 12 months after treatment showing the fruiting bodies of the basidiomycetous *C. purpureum* (picture credit: Graeme Bourdôt, AgResearch Limited, Lincoln, New Zealand)

Following the confirmation of specificity of this pathogen to *C. hirta*, permission was granted in 1986 by the Hawaii Board of Agriculture to use *C. gloeosporioides* f. sp. *clidemiae* to control clidemia. Repeated field inoculations undertaken from 1986 until 1992 at Aiea State Park, Oahu resulted in significant control of the weed

(Trujillo 2005). After several spore formulations were tested for field applications, a 5×10^5 per mL spore suspension amended with 0.5% gelatin and 2.0% sucrose that caused severe defoliation was selected. Annual sprays directed at clidemia in Aiea State Park, Oahu using this spore formulation resulted in effective control of the weed and in the regeneration of several native species such as *Acacia koa* and fern species (Trujillo 2005).

10.3.3 Hedychium gardnerianum–Ralstonia solanacearum: Use of a Bacterial Pathogen

Native to India, *Hedychium gardnerianum* Sheppard ex Ker Gawl. (wild ginger or kahili ginger, Zingiberaceae) is widespread throughout the tropics and invasive in many forest ecosystems including Hawaii, New Zealand, Reunion, South Africa, and Jamaica (Anderson and Gardner 1999). In 1954, *H. gardnerianum* was brought into Hawaii as an ornamental plant where it subsequently escaped cultivation and is currently considered a naturalized plant species. According to Anderson and Gardner (1999), kahili ginger has invaded 500 ha of Hawaii Volcanoes National Park forests (1,000–1,300-m elevation) and is found on all islands in Hawaii (Smith 1985). Displacing native plants in forest ecosystems, kahili ginger forms dense stands that smother the native understory. Further spread is facilitated by the rhizomes making this weed difficult to control (ISSG Database 2006). Because of the fact that kahili ginger is widely distributed throughout the Hawaiian national parks, environmental concerns regarding herbicide use to control kahili ginger have resulted in biological control being considered as the only practical approach for long-term management of this invasive weed in native forests.

A Hawaiian *Ralstonia* (=*Pseudomonas*) *solanacerum* (Smith) Yabuuchi et al. strain was isolated from both edible (*Zingiber officinale* Roscoe) and ornamental gingers. While the isolate from edible ginger was less virulent than the isolate from ornamental ginger, the two isolates were proposed to be one strain based on similar cultural and pathogenicity characteristics (Anderson and Gardner 1999). *Ralstonia solanacerum* systemically infects edible and ornamental gingers causing decay and wilting of the infected tissue. Host specificity tests showed that this bacterium did not infect the native and cultivated solanaceous species tested. All inoculated *H. gardnerianum* plants including rhizomes developed symptoms within 3–4 weeks after inoculation, with most inoculated plants completely dead within 4 months (Anderson and Gardner 1999). While limited infection occurred close to the inoculation sites on *H. coronarium* J. Koenig, *Z. zerumbet* (L.) Sm., *Heliconia latispatha* Benth., and *Musa sapientum* L., no further systemic infection was observed in these species and the plants continued to grow normally (Anderson and Gardner 1999). Despite concerns about the potential negative impacts of the bacterium on the edible ginger, *Z. officinale*, Anderson and Gardner (1999) concluded that kahili ginger infestations were often remote enough to limit the risk of contamination to edible ginger plants. The use of *R. solanacerum* to control of *H. gardnerianum* in Hawaii appears to be successful; however, possible contamination of agricultural lands from runoff water from treatment areas is a concern that should be addressed when using the *R. solanacearum* as a biocontrol agent (Anderson and Gardner 1999).

10.3.4 *Eichhornia crassipes (Mart.) Solms-Laub. (Water hyacinth)–Insects and Pathogens*

The Neotropical invasive aquatic plant water hyacinth (*Eichhornia crassipes* Mart.-Solms-Laub., Pontederiaceae), also called the blue devil, continues to plague many regions of Africa, Asia, Central and South America, and Pacific islands, including many managed and natural forests where human activities are responsible for the weed's establishment and proliferation. Mechanical removal, biological control agents, and chemical herbicides, used alone or in integrated systems, have been reasonably effective for management of water hyacinth on small scales and for short durations at different sites and regions, but the weed remains a recurrent problem throughout the world. Biological control by using insect herbivores (most notably two weevils: *Neochetina bruchi* Hustache and *N. eichhorniae* Warner), naturally occurring pathogens (*Acremonium zonatum* (Saw.) Gams, *Alternaria eichhorniae* Nag Raj & Ponnapa, *Cercospora piaropi* Tharp emend. Conway emend. Tessmann et al.,) and others, and generalist microbial colonizers used in judicious combinations with chemical herbicides and mechanical methods appears to offer a sustainable, long-term solution to the water hyacinth problem (Charudattan 2001). To be effective, the integrated management system should include all appropriate and effective methods of control with biological control serving as the centerpiece (Charudattan 1986). Improvements in the effectiveness of biological control beyond what has been achieved thus far in different parts of the world appear likely through deployment of additional insect and microbial agents (Julien 2001; Evans and Reeder 2001). Although addition of agents to the integrated water hyacinth management system may result in negative rather than additive or synergistic effects, in the case of *Eccritotarsus catarinensis* (Carvalho) (Heteroptera: Miridae), introduced in South Africa in 1996, the effects appear to be generally complementary to those of five previously established arthropods (Coetzee et al. 2005).

There are several reviews on pathogens of water hyacinth and their development and use as bioherbicides (Bateman 2001; Charudattan 2001; Shabana et al. 1997). Among the pathogens that have been studied extensively are *Acremonium zonatum* (Jiménez and Balandra 2007), *Alternaria eichhorniae* (Shabana et al. 1995, 1997), and *Cercospora piaropi* (Charudattan et al. 1985; Tessmann et al. 2001, 2008). Although some recent surveys have been done to find new pathogens (Evans and Reeder 2001; Jiménez and Charudattan 1998), no new candidate has emerged that is better than the known pathogens. While the possibility of finding a previously unknown pathogen with great potential cannot be dismissed, any new attempt at developing a bioherbicide formulation has to rely on one of the three pathogens mentioned here. In this regard, *C. piaropi* is still the leading candidate in our estimation and one of the aims of a bioherbicidal formulation of *C. piaropi* should be as a synergist that magnifies the effects of the arthropods and intensifies the biotic stress. Attempts should be made to use mixtures of pathogens, as has been attempted by Den Breeÿen (1999) and Jiménez and Balandra (2007), as well as strains of *C. piaropi* with a higher level of aggressiveness, higher phytotoxin, and better fitness than previously used (Tessmann et al. 2008).

10.4 Work in Progress: Bioherbicides for Cogongrass and Tropical Soda Apple

10.4.1 Solanum viarum–Tobacco Mild Green Mosaic Tobamovirus: Use of a Plant Virus

Tropical soda apple (*Solanum viarum* Dunal, TSA, *Solanaceae*) is an invasive weed in Florida and several southeastern states in the USA where it is arguably the number one invasive plant species in livestock pastures and surrounding natural areas. Currently, more than a million acres of ranch and natural lands are estimated to be infested with TSA. Native to Argentina, Brazil, and Paraguay, TSA is known to occur in several countries in Central and South Americas and Africa, the Caribbean, and the Indian subcontinent (Mullahey et al. 1998). It is designated a noxious weed in the USA under the US statutes. It is a perennial shrub that has been dubbed *the plant from hell* to aptly describe its thorny and harmful nature, invasiveness, a propensity to form impenetrable thickets, and the difficulty in controlling it (Fig. 10.4). Aided by cattle, wildlife, and birds that eat TSA fruit and spread the seed, TSA has the potential to spread throughout the continental USA (Patterson 1996).

TSA is highly susceptible to a plant virus, *Tobacco mild green mosaic tobamovirus* (TMGMV) and is killed from a hypersensitive reaction to the virus infection (Charudattan et al. 2004; Charudattan and Hiebert 2007). TSA plants are killed completely and quickly; plants of all ages from the seedling to the mature stages are killed in about 21–42 days following inoculation. Younger plants are killed sooner than older plants but the final level of weed kill is generally the same.

TMGMV is a member of the plant virus genus *Tobamovirus* and occurs naturally in Florida and other regions of the USA and the world in about 25 different plant species. It is worldwide in distribution on susceptible *Nicotiana* species (wild and cultivated tobaccos) but is not known to cause any significant economic losses to

Fig. 10.4 Control of tropical soda apple by using the plant virus *Tobacco mild green mosaic tobamovirus* (TMGMV). *Top* row: *Left*: TSA foliage is covered with thorns, making the plant harmful and unpalatable to cattle. *Middle* top and *bottom*: TSA flowers and fruits that are green when immature and yellow when mature. *Right*: Vertical (*top*) and horizontal cross sections of fruits revealing abundant seeds. Second row from top: TSA infestation around trees (*left*) and in

an open field. Third row from *top*: Local lesions (a resistance response) elicited by TMGMV in a TSA leaf (*left*) and TSA plants that have been killed following manual inoculation with TMGM. *Bottom row*: A large TSA plant in the field at the time of inoculation with TMGMV (*left*) and the same plant severely wilted and dying from TMGMV crops. Through extensive research, our group has established that TMGMV could be used safely and effectively as a bioherbicide for TSA (Charudattan and Hiebert 2007; Charudattan et al. 2004). When properly used, TMGMV can provide nearly 100% TSA control, but repeated applications may be necessary to treat plants that emerge after the initial treatment and plants that are missed in the first treatment.

TMGMV could be easily multiplied in the tobacco (*Nicotiana tabacum*) variety "Samsun" *nn*, in which it elicits a mild, nonlethal, systemic mosaic disease and accumulates to a high titer of around 1–2 mg/g of fresh leaf tissue. Taking advantage of this tolerant host–virus interaction, we have developed an industrial process to mass-produce TMGMV and formulate it into two commercial formulations, *SolviNix* LC and *SolviNix* WP. *SolviNix* (*Solvi* from *Solanum viarum* and *Nix* from *put a stop to*) is undergoing large-scale precommercial field testing under an US EPA-issued Experimental Use Permit (EUP). In these trials, *SolviNix* is applied as a postemergent aqueous spray using a boom sprayer that is modified to abrade and spray simultaneously or with a high-pressure sprayer. The former is used for large, open areas while the latter is used for spot applications in wooded areas, under trees, and other areas inaccessible to the spray boom.

We have addressed the risks to nontarget plants, animals, and the environment through a comprehensive risk-analysis study (Charudattan et al., unpublished report; Charudattan and Hiebert 2007). Being a plant virus, TMGMV does not pose risks to any life form besides a small number of plants. Through an exhaustive host-range study involving more than 400 plant species in 58 plant families, we have confirmed that TMGMV is a pathogen adapted to plants in the *Solanaceae*. However, the lethal hypersensitive reaction, as seen in TSA, is restricted to species and cultivars of three genera: *Capsicum* (pepper), *Nicotiana* (tobacco), and *Physalis* (tomatillo). A mechanically transmitted virus that is not insect- or nematode transmitted, TMGMV could be used safely without endangering nontarget plants. It is anticipated that *SolviNix* will be registered in 2009 as the world's first virus-based bioherbicide.

10.5 *Imperata cylindrica* (Cogongrass) – Combined Classical and Bioherbicide Strategies Using Fungal Pathogens

Cogongrass (*Imperata cylindrica*, (L.) P. Beauv., Poaceae), ranked as one of the world's top ten invasive weeds (Holm et al. 1977), infests over 500 million ha worldwide including 200 million ha in Asia and an estimated 100,000 ha in south eastern USA (Schmitz and Brown 1994). A warm-season, rhizomatous, perennial grass species, cogongrass is found throughout tropical and subtropical regions of the world (Holm et al. 1977). It is a serious weed in natural forests and pine plantations in the Southeastern USA (Ramsey et al. 2003). It employs several survival strategies that include having an extensive rhizome system, adaptation to poor soils, drought tolerance, adaptability to fire, wind-disseminated seeds, and high genetic plasticity (MacDonald 2004). Despite the importance of the problems caused by cogongrass throughout the tropical areas of the world, biological control efforts have been rather limited (Caunter 1996). This has paralleled the general absence of attempted, and thus of successful, biological control projects against grasses in general (Waterhouse 1999). Other factors that complicate biocontrol of cogongrass include the existence of closely related grasses of economic or ecologic value (Holm et al. 1977) and potential conflict of interest with groups that value cogongrass (Evans 1991). Similarly, little information exists on pathogens of cogongrass and their potential as biological control agents (Evans 1991). It is likely that fungi associated with cogongrass are more diverse and abundant than indicated by herbarium records (Evans 1991; Charudattan 1997; Minno and Minno 2000). Evans (1987, 1991) suggested that some of the known fungal pathogens of cogongrass should be considered for introduction as classical biological control agents on this invasive weed. Promising species include three rust fungi, *Puccinia fragosoana* Beltrán, *P. rufipes* Dietel, and *P. imperatae* (Magnus) G. Poirault (Evans 1987) and a smut fungus, *Sporisorium (=Sphacelotheca) schweinfurthiana* (Evans 1987), which are well represented on cogongrass in the Old World, mainly Africa (Evans 1991), and the hemibiotrophic fungus *Colletotrichum caudatum* (Caunter 1996). It is interesting to note that these pathogens are common in the Mediterranean region where *I. cylindrica* is not a serious weed (Evans 1991; Holm et al. 1977). The aecial stages of *P. imperatae* and *P. fragosoana* are unknown while *P. rufipes* has a known aecial stage on *Thunbergia*, an alternate host (Cummins 1971) potentially excluding it from further evaluation as a biocontrol agent due to potential nontarget-host effects. Caunter (1996) cited that *P. rufipes*, although present in Malaysia, was having little effect on the cogongrass. The smut fungus, *Sporisorium (=Sphacelotheca) schweinfurthianum* (Thümen) Sacc. (Evans 1987) is well represented on cogongrass in the Old World, mainly Africa (Evans 1991). The host range of the genus *Sporisorium* is restricted to the *Poaceae.* A single cogongrass plant can produce up to 3,000 seeds per inflorescence; implementing a classical biological control program with a smut fungus could effectively reduce the number of seeds produced and ultimately reduce cogongrass density. Determining the value of implementing *S. schweinfurthianum* as an inundative biocontrol agent and integrating a mycoherbicide with the two classical biocontrol rust pathogens would be invaluable to the success of biocontrol of this invasive weed.

Biological control using plant pathogens to manage cogongrass infestations was considered to have potential in the USA (Van Loan et al. 2002). In the southeastern USA, two fungi, *Bipolaris sacchari* (Breda de Haan) Subram., isolated from cogongrass, and *Drechslera gigantea* (Heald & F.A. Wolf) S. Ito, isolated from large crabgrass [*Digitaria sanguinalis* (L.) Scop.], were shown to be pathogenic on a Florida *Imperata cylindrica* accession from Lake Alice, University of Florida by Yandoc (2001). Yandoc et al. (2005) evaluated the efficacy of *B.sacchari* and *D. gigantea* in greenhouse and miniplot trials. Amended spore suspensions of *B. sacchari* and *D. gigantea*, containing 10^5 spores per mL in a 1% aqueous gelatin solution, caused disease symptoms on cogongrass under greenhouse conditions with disease severity ranging from 42 to 49% (Yandoc et al. 2005). An amended spore and oil emulsion suspension (4% horticultural oil, 10% light mineral oil, and 86% water) resulted in higher levels of disease severity than the 1% gelatin. Higher levels of disease severity were achieved when a mixture of both these pathogens was applied under field conditions. Foliar injury ranged from 30 to 86% for *B. sacchari* and from 9 to 70% for *D. gigantea*. Despite the promising levels of disease severity and weed mortality recorded for these fungi, the regenerative potential of the cogongrass rhizomes allowed the plants to outgrow the effects of both pathogens.

In an attempt to explore the feasibility of using fungal pathogens that are effective against different geographic populations of cogongrass, a limited study was conducted by examining the genetic diversity between the USA and West African cogongrass by Den Breeÿen (Den Breeÿen 2007). Inter simple sequence repeat markers (ISSR) were used as a novel approach to understand the relationship between cogongrass accessions from these two regions. Included in the study were three closely related *Imperata* species: *Imperata brasiliensis* Trin., *I. brevifolia* Vasey, and *I. cylindrica* var. *rubra. Imperata brasiliensis* is native to North, Central, and South America and overlaps in its distribution with cogongrass in Florida and possibly elsewhere in the US Southeast, and is morphologically and genetically very similar. *Imperata brevifolia* is native to California, and *I. cylindrica* var. *rubra* is an ornamental variety sold in nurseries throughout the USA. The US and West African *I. cylindrica* accessions were found to be geographically and genetically distinct. Within the USA cogongrass, there was a distinction between cogongrass accessions collected throughout the state of Florida, which confirmed that cogongrass was introduced into the USA multiple times. *Imperata brasiliensis* was not genetically distinct from the USA *I. cylindrica* population forming a sister species to the Florida *I. cylindrica* accessions. In addition, *Imperata cylindrica* var. *rubra* was more closely related to the African accessions. *Imperata brevifolia* was found to be genetically distinct from all the *Imperata* accessions. In addition, *B. sacchari* and *D. gigantea* isolates from the USA and West Africa were evaluated in greenhouse trials to determine their efficacy as biological control agents on the West African *I. cylindrica*. There were no significant differences in disease incidence and disease severity between the Florida and Benin isolates inoculated on the Benin cogongrass. These findings have implications for implementing a biological control approach to manage cogongrass. The hypothesis that fungal biological control within the USA and African cogongrass would result in differential responses across the different accessions was not supported by this study as the West African cogongrass was found to be equally susceptible to the Florida and Benin isolates of *B. sacchari* and *D. gigantea* (Den Breeÿen 2007).

10.6 Conclusions

Here, we have presented a brief overview of several biological control programs using microbial agents against invasive forest weeds. Our emphasis has been on successful projects as well as those that appear to be heading for success. For examples of *unsuccessful* or *marginally successful* projects, the readers are referred to recent reviews by Charudattan (Charudattan 2001a, 2005).

The examples discussed in this chapter highlight the utility of biological control in forest management practices. Biological control offers an alternative, in terms of longevity and self-dispersal (classical biocontrol) to other weed control strategies including chemical and mechanical weed control. From an economic and environmental standpoint, biological control strategies employing plant pathogens should form an important component of integrated forest vegetation management. Further research on forest weed biocontrol should yield several improvements in forest management, with new commercial products and more widely acceptable approaches to forest management.

References

Anderson RC, Gardner ED (1999) An evaluation of the wilt-causing bacterium *Ralstonia solanacearum* as a potential biological control agent for the alien kahili ginger (*Hedychium gardnerianum*) in Hawaiian forests. Biol Control 15:89–96

Anikster Y (1984) The formae speciales. In: Bushnell WR, Roelfs AP (eds) The Cereal Rusts, Vol. I. Academic Press, Orlando, FL, pp. 115–130

Ayres P, Paul N (1990) Weeding with fungi. New Sci 1990:36–39

Barton J (2004) How good are we at predicting the field host-range of fungal pathogens used for classical biological control of weeds? Biol Control 31:99–122

Barton J, Fowler SV, Gianotti AF, Winks CJ, de Beurs M, Arnold GC, Forrester G (2007) Successful biological control of mist flower (*Ageratina riparia*) in New Zealand: agent establishment, impact and benefits to the native flora. Biol Control 40:370–385

Barreto RW, Evans HC (1988) Taxonomy of a fungus introduced into Hawaii for biological control of *Ageratina riparia* (*Eupatorieae*; *Compositae*), with observations on related weed pathogens. Trans Br Mycol Soc 91:81–97

Bateman R (2001) IMPECCA: an international, collaborative program to investigate the development of a mycoherbicide for use against water hyacinth in Africa. In: Hill MP, Julien MH, Center T (eds) Proceedings of the First IOBC Global Working Group Meeting for the Biological and Integrated Control of Water Hyacinth, 16–19 November 1998, St Lucia Park Hotel, Harare, Zimbabwe. Plant Protection Research Institute, Pretoria, pp. 57–61

Bourdôt G (2007) Biological control of gorse (*Ulex europaeus*). IBG News, December 2007, International Bioherbicide Group, pp. 5–6. http://muffin.area.ba.cnr.it/mailman/listinfo/ibg-news

Bourdôt G, Barton J, Hurrell G, Gianotti A, Saville D (2006) *Chondrostereum purpureum* and *Fusarium tumidum* independently reduce regrowth in gorse (*Ulex europaeus*). Biocontrol Sci Technol 16:307–327

Briese DT (2000) Classical biological control. In: Sindel BM (ed) Australian Weed Management Systems. R.G. & F.J. Richardson, Melbourne, VIC, pp. 161–192

Campbell FT (1998) Worst Invasive Plant Species in the Conterminous United States. Report, Western Ancient Forest Campaign, Springfield, VA

Caunter IG (1996) *Colletotrichum caudatum*, a potential bioherbicide for control of *Imperata cylindrica*. In: Moran VC, Hoffman JH (eds) Proceedings of the IX International Symposium on Biological Control of Weeds, University of Cape Town, Rondebosch, South Africa, pp. 525–527

Center TD, Van TK, Rayachhetry M, Buckingham GR, Dray FA, Wineriter S, Purcell MF, Pratt PD (2000) Field colonization of the Melaleuca snout beetle (*Oxyops vitiosa*) in South Florida. Biol Control 19:112–123

Charudattan R (1986) Integrated control of water hyacinth (*Eichhornia crassipes*) with a pathogen, insects, and herbicides. Weed Sci 34 (Suppl 1):26–30

Charudattan R (1988) Inundative control of weeds with indigenous fungal pathogens. In: Burge MN (ed), Fungi in Biological Control Systems. Manchester University Press, Manchester, England, pp. 86–110

Charudattan R (1997) Development of Biological Control for Noxious Plant Species – Progress Report May 15 – November 1997. University of Florida, Gainesville, FL

Charudattan R (2001a) Biological control of weeds by means of plant pathogens: significance for integrated weed management in modern agro-ecology. BioControl 46:229–260

Charudattan R (2001b) Biological control of water hyacinth by using pathogens: opportunities, challenges, and recent developments. In: Julien MH, Hill MP, Center TD, Ding J (eds) Biological and Integrated Control of Water Hyacinth, *Eichhornia* crassipes. Proceedings of the Second Meeting of the Global Working Group for the Biological and Integrated Control of Water Hyacinth, Beijing, China, 9–12 October, 2000. ACIAR Proceedings No. 102. Australian Centre for International Agricultural Research, Canberra, pp. 21–28

Charudattan R (2005) Ecological, practical, and political inputs into selection of weed targets: what makes a good biological control target? Biol Control 35:183–196

Charudattan R, Hiebert E (2007) A plant virus as a bioherbicide for tropical soda apple, *Solanum viarum*. Outlooks Pest Manage 18:167–171

Charudattan R, Linda SB, Kluepfel M, Osman YA (1985) Biocontrol efficacy of *Cercospora rodmanii* on water hyacinth. Phytopathology 75:1263–1269

Charudattan R, Pettersen MS, Hiebert E (2004) Use of *tobacco mild green mosaic virus* (TMGMV)-mediated lethal hypersensitive response (HR) as a novel method of weed control. U.S. Patent No. 6,689,718 B2, 10 February 2004

Coetzee JA, Center TD, Byrne MJ, Hill MP (2005) Impact of the biocontrol agent *Eccritotarsus catarinensis*, a sap-feeding mirid, on the competitive performance of water hyacinth, *Eichhornia crassipes*. Biol Control 32:90–96

Cummins GB (1971) The Rust Fungi of Cereals, Grasses, and Bamboos. Springer, New York

De Jong M (2000) The BioChon story: deployment of *Chondrostereum purpureum* to suppress stump sprouting in hardwoods. Mycologist 14:58–62

De Jong M, Scheepens PC, Zadoks JC (1990) Risk analysis for biological control: a Dutch case study in biocontrol of *Prunus serotina* by the fungus *Chondrostereum purpureum*. Plant Dis 74:189–194

De Jong M, Scheepens P, De Voogd B (2007) Berlin foresters make use of Dutch invention: control of *Prunus serotina* in forests with the fungus *Chondrostereum purpureum*. IBG News, December 2007, International Bioherbicide Group, pp. 7–8. http://muffin.area.ba.cnr.it/mailman/listinfo/ibg-news

Den Breeÿen A (1999) Biological control of water hyacinth using plant pathogens: dual pathogenicity and insect interactions. In: Hill MP, Julien MH, Center T (eds) Proceedings of the First IOBC Global Working Group Meeting for the Biological and Integrated Control of Water Hyacinth, 16–19 November 1998, St Lucia Park Hotel, Harare, Zimbabwe. Plant Protection Research Institute, Pretoria, pp. 75–79

Den Breeÿen A (2007) Biological Control of *Imperata cylindrica* in West Africa Using Fungal Pathogens, PhD Dissertation. University of Florida, Gainesville, FL

Evans HC (1987) Fungal pathogens of some subtropical and tropical weeds and the possibilities for biological control. Biocontrol News Inf 8:7–30

Evans HC (1991) Biological control of tropical grassy weeds. In: Baker FWG, Terry PJ (eds) Tropical Grassy Weeds. CAB International, Wallingford, UK, pp. 52–72

Evans HC, Reeder RH (2001) Fungi associated with *Eichhornia crassipes* (water hyacinth) in the Upper Amazon Basin and prospects for their use in biological control. In: Julien MH, Hill MP, Center TD, Ding J (eds), Biological and Integrated Control of Water Hyacinth, *Eichhornia* crassipes. Proceedings of the Second Meeting of the Global Working Group for the Biological and Integrated Control of Water Hyacinth, Beijing, China, 9–12 October 2000. ACIAR Proceedings No. 102. Australian Centre for International Agricultural Research, Canberra, pp. 62–70

Gardner DE (1992) Plant pathogens as biocontrol agents in native Hawaiian ecosystems. In: Stone CP, Smith CW, Tunison JT (eds) Alien Plant Invasions in Native Ecosystems of Hawaii: Management and Research. Cooperative National Park Resources Studies Unit, University of Hawaii. Honolulu, pp. 432–451

Gerlach J (2006) Invasive Species Specialist Group Database, Ecology of *Clidemia hirta*. http://www.issg.org/database/species/ecology.asp. Accessed 24 July 2006

Green S (2003) A review of the potential use of bioherbicides to control forest weeds in the UK. Forestry 76(3):285–298

Holm LG, Plucknett DL, Pancho JV, Herberger JP (1977) The World's Worst Weeds: Distribution and Biology. University Press of Hawaii, Honolulu, p. 609

ISSG [Invasive Species Specialist Group Database] (2005) Ecology of *Passiflora tarminiana*. http://www.issg.org/database/species/ecology.asp. Accessed 13 July 2005

ISSG [Invasive Species Specialist Group Database] (2006) Ecology of *Hedychium gardnerianum*. http://www.issg.org/database/species/ecology.asp. Accessed 24 July 2006

Julien MH (2001) Biological control of water hyacinth with arthropods: a review to 2000. In: Julien MH, Hill MP, Center TD, Ding J (eds) Biological and Integrated Control of Water Hyacinth, *Eichhornia* crassipes. Proceedings of the Second Meeting of the Global Working Group for the Biological and Integrated Control of Water Hyacinth, Beijing, China, 9–12 October 2000. ACIAR Proceedings No. 102. Australian Centre for International Agricultural Research, Canberra, pp. 8–20

MacDonald GE (2004) Cogongrass (*Imperata cylindrica*) – biology, ecology, and management. Crit Rev Plant Sci 23:367–380

Jiménez MM, Charudattan R (1998) Survey and evaluation of Mexican native fungi for biocontrol of water hyacinth. J Aquat Plant Manage 36:145–148

Jiménez MM, Balandra G (2007) Integrated control of *Eichhornia crassipes* by using insects and plant pathogens in Mexico. Crop Prot 26:1234–1238

Minno MC, Minno M (2000) Natural Enemies of Cogongrass in the Southeastern United States. Report to D. G. Shilling, Mid-Florida Research and Education Center, University of Florida, Apopka, FL

Morin L, Hill RL, Matayoshi S (1997) Hawaii's successful biological control strategy for mist flower (*Ageratina riparia*) – can it be transferred to New Zealand? Biocontrol News Info 18:77N–88N

Morris MJ (1987) Biology of the *Acacia* gall rust, *Uromycladium tepperianum*. Plant Pathol 36:100–106

Morris MJ (1991) The use of plant pathogens for biological weed control in South Africa. Agric Ecosyst Environ 37:239–255

Morris MJ (1997) Impact of the gall-forming rust fungus *Uromycladium tepperianum* on the invasive tree *Acacia saligna* in South Africa. Biol Control 10:75–82

Morris MJ, Wood AR, Den Breeÿen A (1999) Plant pathogens and biological control of weeds in South Africa: a review of projects and progress during the last decade. Afr Entomol Mem 1:129–137

Mullahey JJ, Shilling DG, Mislevy P, Akanda RA (1998) Invasion of tropical soda apple (*Solanum viarum*) into the U.S.: lessons learned. Weed Technol 12:733–736

Norris RF (1999) Ecological implications of using thresholds for weed management. J Crop Prod 2:31–58

Patterson DT (1996) Environmental limits to the distribution of tropical soda apple (*Solanum viarum* Dunal) in the United States. WSSA Abstracts 36:31.

Peters HA (2001) *Clidemia hirta* invasion at the Pasoh Forest Reserve: an unexpected plant invasion in an undisturbed tropical forest. Biotropica 33:60–68

Pimentel D, Zuniga R, Morrison D (2005). Update on the environmental and economic costs associated with alien-invasive species in the United States. Ecol Econ 52:273–288.

Pratt PD, Wineriter S, Center TD, Rayamajhi MB, Van TK (2004) *Boreioglycaspis melaleucae*. In: Coombs EM, Clark JK, Piper GL, Cofrancesco AF Jr (eds) Biological Control of Invasive Plants in the United States. Oregon State University Press, Corvallis, OR, pp. 273–275

Ramsey CL, Jose S, Miller DL, Cox J, Portier KM, Schilling DG, Merritt S (2003) Cogongrass [*Imperata cylindrica* (L.) Beauv.] response to herbicides and disking on a cutover site and in a mid-rotation pine plantation in southern USA. For Ecol Manage 179:195–207

Randall JM (1996) Weed control for the preservation of biological diversity. Weed Technol 10:370–383

Schmitz DC, Brown TC (1994) An Assessment of Invasive Non-Indigenous Species in Florida's Public Lands. Technical Report No. TSS-94–100, Florida Department of Environmental Protection, Tallahassee, FL

Setliff EC (2002) The wound pathogen *Chondrostereum purpureum*, its history and incidence on tress in North America. Aust J Bot 50:645–651

Shabana YM, Charudattan R, ElWakil MA (1995) Evaluation of *Alternaria eichhorniae* as a bioherbicide for water hyacinth (*Eichhornia crassipes*) in greenhouse trials. Biol Control 5:136–144

Shabana YM, Baka ZA, Abdel-Fattah GM (1997). *Alternaria eichhorniae*, a biological control agent for water hyacinth: mycoherbicidal formulation and physiological and ultrastructural host responses. Eur J Plant Pathol 103:99–111

Simberloff D (1997) The biology of invasions. In: Simberloff D, Schmitz DC, Brown TC (eds) Strangers in Paradise. Island Press, St. Louis, MO, pp. 3–17

Smith CW (1985) Impact of alien plants on Hawaii's native biota. In: Stone CP, Scott JM (eds) Hawaii's Terrestrial Ecosystems: Preservation and Management. University of Hawaii Press, Honolulu, pp. 180–250

TeBeest DO, Yang X, Cisar CR (1992) The status of biological control of weeds with fungal pathogens. Annu Rev Phytopathol 30:637–657

Templeton GE (1982) Status of weed control with plant pathogens. In: Charudattan R, Walker HL (eds) Biological Control of Weeds with Pathogens. Wiley, New York, pp. 29–44

Tessmann DJ, Charudattan R, Kistler HC, Rosskopf EN (2001) A molecular characterization of *Cercospora* species pathogenic to water hyacinth and emendation of *C. piaropi*. Mycologia 93:323–334

Tessmann DJ, Charudattan R, Preston JF (2008) Variability in aggressiveness, cultural characteristics, cercosporin production and fatty acid profile of *Cercospora piaropi*, a biocontrol agent of water hyacinth. Plant Pathology 57:957–966

Trujillo EE (1985) Biological control of Hamakua pa-makani with *Cercosporella* sp. in Hawaii. In: Delfosse ES (ed) Proceedings of the VI International Symposium on Biological Control of Weeds, 19–25 August 1984, Vancouver, Canada. Agriculture Canada, Ottawa, Canada, pp. 661–671

Trujillo EE (2005) History and success of plant pathogens for biological control of introduced weeds in Hawaii. Biol Control 33:113–122

Trujillo EE, Norman DJ, Killgore EM (1994) Septoria leaf spot, a potential biocontrol control of banana poke vine in forests of Hawaii. Plant Dis 78:883–885

Van Loan AN, Meeker JR, Minno MC (2002) Cogon grass. In: Van Driesche R, Blossey B, Hoddle M, Lyon S, Reardon R (eds) Biological Control of Invasive Plants in the Eastern United States. USDA Forest Service Publication FHTET-2002–04, p. 413

Vitousek PM (1988) Diversity and biological invasions of Oceanic Islands. In: Wilson EO, Peter FM (eds) Biodiversity. National Academy of Sciences, Washington, DC, pp. 181–189

Waterhouse DF (1998). Foreword. In: Julien MH, Griffiths MW (eds) Biological Control of Weeds—A World Catalogue of Agents and Their Target Weeds, 4th edition. CAB International, Wallingford, UK

Waterhouse DF (1999). When is classical biological control the preferred option for exotic weeds? In: Proceedings of the Symposium of Biological Control in the Tropics, MARDI Training Centre, Serdang, Selangor, 18–19 March 1999. Malaysia CABI, Kuala Lumpur, Malaysia, pp. 52–53

Wester LL, Wood HB (1977) Koster's curse (*Clidemia hirta*), a weed pest in Hawaiian forests. Environ Conserv 4(1): 35–41

Wood AW, Morris MJ (2007) Impact of the gall-forming rust fungus *Uromycladium tepperianum* on the invasive tree *Acacia saligna* in South Africa: 15 years of monitoring. Biol Control 41:68–77

Yandoc CB (2001) Biological control of cogongrass, *Imperata cylindrica* (L.) Beauv. PhD Dissertation, University of Florida, Gainesville, FL, p. 194

Yandoc CB, Charudattan R, Shilling DG (2005) Evaluation of fungal pathogens as biological control agents for cogongrass (*Imperata cylindrica*). Weed Technol 19:19–26

Yandoc-Ables CB, Rosskopf EN, Charudattan R (2006a) Plant Pathogens at Work: Progress and Possibilities for Weed Biocontrol, Part 1: Classical Vs. Bioherbicidal Approach. APSnet Feature Story August 2006. http://www.apsnet.org/online/feature/weed1/

Yandoc-Ables CB, Rosskopf EN, Charudattan R (2006b) Plant Pathogens at Work: Progress and Possibilities for Weed Biocontrol, Part 2: Improving Weed Control Efficacy. APSnet Feature Story August 2006. http://www.apsnet.org/online/feature/weed2/

Chapter 11
Sustainable Control of Spotted Knapweed (*Centaurea stoebe*)

D.G. Knochel and T.R. Seastedt

Abstract Spotted knapweed is native to Eastern Europe, with a locally scarce but widespread distribution from the Mediterranean to the eastern region of Russia. The plant is one of over a dozen *Centaurea* species that were accidentally introduced into North America and now is found in over 1 million ha of rangeland in the USA and Canada. Land managers spend millions of dollars annually in an attempt to control spotted knapweed and recover lost forage production, and meanwhile the plant perseveres as a detriment to native biodiversity and soil stability. These ecological concerns have motivated intense scientific inquiry in an attempt to understand the important factors explaining the unusual dominance of this species. Substantial uncertainty remains about cause–effect relationships of plant dominance, and sustainable methods to control the plant remain largely unidentified or controversial. Here, we attempt to resolve some of the controversies surrounding spotted knapweed's ability to dominate invaded communities, and focus on what we believe is a sustainable approach to the management of this species in grasslands, rangelands, and forests. Application of both cultural and biological control tools, particularly the concurrent use of foliage, seed, and root feeding insects, is believed sufficient to decrease densities of spotted knapweed in most areas to levels where the species is no longer a significant ecological or economic concern.

Keywords Biological control • Biological invasions • *Centaurea stoebe* L. ssp *micranthos* • *Centaurea maculosa* • Knapweed • Sustainable management

11.1 Introduction

Knapweeds and yellow starthistle, plants belonging to the genus *Centaurea* and the closely related genus *Acroptilon*, are members of the Asteraceae that were accidentally introduced into North America from Eurasia over a century ago. These species

D.G. Knochel (✉) and T.R. Seastedt
Department of Ecology and Evolutionary Biology and INSTAAR, an Earth Systems Institute, University of Colorado, Boulder, CO 80309, USA
david.knochel@colorado.edu

Inderjit (ed.), *Management of Invasive Weeds*,

have been identified as major problems for rangelands and forests of the western USA and Canada, occupying millions of hectares and causing millions of dollars of control costs and forage production losses (Duncan et al. 2004; Smith 2004). These species have also been identified as a threat to native biodiversity (Ortega et al. 2006). Sustainable, cost-effective management strategies to mitigate and diminish the negative effects of these knapweeds represent a priority activity desired by a broad spectrum of society, including ranchers, public land managers, and conservation biologists.

Smith et al. (2001) listed 495 research publications on various aspects of knapweed ecology, with most of the research focused at perhaps the most aggressive and widespread of these species, spotted knapweed (*C. stoebe micranthos*, also identified as *C. maculosa* and *C. biebersteinii*; see Ochsmann 2001; Hufbauer and Sforza 2008). A comprehensive review on management of spotted knapweed was provided in the late 1990s (Sheley et al. 1999); however, approximately 170 research articles on spotted knapweed have been published since then (Web of Science search by the authors, August 2007). In spite of this effort, our scientific understanding about cause–effect relationships of plant dominance has only slightly improved, and the important factors explaining the unusual dominance of this species remain largely unidentified or controversial. Sustainable management techniques are clearly desired, as *Centaurea* ranked as the most commonly cited noxious weed genus on government lists in the USA and Canada (Skinner et al 2000). Here, we define sustainable control as the process of using cost-effective management tools that cause a long-term reduction of the target weed to lower densities at which plants persist yet are no longer of ecological nor of economic concern. Ideally, such a sustainable management effort would employ both direct (reduction of plant fecundity and fitness via top–down controls by pathogens and herbivores) and indirect (bottom–up reduction in the available resources for the target plant via cultural methods that enhance native plant competition) methods that require minimal management inputs and reduce susceptibility to reinvasion. Here, we attempt to resolve some of the controversies surrounding spotted knapweed's ability to dominate invaded communities, and focus on what we believe is a sustainable approach to the management of spotted knapweed in grasslands, rangelands, and forests.

11.2 Life History Information on Spotted Knapweed

Spotted knapweed is native to Eastern Europe, with a locally scarce but widespread distribution from the Mediterranean to the eastern region of Russia (Hufbauer and Sforza 2008). The plant is one of over a dozen *Centaurea* species that were accidentally introduced into the North America and now is found in 45 states within the USA. It occupies over 1 million ha of rangeland in the western USA and occurs in over 60,000 ha of Canada (Story 1992). Spotted knapweed is a C_3 perennial forb with a central taproot that grows as a rosette the first year and forms between one and ten flowering stems per year throughout its lifespan (Story et al. 2001). A review by

Sheley et al. (1999) indicated that the plant may live up to 9 years, and is capable of producing 5,000–40,000 seeds per m^2 per year. Schirman (1981) reported that spotted knapweed produced an average of 24–33 seeds per flower head during a 4-year interval, and over the same time period estimated seed production at 11,300–29,600 seeds per m^2. The large seed bank that this species can create within the soil is also persistent, with 25% of seed viable after 8 years (Davis et al. 1993). Given such persistence, any management activities used to reduce densities of this weed must be sustained for a substantial period of time to prevent reestablishment from the seed bank.

The plant has some nutritional value to wildlife, but tends to be an inferior food source for most generalist native species as well as to cattle. While sheep consume this species as a forage crop, sociological and economic issues have not favored the use of sheep in control activities or as a logical alternative to grazing by cattle (Alper 2004).

11.3 Dominance of Knapweed Species

As previously mentioned, the mechanism(s) that allow *Centaurea* spp to achieve a high degree of dominance in many plant communities remain poorly identified and controversial. The ability of *top–down controls* (herbivores) to suppress this species has been debated for over a decade (e.g., Müller-Schärer and Schoeder 1993; Callaway and Ridenour 2004), and the competitive responses of this species across gradients of plant competition and soil resources (e.g., Maron and Marler 2007), and effects on those resources (e.g., Hook et al. 2004) have likewise been a major research topic. Allelopathy has been identified as a potentially significant competitive mechanism. Callaway and Ridenour (2004) and Callaway et al. (2005) argued that the relatively high levels of allelopathic compounds produced by the knapweeds and released upon a plant community that had not evolved tolerance to these chemicals could explain the dominance of invasive *Centaurea* species in North America. Alternatively, a reduction in soil pathogens and/or enhanced positive feedbacks from microflora, including mycorrhizae, encountered in the new environments could also explain this dominance (e.g., Mitchell and Power 2003; Mitchell et al. 2006). The possibility also exists that both allelopathy and positive feedbacks from microbial communities contribute to the dominance and abundance of this species.

Additional mechanisms and alternative hypotheses explaining *Centaurea* abundance have been proposed. Gerlach and Rice (2003) indicated that *C. solstitialis* was successful as an invader due to its abilities to persist within a community and exploit resource opportunities, a characteristic also shared by diffuse knapweed (LeJeune et al. 2006; Seastedt and Suding, 2007). Suding et al. (2004) demonstrated that rosettes of diffuse knapweed were strong competitors under ambient nutrient conditions, but were less competitive under lower nutrient conditions that may have characterized North American grasslands until recently. Further, the superior ability of knapweeds to access soil water beyond the reach of native plants was suggested to allow greater photosynthetic-related advantages (Hill et al. 2006). Elsewhere, experiments on soil biota have tested the collective benefits provided by mycorrhizae and the absence of soil pathogens in the success of invasive species of *Centaurea*. Mycorrhizae fungi provided a competitive advantage to spotted knapweed (Marler et al 1999; Callaway et al. 2004), but in another study were not beneficial to plants losing tissue due to simulated herbivory (Walling & Zabinski 2006). However, Callaway et al. (2001) demonstrated that mycorrhizal interactions with *C. melitensis* allowed the plant to exhibit compensation to grazing damage.

Somewhat surprisingly, low to intermediate levels of herbivory can apparently benefit knapweeds (Callaway et al. 1999; 2006; Thelen et al. 2005; Newingham and Callaway 2006). Herbivores appear to stimulate the competitive ability of the plant, perhaps by enhancing root exudates that stimulate mycorrhizal activity or the production of allelopathic chemicals. Alternatively, Newingham et al. (2007) found that spotted knapweed benefits from root herbivory by shifting overall nitrogen allocation from the damaged root to aboveground stems and reproductive tissues. The combined effects of allelopathy and its reported response to herbivory appear to make spotted knapweed a superinvader. Accordingly, this species was given the title, "The wicked weed of the West" (Alper 2004). As of 2007, most species of *Centaurea* and particularly spotted knapweed were widely perceived as invasive plant species that lack sustainable control measures.

Early attempts to control spotted and diffuse knapweed added a human health concern to land management issues. Two species of gall flies (*Urophora* species) were released on spotted and diffuse knapweed in 1970 (Harris 1980). The insects greatly increased in numbers, but failed to control plant densities. The gall flies provided a substantial amount of biomass available to native predators, and the weed presented an abundant but unused resource by native herbivores, thus greatly altering native food webs (Pearson et al. 2000). Among the native predators that consumed the gall flies were deer mice (*Peromyscus maniculatus*) that were carriers of a virus known to cause the potentially fatal human disease, hantavirus (Pearson and Callaway 2003, 2005, 2006; Ortega et al. 2004). A single broadcast herbicidal application was shown to decrease these impacts on deer mice populations by reducing knapweed abundance and the food source for *Urophora* (Pearson and Fletcher 2008). However, broadcast applications of herbicides intended to temporarily decrease spotted knapweed densities were also meanwhile shown to increase the abundance of another unwanted invasive plant, *Bromus tectorum*, which replaced spotted knapweed as the dominant invader (Y. Ortega, USFWS, personal communication to D.G.K). Alternatively at our site, some of the indirect negative food web effects caused by *Urophora* abundance have already been reduced by the presence and consumption of *Urophora* by another biological control insect, *Larinus minutus* (Seastedt et al. 2007). Because *Larinus* has a large overwintering biomass in the soil, the species may be causing other unknown alterations in food webs.

Recently, a second set of studies has appeared to suggest that spotted knapweed is not invincible in its introduced environment. The identified allelopathic agent for spotted knapweed may not occur in high enough concentrations to have impacts on other plant species (Blair et al. 2005, 2006). A similar finding for diffuse knapweed suggests that this species also lacks sufficient quantities of its putative reported allelopathic chemical to be capable of reducing plant competition as well (Norton et al. 2008). Significant reductions in the densities of spotted knapweed (Corn et al. 2006; Story et al. 2006) or large reductions in seed production by this species (Seastedt et al. 2007) suggested that classical biological control by insect herbivores may in fact be effective. Additional studies suggest that insect herbivory (Jacobs et al. 2006) and competition from native plant species (Pokorny et al. 2005, Rinella et al. 2007) have significant negative effects on the densities of this species. These studies suggest that disturbance events, which lower the productivity of an invaded plant community, may be essential to successful spotted knapweed invasion. Collectively, these results appear to refute much of the evidence presented in previous paragraphs suggesting novel allopathic mechanisms, superior resource use, or the ability of biological control herbivory to improve plant fitness under most field conditions.

11.4 Management of Spotted Knapweed in Natural Areas

Spotted knapweed appears to flourish in many human-generated habitats such as roadsides and areas of soil disturbance, especially where competition from other plants is absent. Because the weed is adapted to such sites, control in these areas without modifying land use may be difficult, and can perhaps be accomplished only with repeated control activities. Such habitats could function as source habitats to disperse seeds into other, less-disturbed plant communities used for grazing or conservation purposes. If, as we suggest later, sustainable control of the weed is possible within relatively *nondisturbed* areas, then the compelling reasons to control the weed in disturbed sites are reduced. In fact, persistence of the weed may be essential in maintaining refugia for biological control agents that provide sustainable control of this species in more intact vegetated areas.

Classical weed management usually attempts to categorize effects into what *tools* work to control a weed species. Using this approach, choices for grasslands and rangelands include *cultural* (enhancing plant competition or preventing weed introduction), *biological* (classical biological control, adding herbivores or pathogens), *mechanical* (tillage, mowing, or fire), and *chemical* (herbicides). Readers are referred to Sheley et al. (1999) for an overview of all management techniques, including those herbicides used to kill the plant. However, as we note later, routine use of herbicides and particularly broadcast applications of herbicides appear insufficient to reduce the long-term dominance characteristics of this species.

While herbicides can provide temporary reductions in densities, the treatment rarely if ever provides long-term benefits, and the weed is too widespread for this to be a viable management option over a large area. In addition, herbicide treatments have been shown to reduce native plant diversity while failing to reduce long-term density of the target plants (Rinella et al. 2008). A single or infrequent application of herbicides could actually increase the subsequent densities of the target plant by reducing plant competition. Prescribed summer fire applied on an annual basis can reduce densities of spotted knapweed (Emery and Gross 2005; MacDonald et al. 2007), but like the use of chemicals, the cost and nontarget effects limit its use. Given the problem with chemical and mechanical tools, the biological and cultural tools would therefore encompass the only viable activities for control of a regionally abundant species such as spotted knapweed. However, some evidence supports the hypotheses that spotted knapweed is a superior competitor in North America via allelopathy, mycorrhizae, or other soil microbial feedback mechanisms, and if this is the case then enhancing plant competition may not be effective. As mentioned previously, the remaining biological tools have begun to show success in some areas. However, if spotted knapweed is not negatively impacted by herbivores or can benefit from herbivory in some situations, then the current suite of biological control agents may only exacerbate the overall problems caused by spotted knapweed (Ortega et al. 2004). Until now, our current science has provided managers with a confusing, mixed message about controls for spotted knapweed.

We argue that the studies that offer the most useful insights into sustainable management are those that provide a long-term assessment of knapweed densities, and are conducted within a whole system perspective. While individual experiments on plant–soil, plant–plant, or plant–herbivore interactions may be perfectly valid given the context of the experiment, interpretations and conclusions may change when the full suite of direct and indirect interaction effects of the activity is imposed. Management decisions should be based on those long-term studies that measure changes in spotted knapweed abundance within the affected communities. Considering that this plant grows in a variety of habitats and climates across much of North America, experiments that assess control tools under a variety of resource conditions are also the most useful for application to management plans.

11.4.1 Classical Biological Control

As previously discussed, biological control agents were first released on spotted and diffuse knapweed in 1970. By 2000, 13 species of insects had been released in an attempt to reduce the abundance of these plants (Story and Piper 2001). A number of these had established on knapweed; however, no significant reductions in the abundance of knapweed had been documented prior to 2006. Given the previously described concern about hantavirus and the influence of knapweed on human health (Pearson and Callaway 2006), the benefits of identifying sustainable management controls of knapweed include not only the need to reduce the target plant's effects on forage loss and biodiversity, but also to reduce problems associated with nontarget effects caused by this biological control insect food-web alteration.

Biological control agent effects on spotted knapweed have been increasingly studied in the last two decades. In a pot study, Steinger and Müller-Schärer (1992) demonstrated that the growth response of spotted knapweed to root herbivory was mediated by soil resource abundance. Since then, experimental studies using actual herbivores or clipping manipulations suggested that *Centaurea* was capable of equal or overcompensation to tissue loss by herbivory. For a species capable of producing allelopathic chemicals, compensation to low levels of root herbivory resulted in a chemical release or changes in root exudates that suppressed competing vegetation (Callaway et al. 1999; Thelen et al. 2005). At a minimum, these experiments suggest that the relative fitness of *Centaurea* is increased when the fitness of competing vegetation is reduced. However, the ability of plants to compensate for herbivory is mediated through relative plant growth rate (RGR) responses (Hilbert et al. 1981). When a plant's RGR is constrained by resource availability, caused either by direct resource limitations or indirect limitations generated by biotic effects (competition or additional herbivory), overcompensation cannot occur. Under reduced resource conditions, the removal of tissue cannot stimulate increased photosynthesis or materials allocation to damaged areas. Further, the particular focal resource limiting plant growth, and information on how herbivory affects a plant's ability to use that resource, may be necessary for understanding these compensation responses under different conditions (Wise and Abrahamson 2007). A study by Steinger and Müller-Schärer (1992) demonstrated that spotted knapweed can compensate for root herbivory by *Agapeta zoegana* (L.) (Lepidoptera: Cochylidae) moth larvae, but not for herbivory by *Cyphocleonus achates* (Coleoptera: Curculionidae) weevil larvae, and that the ability to compensate increased under higher plant resource conditions. A second study confirmed this compensatory response; Newingham et al. (2007) found increased nitrogen allocation to aboveground tissues after root damage from the *Agapeta* moth larvae. These two stud ies test compensation responses by spotted knapweed to herbivory; however, experiments that demonstrate the possible degrees of compensation to both belowground and aboveground herbivory under variable resource conditions have not been conducted.

A large body of evidence obtained by researchers and managers indicates that the closely related species, diffuse knapweed, is controlled by insect herbivory across much (and perhaps all) of its introduced range (Seastedt et al. 2003, 2005, 2007; Coombs et al. 2004; Myers 2004; Smith 2004). Reports from Montana (Jacobs et al. 2006; Story et al. 2006; Corn et al. 2006) suggest that a similar response may be occurring for spotted knapweed in the same region where allelopathy was reported as being an important mechanism for the dominance of this species. There, the declines are attributed to large numbers of two species of gall flies (*Urophora* spp), and the root-feeding weevil, *Cyphocleonus achates*, that were released at those sites in 1988. While reductions in densities of knapweed were being demonstrated in Montana, land managers and researchers working in Minnesota and Colorado reported a similar result using a suite of all available herbivore insects. Among those were *Urophora* spp, *Larinus minutus* Gyllenhal (Coleoptera: Curculionidae), and *Cyphocleonus achates* (Cortilet and Northrop 2006; Michels et al. 2007) (Fig. 11.1).

A very reasonable question to ask is, if the suite of biological control agents that currently exist in the USA have all been approved and released since 1991, why have not we seen more evidence for a negative effect on spotted knapweed? Given the response we have observed of diffuse knapweed to insect herbivores, we suggest that successful biological control of spotted knapweed has been similarly slow

Fig. 11.1 Biological control fly, *Urophora affinis*, and the weevil *Larinus minutus* ovipositing into spotted knapweed flower heads. On the right is *Cyphocleonus achates* on a diffuse knapweed plant in a prairie near the Colorado foothills (USA) (photo courtesy of David G. Knochel)

to manifest itself due to (1) the failure to find the right combination of control insects (Müller-Schärer and Schroeder 1993), (2) the tendency of insects to disperse or require a substantial time period before increasing to high densities at individual sites, (3) other management activities have interfered with the efficacy of the insects, and (4) lag effects were generated by the availability of a large and persistent seed bank established prior to the increase in herbivore release and activity.

At our research site, we released the insects discussed here against a population of spotted knapweed and began monitoring indices of spotted knapweed abundance in 2002, a year when our site was impacted by a severe drought (Seastedt et al. 2007). We have followed stem densities and seed production of this population through the present (Table 11.1). Knapweed seed production declined since 2003 and we interpreted this to suggest that insects released in 2001 were initiating top–down constraints, similar to that documented for diffuse knapweed. Our highest seed production numbers in 2004 and 2007 are still well below those reported by Schirman (1981), where spotted knapweed grew in areas lacking top–down pressure from herbivory. The damage by *Urophora spp.* and *Larinus minutus* consuming seed and aboveground tissues from the majority of the plants at out site, as well as the effects of the *Cyphocleonus achates* weevil feeding on root tissue, suggest that this combination of insects may provide success in biological control. The abundance of all released insects has increased since 2001, and experiments are under way to determine if the insects reach high enough densities to continue to push the spotted knapweed population into steady decline. Blair et al. (2008) compared herbivore loads on both *C. diffusa* and *C. stoebe* in North America and Europe, and found that *C. stoebe* has largely escaped root herbivores but not seedhead feeders in the introduced range. The root herbivores may thus simply require greater time and assistance to become widely established or effective on spotted knapweed populations. In a study by Pokorny et al. (2005), 2,000 spotted knapweed seeds per m^2 were added to areas of manipulated

Table 11.1 Seeds, *Urophora*, and *Larinus* found in flower heads, and estimates of seed production for *C. stoebe* in Colorado (updated from Seastedt et al. 2007)

Year	Sample Size	Number per flower head[a] Seeds	Urophora	Larinus	Seed production (number m^{-2})
2002	294	1.46 (0.180)	0.30 (0.041)	0.89 (0.025)	No data
2003	311	4.26 (0.335)	0.43 (0.053)	0.63 (0.036)	4,600
2004	429	3.06 (0.250)	1.15 (0.073)	0.10 (0.014)	1,030
2005	540	2.15 (0.151)	1.81[b] (0.065)	0.74 (0.018)	260
2006	288	1.80 (0.215)	0.57[b] (0.064)	0.57 (0.032)	0
2007	227	6.01 (0.441)	0.14[b] (0.031)	0.80 (0.042)	2,970

[a]Values are means and SE from flower heads from a single population

[b]*Urophora* values after 2004 include estimates of *Urophora* present but consumed by *Larinus*. Results obtained before 2005 are based on live *Urophora* or pupae fragments

plant competition. The study documented lower abundance of the weed after several years in plots where competing vegetation remained intact, and using their estimates we predict that the seed production and competition at our site will eventually result in knapweed densities reduced to low levels of persistence.

While we are not certain if foliage herbivory reduces survivorship of immature plants, the combined impact of root feeders in addition to seed predators best explains results of our long-term monitoring efforts. Corn et al. (2007) suggest that insect effects are independent of climate. At our site, we have emphasized measurements on seed production. Our results appear to parallel those of Schirman (1981) that suggest that plants produce more seeds in wetter years, or in years that follow drought years. Plant size, seeds produced per flower head, and seeds produced per unit area appear to be influenced by the amounts and seasonality of precipitation. Both 2003 and 2007 were years of high spring moisture that followed dry years, and both years resulted in pulses of plant growth and seed production. The ability of this species to exploit spring rainfall and increase seed production following drought supports the notion that spotted knapweed is an effective opportunist (Hill et al. 2006), similar to diffuse knapweed (LeJeune et al. 2006; Seastedt and Suding 2007) and yellow star thistle (Gerlach and Rice 2003).

11.5 Resolving Research Contradictions and Developing a Management Framework

While factors identified to be responsible for the dominance and persistence of *Centaurea* spp. often seem contradictory, we believe that the collective findings reported here can be reconciled to a conceptual model for *Centaurea*. This model recognizes its dominance as the net result of concurrent top–down (herbivory) and bottom–up (plant-mediated resource abundance) effects on weed abundance (Fig. 11.2). Up until the mid 1990s, the magnitude of top–down controls was inadequate to reduce seed production to a level where additional seedling mortality caused by plant competition and resource limitation could prevent dominance by the species. In addition, the abundance of biological control insects may not have increased to high enough levels at individual sites to warrant a population decline. In some areas, these top–down controls are now in place, but the large seed bank produced by the weed during preinsect control years continues to subsidize knapweed densities. Uncertainty exists regarding the intensity and regional extent of the various species characteristics such as allelopathy, or the microbial community feedbacks on the plant.

We also note that it is important to recognize the unique interactions between plants, insects, and soil biota that are represented by *Centaurea* populations in the invaded range. The *non-coevolved* interactions (e.g., unusually strong top–down controls imposed by biological control insects) in areas outside the plant's native range have the potential to neutralize or amplify other non-coevolved interactions (e.g., unusually strong competitive effects due to plant characteristics; also see Pearson and Callaway 2005). The non-coevolved competitive interactions have led to high levels of plant dominance by spotted knapweed in North America, occurring in densities that have never been observed in its home range. However, sustainable management activities employing novel combinations of negative biotic and

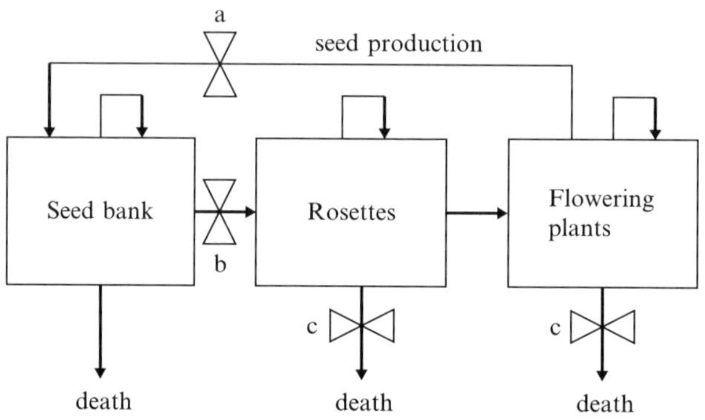

Fig. 11.2 How top–down and bottom–up factors collectively control densities of spotted knapweed. The density of the weed per unit area is composed of seeds in the seed bank, immature plants (seedlings and rosettes), and flowering plants. Each year a portion of the seed bank germinates or dies, rosettes bolt to become flowering plants, and flowering plants produce seed. Key control points include factors affecting seed production (a), rosette establishment (b), and survival of rosettes and adult plants (c). Sustainable control procedures involve reducing seed production by biological (top–down) control by herbivory from multiple insect species, and by decreasing seedling survivorship by increasing plant competition (bottom–up) activities. This model is similar to that proposed by Myers and Risley (2000) for diffuse knapweed, except that here, seed production is reduced to levels where interspecific plant competition affecting seedling survivorship, perhaps mediated by additional herbivory, is capable of controlling densities of the plant

abiotic feedbacks are now believed sufficient to reverse these patterns of dominance.

Many research findings report spotted knapweed to exhibit positive feedback with microbes and mycorrhizae in its introduced habitats, and the plant certainly functions as an aggressive competitor with native plant species. Further, spotted knapweed appears to have at least temporarily escaped specialist herbivores and perhaps some specialist pathogens present in its native country. The species may or may not exhibit allelopathic characteristics that enhance its invasiveness and competitive characteristics, but this fact may be irrelevant given controls on seed production and seedling survivorship. We find no fault with those studies that have found that single species of herbivores can increase the competitive abilities of spotted knapweed under a specified set of resource conditions (Callaway et al. 1999; Thelen et al. 2005). However, we do find that the experimental findings of Pokorny et al. (2005) and Jacobs et al. (2006) are much more consistent with our observations regarding the relative inability for individual spotted knapweed seedlings to become established or persist in intact communities when multiple specialist herbivore insects are present. Further, the herbivore studies by Corn et al. (2006) and Story et al. (2006) match the management findings of Cortilet and Northop (2006), and monitoring studies of Michels et al. (2007), documenting spotted knapweed decline due to biological control insects.

Our studies of herbivore impacts in the presence of strong interspecific competition and moderate to low soil resources also suggest similar declines (Seastedt et al. 2007 and unpublished results). Collectively, these studies argue that insect herbivory is effective at controlling densities of spotted knapweed, and that the degree of control will vary depending on site characteristics. Thus, we believe that (1) multiple herbivores appear to prevent spotted knapweed from exhibiting compensatory responses to tissue removal and seed destruction by herbivory, and (2) that the long-term effect of seed reduction in the presence of competing species is sufficient to reduce the densities of spotted knapweed. What we do not know, however, and what remains an open research issue, is the range of habitats in which this response is likely to occur.

We predict that the suite of insects released against spotted knapweed will eventually be effective across at least a substantial portion of the introduced range of this species. This prediction argues that many areas may not require additional management activities beyond monitoring plant and insect abundance. However, studies also indicate that the reduction in densities of spotted knapweed may be a slow process, and one that might be facilitated by cultural techniques (e.g., adding desirable plant seeds; Jacobs et al. 2006). Use of biological control agents does not preclude the use of other proactive control measures for spotted knapweed, but the evidence for the efficacy of biological control insects published since 2006 reduces the urgency to proactively manage this plant species with other mechanical or chemical techniques. Accordingly, the use of both cultural and biological control tools in natural areas will likely decrease spotted knapweed densities to a level where the plant persists as neither an ecological nor as an economic concern. We believe that the moniker, "the wicked weed of the West" (Alper 2004) may now be transferred to another invasive plant species.

Acknowledgments We thank Monika Chandler, Minnesota Department of Agriculture, for information and discussions on the biocontrol program of spotted knapweed conducted in Minnesota, and Dr. Jerry Michels for information on Texas A&M studies. Our research on spotted knapweed has been funded by the National Research Initiative of the USDA Cooperative State Research, Education and Extension Service, grant number 06-03618.

References

Alper J (2004) The wicked weed of the West. Smithsonian Mag 35(7): 33–36

Bais HP, Vepachedu R, Gilroy S, Callaway RM, Vivanco JM (2003) Allelopathy and exotic plant invasions: from molecules and genes to species interactions. Science 301: 1377–1380

Blair AC, Hanson BD, Brunk GR, Marrs RA, Westra P, Nissen SJ, Hufbauer RA (2005) New techniques and findings in the study of a candidate allelochemical implicated in invasion success. Ecol Lett 8: 1039–1047

Blair AC, Nissen SJ, Brunk GR, Hufbauer RA (2006) A lack of evidence for an ecological role of the putative allelochemical (±)-catechin in spotted knapweed invasion success. J Chem Ecol 32(10): 2327–2331

Blair AC, Schaffner U, Häfliger P, Meyer SK, Hufbauer RA (2008) How do biological control and hybridization affect enemy escape? Biol Control doi:10.1016/j.biocontrol.2008.04.014

Callaway RM, Ridenour WM (2004) Novel weapons: invasive success and the evolution of increased competitive ability. Front Ecol Environ 2(8): 436–443

Callaway RM, DeLuca TH, Belliveau WM (1999) Biological-control herbivores may increase competitive ability of the noxious weed *Centaurea maculosa*. Ecology 80(4): 1196–1201

Callaway RM, Newingham B, Zabinski CA, Mahall BE (2001) Compensatory growth and competitive ability of an invasive weed are enhanced by soil fungi and native neighbours. Ecol Lett 4(5): 429–433

Callaway RM, Thelen GC, Barth S, Ramsey PW, Gannon JE (2004) Soil fungi alter interactions between the invader *Centaurea maculosa* and North American natives. Ecology 85(4): 1062–1071

Callaway RM, Ridenour WM, Laboski T, Weir T, Vivanco JM (2005) Natural selection for resistance to the allelopathic effects of invasive plants. J Ecol 93: 576–583

Callaway RM, Kim J, Mahall BE (2006) Defoliation of *Centaurea solstitialis* stimulates compensatory growth and intensifies negative effects on neighbors. Biol Invasions 8(6): 1389–1397

Coombs EM, Clark JK, Piper GL, Cofrancesco AF Jr (2004) Biological control of invasive plants in the United States. Oregon State University Press, Corvallis, OR

Corn JG, Story J, White LJ (2006) Impacts of the biological control agent *C achates* on spotted knapweed, *Centaurea maculosa*, in experimental plots. Biol Control 37: 75–81

Corn JG, Story JM, White LJ (2007) Effect of summer drought relief on the impact of the root weevil *Cyphocleonus achates* on spotted knapweed. Environ Entomol 36(4): 858–863

Cortilet AB, Northrop N (2006) Biological control of European buckthorn and spotted knapweed. Minn Dept Agri Final Program Report, St Paul, MN. http://www.mda.state.mn.us/news/publications/pestsplants/weedcontrol/knapweedlcmrfinalreport.pdf. Accessed 27 May 2008.

Davis ES, Fay PK, Chicoine TK, Lacey CA (1993) Persistence of spotted knapweed (*Centaurea-maculosa*) seed in soil. Weed Sci 41(1): 57–61

Duncan CA, Jachetta JJ, Brown ML, Carrithers VF, Clark JK, DiTomaso JM, Lym RG, McDaniel KC, Renz MJ, Rice PM (2004) Assessing the economic, environmental, and societal losses from invasive plants on rangeland and wildlands. Weed Technol 18: 1411–1416

Emery SM, Gross KL (2005) Effects of timing of prescribed fire on the demography of an invasive plant, spotted knapweed *Centaurea maculosa*. J Appl Ecol 42: 60–69

Gerlach JD, Rice KJ (2003) Testing life history correlates of invasiveness using congeneric plant species. Ecol Appl 13(1): 167–179

Harris P (1980) Effects of *Urophora affinis* Frfld and *U quadrifasciata* (Meig) (Diptera: Tephritidae) on *Centaurea diffusa* Lam and *Centaurea maculosa* Lam (Compositae). Zeitschrigt fur Angewandte Entomologie 90: 190–210

Hilbert DW, Swift DM, Detling JK, Dyer MI (1981) Relative growth rates and the grazing optimization hypothesis. Oecologia 51: 14–18

Hill JP, Germino MJ, Wraith JM, Olsen BE, Swan MB (2006) Advantages in water relations contribute to greater photosynthesis in *Centaurea maculosa* compared with established grasses. Int J Plant Sci 167(2): 269–277

Hook PB, Olson BE, Wraith JM (2004) Effects of the invasive forb *Centaurea maculosa* on grassland carbon and nitrogen pools in Montana, USA. Ecosystems 7(6): 686–694

Hufbauer RA, Sforza R (2008) Multiple introductions of two invasive *Centaurea* taxa inferred from cpDNA haplotypes. Divers Distrib 14: 252–261.

Jacobs JS, Sing SE, Martin JM (2006) Influence of herbivory and competition on invasive weed fitness: observed effects of *Cyphocleonus achates* (Coleoptera: Curculionidae) and grass-seeding treatments on spotted knapweed performance. Environ Entomol 35: 1590–1596

LeJeune KD, Suding KN, Seastedt TR (2006) Nutrient availability does not explain invasion and dominance of a mixed grass prairie by the exotic forb *Centaurea diffusa* Lam. Appl Soil Ecol 32 (1): 98–110

Maron J, Marler M (2007) Native plant diversity resists invasion at both low and high resource levels. Ecology 88(10): 2651–2661

Marler MJ, Zabinski CA, Callaway RM (1999) Mycorrhizae indirectly enhance competitive effects of an invasive forb on a native bunchgrass. Ecology 80(4): 1180–1186

MacDonald NW, Scull BT, Abella SR (2007) Mid-spring burning reduces spotted knapweed and increases native grasses during a Michigan experimental grassland establishment. Restor Ecol 15(1): 118–128

Michels GJ Jr, Carney VA, Jurovich D, Kassymzhanova-Mirik S, Jones E, Barfot K, Bible J, Mirik M, Best S, Bustos E, Karl B, Jimenez D (2007) Biological control of noxious weeds on federal installations in Colorado and Wyoming. Texas Agric Experiment Station 2006 Annual Report, 222 pp. http://amarillotamuedu/programs/entotaes/CNWB%20Annual%20Reports.htm. Accessed 27 May 2008

Mitchell CE, Power AG (2003) Release of invasive plants from fungal and viral pathogens. Nature 421(6923): 625–627

Mitchell CE, Agrawal AA, Bever JD, Gilbert GS, Hufbauer RA, Klironomos JN, Marion JL, Morris WF, Parker IM, Power AG, Seabloom EW, Torchin ME, Vazquez DP (2006) Biotic interactions and plant invasions. Ecol Lett 9: 726–740

Müller-Schärer H, Schroeder D (1993) The biological-control of *Centaurea* spp in North-America – do insects solve the problem? Pestic Sci 37: 343–353

Myers JH (2004) A silver bullet in the biological control of diffuse knapweed. ESA 2004 Annual meeting abstract. http://abstractscoallenpresscom/pweb/esa2004/document/36421. Accessed 27 May 2008

Myers JH, Risley C (2000) Why reduced seed production is not necessarily translated into successful biological weed control. In: Spencer NR (ed) Proceedings of the X international symposium on biological control MSU, Bozeman, Montana, pp. 569–581

Myers JH, Bazely DR (2003) Ecology and control of introduced plants. Cambridge University Press, Cambridge, UK

Newingham BA, Callaway RM (2006) Shoot herbivory on the invasive plant, *Centaurea maculosa*, does not reduce its competitive effects on conspecifics and natives. Oikos 114: 397–406

Newingham BA, Callaway RM, BassiriRad H (2007) Allocating nitrogen away from an herbivore: a novel compensatory response to root herbivory. Oecologia 153: 913–920

Norton AP, Blair AC, Hardin JG, Nissen SJ, Brunk GR (2008) Herbivory and novel weapons: no evidence for enhanced competitive ability or allelopathy induction of *Centaurea diffusa* by biological controls. Biol Invasions 10(1): 79–88

Ochsmann J (2001) On the taxonomy of spotted knapweed. In: Smith L (ed) Proceedings, First international knapweed symposium of the 21st century, Coeur d'Alene, ID USDA-ARS, Albany, CA, pp. 33–41.

Ortega YK, Pearson DE, McKelvey KS (2004) Effects of biological control agents and exotic plant invasion on deer mouse populations. Ecol Appl 14: 241–253

Ortega YK, McKelvey KS, Six DL (2006) Invasion of an exotic forb impacts reproductive success and site fidelity of a migratory songbird. Oecologia 149: 340–351

Pearson DE, Callaway RM (2003) Indirect effects of host-specific biological control agents. Trends Ecol Evol 18: 456–461

Pearson DE, Callaway RM (2005) Indirect non-target effects of host-specific biological control agents: implications for biological control. Biol Control 35: 288–298

Pearson DE, Callaway RM (2006) Biological control agents elevate hantavirus by subsidizing deer mouse populations. Ecol Lett 9: 443–450

Pearson DE, Fletcher RJ (2008) Mitigating exotic impacts: restoring deer mouse populations elevated by an exotic food subsidy. Ecol Appl 18(2): 321–334

Pearson DE, McKelvey KS, Ruggiero LF (2000) Non-target effects of an introduced biological control agent on deer mouse ecology. Oecologia 122(1): 121–128

Pokorny ML, Sheley RL, Zabinski CA, Engel RE, Svejcar TJ, Borkowski JJ (2005) Plant functional group diversity as a mechanism for invasion resistance. Restor Ecol 13(3): 448–459

Rinella MJ, Pokorny ML, Rekaya R (2007) Grassland invader responds to realistic changes in native species richness. Ecol Appl 17(6): 1824–1831

Rinella MJ, Maxwell BD, Fay PK, Weaver T, Sheley RL (2008) Control effort exacerbates invasive species problem. Ecol Appl (in press)

Schirman R (1981) Seed production and spring seedling establishment of diffuse and spotted knapweed. J Range Manage 34(1): 45–47

Seastedt TR, Suding KN (2007) Biotic resistance and nutrient limitation controls the invasion of diffuse knapweed (*Centaurea diffusa*). Oecologia 151: 626–636

Seastedt TR, Gregory N, Buckner D (2003) Reduction of diffuse knapweed by biological control insects in a Colorado grassland. Weed Sci 51: 237–245

Seastedt TR, Suding KN, LeJeune KD (2005) Understanding invasions: the rise and fall of diffuse knapweed (*Centaurea diffusa*) in North America. In: Inderjit (ed) Invasive plants: ecological and agricultural aspects. Birkhouser-Verlag, Basal, Switzerland

Seastedt TR, Knochel DG, Garmoe M, Shosky SA (2007) Interactions and effects of multiple biological control insects on diffuse and spotted knapweed in the Front Range of Colorado. Biol Control 42: 345–354

Sheley RL, Jacobs JS, Carpinelli ML (1999) Spotted knapweed. In: Sheley RL, Petroff JK (eds) Biology and management of noxious rangeland weeds. Oregon State University, Corvallis, OR, pp. 350–361

Skinner K, Smith L, Rice P (2000) Using noxious weed lists to prioritize targets for developing weed management strategies. Weed Sci 48: 640–644

Smith L (2004) Impact of biological control agents on *Centaurea diffusa* (diffuse knapweed) in central Montana. In: Cullen J (ed) I Symposium on biological control of weeds, Canberra, Australia CSIRO, Australia

Smith L, Mayer M (2005) Field cage assessment of interference among insects attacking seed heads of spotted and diffuse knapweed. Biocontrol Sci Technol 15(5): 427–442

Smith L, Story JM, DiTomaso JM (2001) Bibliography of spotted knapweed, yellow starthistle and other weedy knapweeds. In: Smith L (ed) Proceedings of the first international knapweed symposium of the twenty-first century, Coeur d'Alene, Idaho USDA-ARS, Albany, CA

Steinger T, Müller-Schärer H (1992) Physiological and growth-responses of *Centaurea-maculosa* (Asteraceae) to root herbivory under varying levels of interspecific plant competition and soil-nitrogen availability. Oecologia 91(1): 141–149

Story JM, Piper GL (2001) Status of biological control efforts against spotted and diffuse knapweed. In: Smith L (ed) Proceedings of the first international knapweed symposium of the twenty-first century. Coeur d'Alene, Idaho USDA-ARS, Albany, CA, pp. 11–17

Story JM, Boggs KW, Good WR (1992) Voltinism and phenological synchrony of Urophora affinis and U. Quadriasciata (Diptera: Tephritidae), two seed head flies introduced against spotted knapweed in Montana. Env Entomol 21: 1052–1059.

Story JM, Smith L, Good WR (2001) Relationship among growth attributes of spotted knapweed (*Centaurea maculosa*) in western Montana. Weed Technol 15: 750–761

Story JM, Coombs EM, Piper GL (2004) In: Coombs EM, Clark JK, Piper GL, Cofrancesco AF Jr (eds) Biological Control of Invasive Plants in the United States. Oregon State University Press, Corvallis, OR, pp. 204–205

Story JM, Callan W, Corn JG, White LJ (2006) Decline of spotted knapweed density at two sites in western Montana with large populations of the introduced root weevil, *Cyphocleonus achates* (Fahraeus). Biol Control 38: 227–232

Suding KN, LeJeune KD, Seastedt TR (2004) Competitive impacts and responses of an invasive weed: dependencies on nitrogen and phosphorus availability. Oecologia 141(3): 526–535

Thelen GC, Vivanco JM, Newingham G, Good W, Bais HP, Landres P, Caesar A, Callaway RM (2005) Insect herbivory stimulates allelopathic exudation by an invasive plant and the suppression of natives. Ecol Lett 8: 209–217

Vivanco JM, Bais HP, Stermitz FR, Thelen GC, Callaway RM (2004) Biogeographical variation in community response to root allelochemistry: novel weapons and exotic invasion. Ecol Lett 7 (4): 285–292

Walling SZ, Zabinski CA (2006) Defoliation effects on arbuscular mycorrhizae and plant growth of two native bunchgrasses and an invasive forb. Appl Soil Ecol 32: 111–117

Wise MJ, Abrahamson WG (2007) Effects of resource availability on tolerance of herbivory: a review and assessment of three opposing models. Am Nat 169(4): 443–454

Chapter 12
Managing Parthenium Weed Across Diverse Landscapes: Prospects and Limitations

K. Dhileepan

Abstract Parthenium is a weed of global significance affecting many countries in Asia, Africa, and the Pacific Islands. Parthenium causes severe human and animal health problems, agricultural losses as well as serious environmental problems. Management options for parthenium include chemical, physical, legislative, fire, mycoherbicides, agronomic practices, competitive displacement and classical biological control. The ability of parthenium to grow in a wide range of habitats, its persistent seed bank, and its allelopathic potential make its management difficult. No single management option would be adequate to manage parthenium across all habitats, and there is a need to integrate various management options (e.g. grazing management, competitive displacement, cultural practices) with classical biological control as a core management option.

Keywords *Parthenium hysterophorus* • Biological control • Integrated weed management

12.1 Introduction

Parthenium (*Parthenium hysterophorus* L.; Asteraceae) (Fig. 12.1) is an annual herb with a deeply penetrating taproot and an erect shoot. Young plants form a rosette of leaves close to the soil surface. As it matures, the plant develops many branches on its upper half, and may eventually reach a height of up to 2 m (McFadyen 1992). Parthenium grows vigorously in summer, but with good rainfall and warm temperature, it has the ability to germinate and establishes at any time of the year. Flowering usually commences 6–8 weeks after germination, and soil moisture seems to be the major contributing factor to the duration of flowering

K. Dhileepan
Queensland Department of Primary Industries and Fisheries, Alan Fletcher Research Station, Sherwood, QLD 4075, Australia
k.dhileepan@dpi.qld.gov.au

Inderjit (ed.), *Management of Invasive Weeds*,

Fig. 12.1 *Parthenium hysterophorus*

(Navie et al. 1996). Parthenium is a prolific seed producer, and a fully grown plant can produce more than 15,000 seeds in its lifetime (Haseler 1976). Seeds persist and remain viable in soil for reasonably long periods, with nearly 50% of the seed bank viable up to 6 years (Navie et al. 1998a).

Parthenium is a weed of global significance (Adkins et al. 2005). Parthenium occurs in Bangladesh (Navie et al. 1996), India (Mahadevappa and Patil 1997), Israel (Joel and Liston 1986), Pakistan (Javaid and Anjum 2005; Shabbir and Bajwa 2006), Nepal (Adhikari and Tiwari 2004), China (Navie et al. 1996), Sri Lanka (Jayasurya 2005), Taiwan (Peng et al. 1988), and Vietnam (Nath 1981) in Asia, several Pacific islands including New Caledonia, Papua New Guinea, Seychelles,

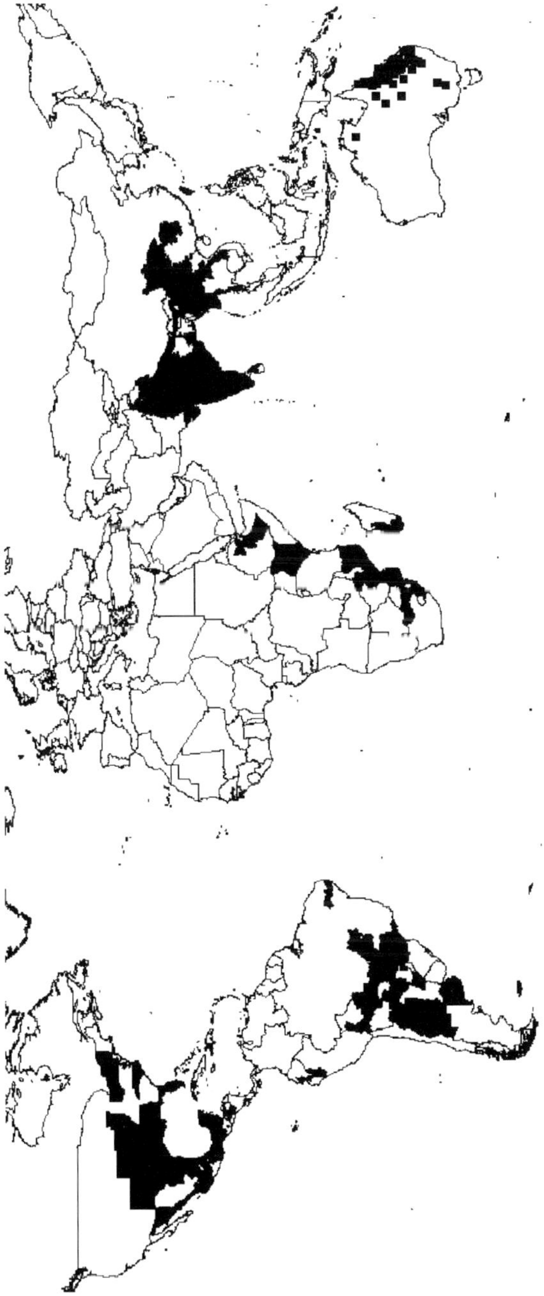

Fig. 12.2 Global distribution of parthenium weed

Vanuatu (Anonymous 2003; Parsons and Cuthbertson 2001), and several African countries including Ethiopia, Kenya, Madagascar, Mozambique, South Africa, Somalia, Swaziland, and Zimbabwe (i.e., Bowen 2001; CABI 2004; Da Silva et al. 2004; Fessehaie et al. 2005; Frew et al. 1996; Hilliard 1977; MacDonald et al. 2003; Nath 1988; Njoroge 1986, 1989, 1991; Strathie et al. 2005; Tamado and Milberg 2000; Tamado et al. 2002a; Taye et al. 2004b) (Fig. 12.2). Parthenium, a major weed in Australia and India, is attaining a major weed status in other countries also.

12.2 History of Parthenium Infestations

Parthenium occurs naturally throughout the tropical and subtropical Americas from southern United States of America (USA) to southern Brazil and northern Argentina (Navie et al. 1996; Towers 1981). Parthenium in Argentina, Bolivia, Chile, Paraguay, Peru, and Uruguay appears different to those in North America with yellow flowers (Dale 1981). Genotypic studies confirmed the existence of distinct North American and Central American populations (Graham and Lang 1998).

Parthenium was accidentally introduced into India in 1956 (Rao 1956), possibly through contaminated seed imports from north America, and has since then spread over most parts of the Indian subcontinent, including Pakistan (Shabbir and Bajwa 2006), Nepal (Adhikari and Tiwari 2004), Sri Lanka (Jayasurya 2005), and Bangladesh (Navie et al. 1996). Sources of parthenium introductions in other Asian countries are not fully known. In Australia, parthenium was first identified in 1955, also possibly along with contaminated seed imports from Texas, USA, and was proclaimed as a noxious plant in 1975 (Auld et al. 1982–1983). Currently, parthenium mainly occurs in Queensland, (Chippendale and Panetta 1994; McFadyen 1992) but has the potential to spread throughout Australia (Adamson 1996). It is believed that parthenium has spread to neighbouring Papua New Guinea and East Timor Islands from Australia. In Africa, invasion of parthenium was reported in Swaziland and Mozambique in mid 1960s, Kenya in mid 1970s, and Ethiopia, South Africa and Zimbabwe in early 1980s. It is suspected that parthenium in Ethiopia may have been introduced through contaminated grains from North America (Fessehaie et al. 2005; Frew et al. 1996), but the mode and source of introduction in other countries in Africa are not fully known. Genetic analysis suggests that parthenium genotypes found in Australia, India, and Africa are possibly originated from southern Texas, USA (Graham and Lang 1998).

12.3 Deleterious Effects

Parthenium causes severe human (Cheney 1998; Kololgi et al. 1997; McFadyen 1995; Rao et al. 1977; Wedner et al. 1986) and animal (Tudor et al. 1982; Kadhane et al. 1992) health problems, agricultural losses (Firehun and Tamado 2006; Navie et al. 1996; Tamado and Milberg 2000; Tomado et al. 2002a) as well as serious environmental problems (Chippendale and Panetta 1994). In Australia, parthenium affected over 170,000 km^2 of grazing land in Queensland (Chippendale and Panetta 1994; McFadyen 1992) and reduced beef production by AUD 16.5 million annually (Chippendale and Panetta 1994). In India, parthenium caused yield losses of up to 40% in crops (Khosla and Sobti 1979) and reduced forage production by up to 90% (Nath 1981). Parthenium has reduced the species richness and species diversity of other plant species (Sridhara et al. 2005) and their seed banks (Navie et al. 2004). Parthenium also acts as a reservoir host for plant pathogens and insect pest of crop plants (e.g. Basappa 2005; Evans 1997a; Govindappa et al. 2005; Navie et al. 1996; Rao et al. 2005; Robertson and Kettle 1994). Parthenium and related genera contain sesquiterpene lactones (Picman and Towers 1982), which induce severe contact dermatitis and other allergic symptoms (Towers 1981). Stock animals, especially horses, suffer from allergic skin reaction while grazing infested paddocks. Parthenium is generally unpalatable and toxic to cattle, buffalo and sheep (e.g. Narasimham et al. 1980; Kadhane et al. 1992). Consumption of large amounts will produce taints in mutton (Tudor et al. 1982) and can even kill livestock.

12.4 Management Options

Chemical control is the first line of defence (Holman 1981), but high costs of herbicides prohibit their long-term use for parthenium management in grazing areas, public and uncultivated areas and forests. To eradicate localised infestations, for roadside infestations or when the weed is a problem in certain crops, control can be achieved by using herbicides (e.g. Holman 1981). In areas where chemical control is not economical, other options such as the use of competitive plants to displace parthenium (e.g. Joshi 1991a, b; O'Donnell and Adkins 2005), fire (Vogler et al. 2002) and other physical methods including mulching green parthenium plants have been suggested as suitable options either individually on in combinations.

12.4.1 Chemical Control

Herbicides, either as pre- or post-emergence application, can provide effective control of parthenium in crops (e.g. Holman 1981; Navie et al. 1996; Dawson and Sarkar 1997), infestations along road side (e.g. Brooks et al. 2004) and wasteland (e.g. Dixit and Bhan 1997; Yadav et al. 1997). The effectiveness of herbicides depends on the timing of application, and often more selective herbicides are preferred to minimise non-target damage. Chemical control is the first line of defence in eradication and containment programs. Chemical control is also the most suitable option for managing parthenium in urban areas, to reduce human and animal health impacts as well as in high-value crops. However, high costs of herbicides prohibit their long-term use for parthenium management in grazing areas, public and uncultivated areas and forests.

12.4.2 Physical/Mechanical Control

Manual removal of parthenium before flowering is often carried out to reduce localised infestations in residential areas and high-value crops, but not in large infestations. However, handling of parthenium is not recommended because of the health risks associated with parthenium weed. Other physical methods such as grading, slashing and ploughing in large infestations can provide some relief over short term, but they are not effective in the long-term management, as they are known to enhance regeneration of parthenium (Kohli et al. 1997). It has been suggested that ploughing in parthenium in the rosette stage before seed-set helps to retain soil moisture, but this practice needs to be followed up by sowing a crop or direct seeding of perennial pasture.

12.4.3 Competitive Displacement

Parthenium is known to be allelopathic, and is capable of reducing growth and germination of crops resulting in reduced crop yield and contaminated crop products. Several beneficial plants are also known to be allelopathic and have the potential to compete and displace parthenium (Table 12.1). Most of the research so far on using beneficial plants to competitively displace parthenium has been restricted to India (Akula and Kondap 1997; Dhawan et al. 1997; Joshi 1991a, b; Kandasmy and Sankaran 1997; Kauraw et al. 1997; Sushilkumar and Bhan 1997; Yaduraju et al. 2005; Gautam et al. 2005a), with only limited studies done in Australia (O'Donnell and Adkins 2005), South Africa (van der Laan 2006) and Pakistan (Anjum and Bajwa 2005; Anjum et al. 2006; Javaid et al. 2005). Potential for the large-scale use of competitive plants to displace parthenium in the field is yet to be studied.

12.4.4 Fire

Fire is commonly used for pasture management and woody weed control in northern Australia (Grice and Brown 1996). Parthenium-dominant pastures will not carry a fire, because of lack of adequate fuel load for an effective fire to occur. In areas with adequate fuel load (i.e. national parks, road side infestation, etc.), more parthenium incidence was observed in burnt areas compared with nearby unburnt areas. On the basis of anecdotal evidence, burning of parthenium-infested areas is

Table 12.1 Plants that have the potential to competitively displace parthenium (see text for references for each country)

Country	Competitive beneficial plants
India	Acanthaceae
	Andrographis paniculata Nees (medicinal)
	Amaranthaceae
	Achyranthus aspera L. (medicinal)
	Alternanthera sessilis (L.) DC. (invasive?)
	Amaranthus spinosus L. (food, dye, etc.)
	Aerva javanica Juss (food)
	Asteraceae
	Tagetus erecta L. (crop, ornamental)
	T. patula L. (ornamental, natural pesticide)
	Capparaceae
	Cleome gynandra L. (medicinal & edible)
	Chenopodiaceae
	Chenopodium album L. (vegetable & poultry feed)
	Euphorbiaceae
	Croton bonplandianum Baill (medicinal)
	Fabaceae
	Cassia auriculata L. (medicinal)
	C. occidentalis L. (medicinal)
	C. sericea SW (food, medicinal)
	C. tora L. (medicinal, natural pesticide, gelling agent)
	Stylosanthes scabra Vogel (non-native pasture)
	Tephrosia purpurea L. (medicinal)
	Lamiaceae
	Hyptis suaveolens (L.) Poit. (non-native, medicinal)
	Ocimum canum Sim. (medicinal)
	Malvaceae
	Abutilon indicum (Linn.) Sweet. (medicinal)
	Malva pusilla Sm. (medicinal)
	Sida acuta Burm.f. (invasive, medicinal)
	S. rhombifolia L. (invasive)
	S. spinosa L. (medicinal)
	Nyctaginaceae
	Mirabilis jalapa L. (non-native, ornamental)
	Poaceae
	Cenchrus ciliaris L. (non-native pasture)
	Panicum maximum Jacq. (non-native pasture)
Australia	Fabaceae
	Clitoria ternatea L. (introduced legume, cattle feed)
	Glycine latifolia Newell & Hymowitz (native pasture legume)
	Macroptilium bracheatum Marechal & Baudet (pasture legume)
	Stylosanthes seabrana B.L. (non-native pasture legume)
	Poaceae
	Bothriochloa insculpa A. Camus (non-native pasture)
	Dicanthium aristatus (Poir.) C.E.Hubb (non-native pasture)
	Cenchrus ciliaris L. (non-native pasture)
South Africa	Poaceae
	Eragrostis curvula (Schrad.) Nees (native grass)
	Panicum maximum Jacq. (non-native pasture)

(continued)

Table 12.1 (continued)

Country	Competitive beneficial plants
	Digitaria eriantha Steud. (native grass)
Pakistan	Poaceae
	Imperata cylindrica (L.) Beauv. (native grass)
	Desmostachya bipinnata (L.) Stapf. (native grass)
	Dichanthium annulatum (Forssk.) Stapf. (native grass)
	Cenchrus pennisetiformis Hochst. & Steud. (native grass)
	Sorghum halepense (L.) Pers. (native grass)

often discouraged, even if there is adequate fuel load. Subsequent field experiments confirmed that fire did not affect the size of parthenium seed banks, nor does smoke from such fires stimulate parthenium seed germination (Vogler et al. 2002). Fire, however, resulted in an increase in parthenium densities, supporting earlier field observations, but the parthenium populations declined after subsequent fires.

12.4.5 Cultural Practices

Cultural practices are often used to minimise parthenium infestations in crops, but their role in other habitats such as forests, wasteland and roadside infestations is yet to be implemented. Agronomic practices, such as manipulation of crop varieties, sowing density and planting date, or the use of cover crops (e.g. cowpea) are widely used to outcompete parthenium in sorghum crops in Ethiopia (Tomado et al. 2002a). In India also, crop rotation incorporating marigold (*Tagetus erecta* Linn.) between regular crops has been suggested to reduce parthenium infestations (e.g. Kauraw et al. 1997). In Australia, ploughing and mulching green parthenium before planting winter crops (e.g. wheat) has been suggested to improve soil moisture retention. However, no scientific studies have been made so far to validate these claims.

12.4.6 Mycoherbicides

Mycoherbicides are fungal-based bioherbicides, and unlike herbicides, will have no or limited non-target damage. Information is widely available on the mycoflora (e.g. *Candidatus Phytoplasma aurantifolia*, *Phoma herbarum* Westend, *Sclerotium rolfsii* Sacc., *Fusarium pallidoroseum* Sacc.) associated with parthenium in India (e.g. Bagyanarayana and Manoharachary 1997; Deshpande et al. 1997; Gayathri and Pandey 1997; Kauraw et al. 1997b; Singh 1997; Manickam et al. 1997a, b; Jeyalakshmi et al. 2005; Kumar and Evans 2005; Vikrant et al. 2007), South Africa

(Wood and Scholler 2002) and Ethiopia (Taye 2005, 2006; Taye et al. 2002, 2004b, c), but studies on their potential role as mycoherbicides in the field, and its economics are not known. Mycoherbicides have a potential to supplement herbicides, especially in crops and urban areas, but may have limited application value in other areas (e.g. forests, waste land and pastures) where their use may not be economical.

12.4.7 Legislative Control

Preventing the spread of parthenium is the most cost-effective management strategy. Parthenium is known to spread through contaminated vehicles, machinery, livestock, grain and other products. In Australia, parthenium is a declared weed in all states, which makes the sale, movement or distribution of parthenium within Australia prohibited (Table 12.2). In Queensland, there are several 'washdown' facilities at strategic points and vehicles travelling from infested areas are required to clear their vehicles before travelling to parthenium-free areas. It is legally mandatory for suppliers of stocks, machinery, soil, water or other products from areas of known parthenium infestations to declare that the material they supply is free of parthenium. The Australian Quarantine and Inspection Service prohibits the introduction of parthenium as a nursery stock or through contaminated seed imports.

Although parthenium is widespread throughout India, it has been declared as a weed only in the Karnataka state (Table 12.2). However, there has been no coordinated nationwide program to prevent the spread and management of parthenium in India. In South Africa and Sri Lanka, parthenium is a declared weed (Table 12.2), where movement of parthenium is prohibited in both rural and urban areas. However, in other countries no such nationwide legislative mechanism exists to prevent further spread of parthenium.

Table 12.2 Declaration status of parthenium as a noxious weed in various countries

Country	Category	Declaration
South Africa	Category 1	Conservation of Agricultural Resources Act (Act No. 43 of 1983)
Australia	Weed of national significance	Australian Weeds Strategy
Queensland, Australia	Class 2	Land Protection (Pest and Stock Route Management) Act 2002
New South Wales, Australia	Class 1	NSW Noxious Weeds Act 1993
Karnataka, India	Noxious weed	Agricultural Pests and Diseases Act, 1968 (on 23 October 1975)
Sri Lanka	Noxious weed	Plant Protection Ordinance No. 35 of 1999

12.4.8 Biological Control

12.4.8.1 Australia

In Australia, biological control of parthenium was initiated in 1977 with surveys conducted in Mexico, USA (Evans 1983, 1997a, b; McClay 1980; McClay et al. 1995), Brazil, Argentina (McFadyen 1976, 1979) and the Caribbean Islands (Bennett 1976). So far, nine species of insects and two rust fungi have been introduced into Australia (Dhileepan and McFadyen 1997; Griffiths and McFadyen 1993; McClay et al. 1990; McFadyen 1985, 1992, 2000; McFadyen and McClay 1981; Parker et al. 1994; Wild et al. 1992; Table 12.3). Among them, seven species of insects and two rust fungi have been successfully established as biological control agents in Australia (Dhileepan et al. 1996; Dhileepan and McFadyen 1997; McFadyen 1992, 2000).

Table 12.3 Introduced parthenium biological control agents (see text for references for each biological control agent in different countries)

Biological control agents	Introduced country	Source country	Year	Establishment status
Lepidoptera: Tortricidae				
Epiblema strenuana Walker	Australia	Mexico	1982	Widespread and abundant
Platphalonidia mystica (Razowski & Becker)	Sri Lanka	Australia	2004	Unknown
Lepidoptera: Sessidae	Australia	Argentina	1992	Unknown
Carmenta nr *ithacae* (Beutenmüller)	Australia	Mexico	1998	Localised
Lepidoptera: Bucculatricidae	Australia	Mexico	1998	Abundant and localised
Bucculatrix parthenica Bradley				
Coleoptera: Chrysomelidae				
Zygogramma bicolorata Pallister	Australia	Mexico	1980	Abundant and localised
Listronotus setosipennis Hustache	India	Mexico	1984	Abundant and widespread
Smicronyx lutulentus Dietz	Pakistan	India	2006	Localised
Coleoptera: Curculionidae	Australia	Argentina and Brazil	1982	Abundant and widespread
Conotrachelus albocinereus Fiedler	Australia	Mexico	1981	Abundant and widespread
	Australia	Argentina	1995	Localised
Homoptera: Delphacidae				
Stobaera concinna (Stål).	Australia	Mexico	1983	Unknown
Basidiomycotina: Uredinales				
Puccinia abrupta parthenii-cola Parmelee	Australia	Mexico	1991	Localised

(continued)

Table 12.3 (continued)

Biological control agents	Introduced country	Source country	Year	Establishment status
Puccinia melampodii Dietel & Holway	Ethiopia	Kenya?	1997	Localised
	India	Unknown	1980?	Unknown
	Kenya	Unknown	Unknown	Localised
	South Africa	Unknown	1995	Widespread
	Australia	Mexico	1999	Unknown

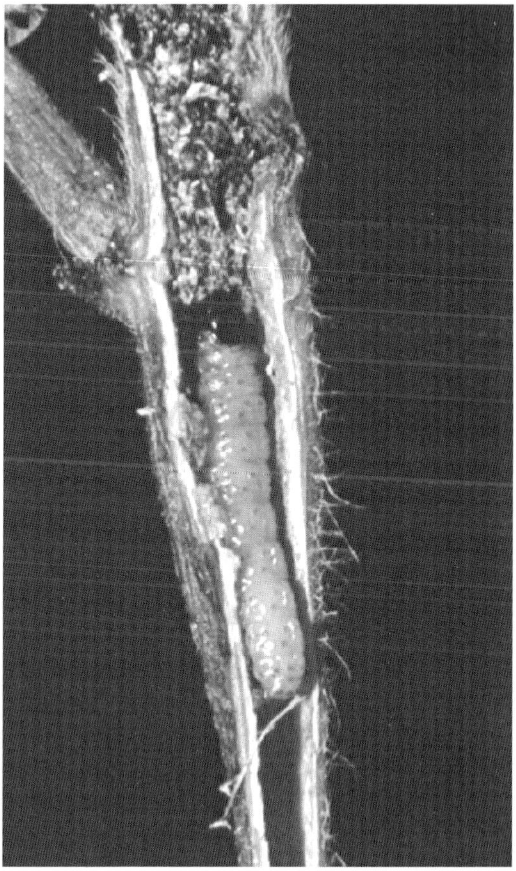

Fig. 12.3 *Epiblema strenuana* larva inside the stem gall

The stem-galling moth *Epiblema strenuana* Walker (Fig. 12.3) was introduced to Australia from Mexico in 1982 (McClay 1987; McFadyen 1987, 1992). It became established and widespread within 2 years of introduction, and now occurs in all parthenium-infested areas. Galling by *E. strenuana* causes serious visible symptoms on parthenium (McFadyen 1992), and the impact on plant height, leaf production, flower production and plant biomass becomes significant when the gall damage is initiated at early stages of plant growth (Dhileepan 2001, 2003b, 2004; Dhileepan and McFadyen 1997, 2001; Navie et al. 1998b). Grass competition significantly increased the effectiveness of the parthenium stem-galling moth *E. strenuana* (Navie et al. 1998b).

The leaf-feeding beetle *Zygogramma bicolorata* Pallister (Fig. 12.4) was introduced from Mexico into Australia in 1980 (McFadyen and McClay 1981). Evidence of *Z. bicolorata* activity on parthenium in Australia was first noticed in 1990 (Dhileepan and McFadyen 1997). Since then, due to both natural spread by the beetle and deliberate spread efforts by farmers, the area with *Z. bicolorata* defoliation has increased to around 12,000 km^2, in central Queensland (Dhileepan et al. 2000). Adult beetles can live up to 2 years and usually spend around 6 months diapausing in the soil during autumn and winter (McFadyen 1992). In central Queensland the leaf-feeding beetle *Z. bicolorata* caused 91–100% defoliation, resulting in reductions in weed density by 32–93%, plant height by 18–65%, plant biomass by 55–89%, flower production by 75–100%, soil seed bank by 13–86% and seedling emergence by 73–90% (Dhileepan et al. 2000).

The stem-boring weevil *Listronotus setosipennis* (Hustache) (Coleoptera: Curculionidae) (Fig. 12.5) was introduced into Australia from northern Argentina and southern Brazil there during 1982–1986 (McFadyen 1985; Wild et al. 1992) after host specificity tests had been conducted in Brazil (Wild 1980) and Australia (Wild et al. 1992). *L. setosipennis* became established in 1983 soon after the first release (Wild et al. 1992) but its field incidence remained low and sporadic (McFadyen 1992; Dhileepan et al. 1996; Dhileepan and McFadyen 1997; Dhileepan 2003a). Adult feeding and oviposition damage is negligible. Larval feeding has the ability to kill or prevent further development of parthenium seedlings (Wild et al. 1992). Incidence of *L. setosipennis* was recorded in 48% of the parthenium-infested

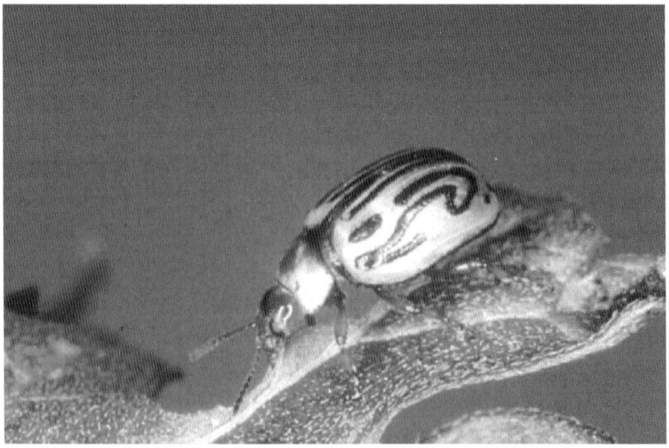

Fig. 12.4 *Zygogramma bicolorata* adult

Fig. 12.5 *Listronotus setosipennis* adult

sites, with 16% of the sites showing high to very high levels (Dhileepan 2003a). The realised impact of *L. setosipennis* damage in the field in Australia has been less than the potential impact as estimated through controlled trials, because of higher number of *L. setosipennis* larvae per plant utilised in the trials (Dhileepan 2003a).

The seed-feeding weevil *Smicronyx lutulentus* Dietz (Coleoptera: Curculionidae) from Mexico (McClay 1979; McFadyen and McClay 1981) was released in Australia (McFadyen and McClay 1981) from 1981 to 1983. Field establishment was confirmed only in 1996 (Dhileepan and McFadyen 1997). Adults feed on young leaves, but cause negligible feeding damage. Larval feeding causes significant reduction in seed output. In Mexico up to 30% seed destruction is attributed to *S. lutulentus* damage (McClay 1985). In Australia, the incidence of *S. lutulentus* is sporadic and localised with limited impact on seed production. With the current low infestation levels it will be difficult to estimate its impact on parthenium under field conditions (Dhileepan et al. 1996).

The stem-galling weevil *Conotrachelus albocinereus* Fiedler (Coleoptera: Curculionidae) was introduced from Argentina and Brazil. Adult feeding damage is not significant, and causes only minor damage to young leaves (McFadyen 2000). Damage is mainly due to larval feeding, which results in the fracturing of the vertical continuity of vascular tissues, thereby disrupting the host plant's overall metabolism (Florentine et al. 2002). Galling often kills axillary shoots, but the main stem remains unaffected. This insect might have established at a few sites in central Queensland, but is not in sufficient numbers to indicate widespread field establishment.

The clear-wing moth *Carmenta* nr *ithacae* (Beutenmüller) (Lepidoptera: Sesiidae) (Fig. 12.6) collected from Mexico is a highly host-specific agent (McFadyen and Withers 1997; Withers et al. 1999) and was released in Australia in 1998. Damage is from larvae that feed on the cortical tissue of the taproot and crown. Larvae are found on all growth stages of parthenium, and heavily infested plants often die. This agent has been recovered from the field only at irrigated

Fig. 12.6 *Carmenta nr.* ithacae adult

parthenium nursery sites in central Queensland recently, but not in sufficient numbers to indicate its widespread field establishment or potential impact.

The leaf-mining moth *Bucculatrix parthenica* Bradley (Lepidoptera: Bucculatricidae), a highly host-specific agent native to Mexico, was released in Australia from 1984 and its field establishment was confirmed in 1987 (McClay et al. 1990). The leaf-mining moth became established widely in both central and north Queensland, but failed to establish in southeast Queensland. Damage is caused by the larval feeding, which is evident on all growth stages of parthenium. The insect is rare in Mexico, but has become abundant in Queensland at some sites.

The winter rust *Puccinia abrupta* Diet. & Holw. var. *partheniicola* (Jackson) Parmelee (Uredinales), collected from the semiarid, upland regions (1400–16,000 m above sea level) of Mexico (Evans 1987, 1997a), was the first pathogen to be released on parthenium in Australia. It is a highly host-specific pathogen (Parker et al. 1994; Taye et al. 2004a; Tomley 1990) and its release in Australia began in 1991 and continued till 1995. The winter rust became established only in a few localised areas in central Queensland (Dhileepan and McFadyen 1997) with long dew periods and cooler temperatures (Fauzi et al. 1999), but its impact on parthenium in these areas appears to be not significant (Dhileepan 2003). This agent did not establish in north Queensland with warmer and drier conditions (Dhileepan et al. 1996).

The summer rust *Puccinia melampodii* Dietel and Holway (Uredinales) (Fig. 12.7) collected from low-altitude regions of Mexico (Evans 1997a; Seier et al. 1997; Tomley 2000) is highly host specific, damaging and adapted to areas with high temperatures and limited periods of humidity (Holden et al. 1995; Seier 1999; Seier and Tomley 2000). However, its incidence in Mexico was highest during the wet season. Field release of the summer rust in Queensland commenced in January 2000 (Dhileepan et al. 2006). Field establishment of the summer rust was evident in 88% of the release sites in Australia (Dhileepan et al. 2006). As predicted, this rust became established immediately, but with higher prevalence and intensity in north Queensland than in central Queensland. However, the impact of summer rust

Fig. 12.7 Summer rust *Puccinia melampodii*

on seedling establishment, plant height, flower production, plant biomass and plant density at the end of the first year was not significant (Dhileepan 2003b).

The biological control agents had a significant negative impact on parthenium at both individual plant and population level, and the impact was more severe in central Queensland than in north Queensland (Dhileepan 2001, 2003b). As a result, there was a significant increase in grass biomass production due to biological control (Dhileepan 2007). In central Queensland, there was a 40% increase in grass biomass in 1997 due to 96% defoliation by the leaf-feeding beetle *Z. bicolorata* and galling in 100% of the plants by the moth *E. strenuana*. In north Queensland, grass biomass increased by 52% in 1998 due to reduced parthenium seedling emergence, and by 45% in 2000 (Dhileepan 2007), due to the combined effects of galling by the moth *E. strenuana* and the establishment of the summer rust *P. melampodii* in 72% of the plants. In economic terms, benefits from increased grass production due to biological control have been estimated to support an additional 0.002 of an animal ha/year, which is equivalent to AUD 1.25/ha/year for buffel grass and AUD 1.19/ha/year for the Queensland blue grass (Adamson and Bray 1999). At today's value (April 2007), the return is much higher, i.e. AUD 2.50/ha/year for buffel grass and AUD 2.40/ha/year for the Queensland blue grass (Dhileepan 2007). With more than 170,000 km^2 of parthenium infestation in Queensland, this will translate to an economic benefit of AUD 37 million annually to the Queensland grazing industry. These benefits are in addition to the saving of AUD 8 million annually in medical costs in treating allergic dermatitis and asthma in property workers from infested areas (Page and Lacey 2006).

12.4.8.2 Ethiopia

The winter rust (*P. abrupta* var. *partheniicola*) was first reported in Ethiopia in 1997, and now it is known to occur there commonly in cool and humid areas at high altitudes (1,500–2,500 m above mean sea level) where rainfall varies from 400 to 700 mm (Taye et al. 2002, 2004a). In Ethiopia, the winter rust significantly reduced the plant height, number of leaves, number of branches and total biomass of parthenium (Taye et al. 2004a). The leaf-feeding beetle Z. *bicolorata* will be imported from South Africa into quarantine in Ethiopia in 2008 for further testing on native and economically important plant species. The stem-galling moth *E. strenuana* is not being considered for introduction into Ethiopia, in view of its potential to feed on niger (*Guizotia abyssinica* (L.f.) Cass.), a major oil seed crop there.

12.4.8.3 India

In India, a biological control program against parthenium was initiated in 1983, and since then only the leaf-feeding beetle *Z. bicolorata* has been introduced. It was introduced from Mexico in 1984 in the Bangalore region, and it became established in the same year (Jayanth 1987a; Jayanth and Bali 1994a; Jayanth and Nagarkatti 1987). However, its population levels attained damaging levels only after 3 years (Jayanth and Visalakshy 1994a, 1996). Field releases continued in 15 states in India (Viraktamath et al. 2004), and now after 20 years the beetle occurs in the majority of areas in India with parthenium infestations, ranging from the tropical south to sub-Himalayan regions in the north (Basappa 1997; Bhatia et al. 2005; Dhiman and Bhargava 2005; Gautam et al. 2005b, 2006; Gupta and Anil Sood 2002, 2005; Gupta et al. 2004; Jadhav and Ashok Varma 2001; Maninder et al. 1998; Pandey et al. 2001; Sarkate and Pawar 2006; Sharma and Sujauddin 2006; Susilkumar and Bhan 1998; Uniyal et al. 2001), but not in the hot and dry arid northwest region (i.e. Rajasthan State). It is unlikely that the leaf-feeding beetle *Z. bicolorata* will survive in this region, as the summer temperature exceeds 45°C, resulting in very high mortality among eggs, larvae and diapausing adults (Jayanth and Bali 1993a). Incidence of the leaf-feeding beetle *Z. bicolorata* has also been reported from the Punjab region in Pakistan (Javaid and Shabbir 2007). The leaf-feeding beetle *Z. bicolorata* caused 85–100% defoliation, resulting in up to 99.5% reduction in the parthenium weed density in Bangalore region (Jayanth and Bali 1994a; Jayanth and Visalakshy

1996). Similar impacts have been observed in other areas in India also (e.g. Sushilkumar 2000; Dhiman and Bhargava 2005; Jaipal 2007). Defoliation by the leaf-feeding beetle *Z. bicolorata* also resulted in the re-establishment of native vegetation (Jayanth and Visalakshy 1996; Sridhara et al. 2005). However, information on the long-term impact of defoliation by the leaf-feeding beetle *Z. bicolorata* in India is lacking. Though non-target feeding by leaf-feeding beetle *Z. bicolorata* on sunflower crop was reported (Kumar 1992; Chakravarthy and Bhat 1994, 1997; Chakravarthy et al. 1994, 1996), later studies (Jayanth et al. 1993, 1997, 1998; Jayanth and Visalakshy 1994b; Swamiappan et al. 1997a, 1997b; Bhumannavar et al. 1998; Withers 1998, 1999; Viraktamath et al. 2004; Patel and Viraktamath 2005) indicated that the chance of the leaf-feeding beetle *Z. bicolorata* becoming a pest of sunflower is negligible. No economic loss due to the leaf-feeding beetle *Z. bicolorata* feeding on sunflower even at higher insect densities was recorded in India (Kulkarani et al. 2000).

The stem-galling moth *E. strenuana* was not approved for field release in India due to oviposition and larval feeding on niger and sunflower (*Helianthus annus* L.) crops (Jayanth 1987b; Singh 1997).

Parthenium winter rust, though not intentionally released as a biological control agent, has been reported in India (Bagyanarayana and Manoharachary 1997; Parker et al. 1994), but the strains occurring in India do not appear to be widespread or aggressive (e.g. Kumar and Evans 1995). Hence, host specificity of a highly virulent isolate of the winter rust *P. abrupta* var. *partheniicola* from Mexico as a biological control agent for parthenium is being explored (Kumar and Evans 1995). However, there are no immediate plans to import this rust to India for further studies.

Inability to establish the seed-feeding weevil *S. lutulentus* in quarantine in India using adults collected in Australia prevented further studies on the host specificity of this agent there.

12.4.8.4 South Africa

A biological control program on parthenium was initiated in 2003 (Strathie 2007). The leaf-feeding beetle *Z. bicolorata*, the stem-boring weevil *L. setosipennis*, the stem-galling moth *E. strenuana* and the summer rust *P. melampodii* (Ntushelo and Wood 2007; Strathie et al. 2005) were prioritised for host-specificity tests, in view of their potential impacts and suitability for local climatic conditions. A colony of the leaf-feeding beetle *Z. bicolorata* was established in quarantine from adults collected in Australia, and host-specificity tests are in progress with results indicating a strong likelihood of release (Strathie et al. 2005). In South Africa, the stem-boring weevil *L. setosipennis* colony was established in quarantine from adults collected from yellow-flowering parthenium in Santiago del Estero and outside Metán in Salta province in northwestern Argentina, and host-specificity tests are in progress (Strathie et al. 2005). An attempt to establish a culture of the stem-galling weevil *E. strenuana* in quarantine did not succeed, likely due to low humidity, but it will be imported from Australia again (Strathie et al. 2005).

The parthenium winter rust was first observed in South Africa, in the town of Brits, in the northwest province (Wood and Scholler 2002) in 1995, and now also occurs in Mpumalanga and KwaZulu-Natal provinces. However, the winter rust was not intentionally released as a biological control agent in South Africa, and also the strain known to occur here does not appear to be widespread or aggressive (e.g. Kumar and Evans 1995).

12.4.8.5 Sri Lanka

Biological control effects were initiated in 2003 with the importation of the stem-galling moth *E. strenuana* and the summer rust *P. melampodii* from Australia (Jayasuriya 2005). Host-specificity tests confirmed the suitability and safety of the stem-galling moth *E. strenuana* as a biological control agent for parthenium in Sri Lanka, and the moth was field released in 2004 (Jayasuriya 2005). The summer rust was imported to Sri Lanka in 2003, and the pathogenicity and host-specificity tests indicate that this rust is suitable for release in Sri Lanka (Jayasuriya 2005). Attempts to establish the leaf-feeding beetle *Z. bicolorata* in quarantine in Sri Lanka have not been successful.

12.4.8.6 Papua New Guinea

Attempts to establish colonies of the leaf-feeding beetle *Z. bicolorata* and the stem-galling moth *E. strenuana* in quarantine in Papua New Guinea so far have failed.

12.5 Managing Parthenium Across Landscapes

12.5.1 Cropping Area

Parthenium is a weed of a wide range of crops in several countries (Table 12.4). In India, parthenium is a problem in cropping areas in majority of the States (Angiras and Saini 1997; Mahadevappa 1997; Patil et al. 1997; Dawson and Sarkar 1997; Sarkar 1997) and causes yield losses of up to 40% in crops (Khosla and Sobti 1979) and reduces forage production by up to 90% (Nath 1981). In Ethiopia also, parthenium is primarily a weed in cropping areas and is ranked as the most serious weed by the farmers (Tamado and Milberg 2000). Hand hoeing and hand pulling are the most common management options, which are effective in controlling parthenium in maize and sorghum (Fessehaie et al. 2005). The use of herbicides by Ethiopian small scale farmers is not economically feasible (Tomado and Milberg 2000, 2004) due to the low-economic value of the crops. In South Africa, parthenium is a problem weed in sugarcane and banana growing areas. Registered herbicides are available to manage parthenium in crops (e.g. Mahadevappa

Table 12.4 Crops affected by parthenium in various countries

Country	Crops
India[1]	*Achras zapota* (sapota)
	Allium cepa L. (onion)
	Allium sativum L. (garlic)
	Anacardium occidentale (cashew)
	Ananas comosus (L.) Merr. (pineapple)
	Arachis hypogaea L. (groundnut)
	Areca catechu L. (arecanut)
	Cajanus cajan L. (Arhar)
	Carica papaya L. (papaya)
	Citrus spp.
	Cocos nucifera L. (coconut)
	Coffea Arabica L.(coffee)
	Glycine max (L.) Merr. (soybean)
	Gossypium hirsutum L. (cotton)
	Lycopersicon esculentum Miller (tomato)
	Mangifera indica L. (mango)
	Musa spp. (banana)
	Oryza sativa L. (upland rice)
	Pennisetum glaucum (L.) R. Br. (pearl millet)
	Piper nigrum L. (black pepper)
	Pisum sativum L. (field pea)
	Psidium guajava L. (guava)
	Saccharum officinarum L.(sugarcane)
	Sesamum indicum L. (sesame)
	Solanum tuberosum L. (potato)
	Sorghum spp. (sorghum)
	Triticum aestivum L. (wheat)
	Vigna radiata (L.) R. Wilcz (green gram)
	Vigna unguiculata, (L.) Walp. (cowpea)
	Vitis vinifera L. (grapes)
	Zea mays L. (maize)
Kenya[2]	*Coffea arabica L.*(coffee)
Ethiopia[3]	*Sorghum bicolour* (L.) Moench (sorghum)
	Saccharum officinarum L. (sugarcane)
	Eragrostis tef (Zucc.) Trotter (tef)
Australia[4]	*Helianthus annuus* L (sunflower)
	Sorghum spp. (sorghum)
Sri Lanka[5]	*Capsicum annum* (chilli)
	Oryza sativa L. (upland rice)
South Africa[6]	*Saccharum officinarum* L. (sugarcane)
	Musa spp. (banana)
Israel[7]	*Lycopersicon esculentum* Miller (tomato)
	Gossypium spp. (cotton)
Pakistan[8]	*Gladiolus spp.* (Sward lily)
	Trifolium alexandrinum L. (Egyptian clover)

[1]Mahadevappa and Patil 1997; Prasad et al. 2005
[2]Njoroge 1986, 1991
[3]Tamado and Milgerg 2000, 2004; Tomado et al. 2002; Fessehaie et al. 2005; Firehun and Tomado 2006
[4]Dhileepan, personal observation
[5]Jayasurya 2005
[6]Strathie et al. 2005
[7]Joel and Liston 1986
[8]Shabbir and Bajwa 2006

Table 12.5 Current (bold lettering) and required management options for parthenium in diverse habitats

	Habitats				
Countries	Crops	Grazing area and pastures	Forest and nature reserve	Wasteland and roadside	Urban area
Australia	M	L + GM + BC + CD	BC	C + BC + CD	C
Ethiopia	M + C + AP	L + GM + BC	GM + BC	BC	BC
India	M + C + AP	CD + GM + BC + L	GM + BC	CD + GM + BC	M + CD + BC
Kenya	C + AP	GM + BC	GM + BC	L + CD + BC	M + CD + BC
Mozambique	C + AP	L + GM + BC + CD	GM + BC	L + CD + BC	CD + BC
Nepal	C + AP	L + BC	BC	BC	BC
Pakistan	M + C + AP	GM + BC + CD	GM + BC	CD + BC + GM	M + CD + BC
South Africa	M + C + AP	GM + CD + BC	GM + CD + BC	CD + BC	CD + BC
Sri Lanka	M + C	L + GM + BC + CD	M + L + GM + BC	M + BC + CD	M + BC
Swaziland	C + AP	L + M + BC + CD	GM + BC	L + M + BC + CD	BC
Zimbabwe	C + AP	GM + BC	BC	BC	BC

M mechanical, *C* chemical, *AP* agronomic practices, *L* legislative, *GM* grazing management, *CD* competitive displacement, *BC* biological control

and Patil 1997; Navie et al. 1996), but their use appears not widespread due to the cost of herbicides. In Kenya, parthenium is a major weed in coffee plantations, where herbicides appear less effective (Njoroge 1991). In Australia, currently parthenium is not a major crop weed, but its incidence has been reported in sugarcane, sunflower and sorghum (Navie et al. 1996, Parsons and Cuthbertson 2001). No detailed information is available of the role of parthenium as a crop weed in other countries (e.g. Bangladesh, East Timor, Mozambique, Nepal, Papua New Guinea, Sri Lanka, Swaziland and Zimbabwe). An integrated approach involving current and additional management tools is recommended for parthenium in crops (Table 12.5).

12.5.2 Grazing Areas

Parthenium is a well-known weed in grazing areas in all countries where they are known to occur. In Australia, parthenium is a serious problem in perennial grasslands in central Queensland, where it reduces beef production by as much as AUD 16.5 million annually (Chippendale and Panetta 1994). Recent study has shown that fire does not significantly reduce the parthenium soil seed bank, nor does smoke from such fires stimulate parthenium seed germination (Vogler et al. 2002). In grazing areas, management of parthenium can be achieved by maintaining good pasture grass growth to maximise competition against the weed. Managing grazing to maintain pasture cover/biomass and desirable pasture composition is potentially the most important factor influencing the amount of parthenium present in native pastures. Classical biological control has been seen as a better alternative to herbicides in perennial grasslands as well as in areas such as wastelands and forest, where the use of herbicides is uneconomical. A combination of grazing management and classical biological control resulted in significant increase in pasture production in Queensland, Australia (Dhileepan 2007). Grazing management is widely adapted in South Africa and Swaziland. A similar approach along with other management options (Table 12.5) is recommended for managing parthenium in pastures and grazing areas.

12.5.3 Forests and Nature Reserves

Parthenium has been reported in various forests and nature reserves in India, Pakistan, South Africa, Zimbabwe and Ethiopia (Table 12.6). So far the only line of management has been to integrate grazing management either with biological control as in Australia or with competitive displacement as in South Africa. Control programmes

Table 12.6 Parthenium incidence in national parks and forests

Country	Region/state/province	National park/reserve
India	Orissa	Kaziranga National Park
	Karnataka	Bandipur National Park
	Chandigarh	Sukhna Wildlife Sanctuary
	Uttaranchal	Mothronwala Swamp
		Rajaji National Park
		Jim Corbett National Park
	Rajasthan	Keoladeo National Park
	Tamil Nadu	Mudumalai Wildlife Sanctuary
		Nilgiri Bioreserve
	Kerala	Chinnar Wildlife Sanctuary
Australia	Queensland	Albinia National Park
		Mazeppa National Park
South Africa	Mpumalanga province	Kruger National Park
	KwaZulu-Natal	Ndumo Park
		Tembe Park
		Hluhluwe-iMfolozi Park
Swaziland	Lebombo mountain	Lebombo Conservancy
	Northeastern Swaziland	Mbuluzi Game Reserve
		Mlawula Nature Reserve
Zimbabwe	Bulawayo	Chipangali Wildlife Sanctuary
Ethiopia	Oromia	Awash National Park
Pakistan	Punjab	Chhanga Manga Forest

utilising cutting and burning to control alien plants including parthenium are being practiced in Swaziland. However, long-term effectiveness of such methods in reducing parthenium abundance is unknown. Management practices were successfully implemented in Mlawula Reserve in northeastern Swaziland to improve the condition of the vegetation in overgrazed areas with dense infestations of parthenium by lowering game levels and therefore reducing overgrazing and the incidence of parthenium (Bowen 2001). In protected Nature Reserves, use of more aggressive and or non-native plants to displace parthenium is often not encouraged. An approach incorporating classical biological control and grazing management (Table 12.5) is more suitable for managing parthenium in protected forests and nature reserves.

12.5.4 Wasteland and Roadside

In all countries where parthenium is known to occur, it is widespread in wasteland, fallow and roadside areas where control options often do not provide any economic returns. Hence, in many countries, no concerted efforts are being made to manage parthenium in these areas. However, parthenium infestations in these areas appear responsible for the spread of the weed to new areas, and hence managing parthenium in these areas would greatly reduce the chances of further spread. In Australia, a robust roadside management program using herbicides is in operation in majority of the local Shires and Councils where parthenium is known to occur. Research is in progress in Australia to identify suitable native plants to competitively displace parthenium in these areas. In India, competitive displacement of parthenium by other beneficial plants has been widely attempted in several states (Gautam et al. 2005a; Joshi, 1991a, 1991b; Kandasamy and Sankaran 1997; Kauraw et al. 1997; Sushilkumar and Bhan 1997). However, the long-term impact of such programmes is not fully known. Required management options for parthenium differ markedly between countries, but essentially include biological control as one of the options in all countries (Table 12.5).

12.5.5 Urban Areas

Parthenium is prominent in majority of the towns and cities in India, and it has the potential to become a major urban weed in other countries also (e.g. Bangladesh, Ethiopia, Kenya, Mozambique, Pakistan, Sri Lanka and Zimbabwe). In India, parthenium infestation in urban areas is responsible for causing severe human health problems such as contact dermatitis and allergic rhinitis (e.g. Kololgi et al. 1997; Handa et al. 2001; Sharma et al. 2005; Sashidhar et al. 1997; Towers and Rao 1992). But no management program is in place in most of the towns. A similar situation appears to exist in other countries also. In Australia, though parthenium is not a problem weed in urban areas, there are ongoing campaigns through television and other print media to increase public awareness. It is the responsibility of the local Shires and Council to eradicate, using herbicides, any parthenium infestations in urban and residential areas. Though manual control, including hand pulling appear the cheapest option, in view of the health risk, a chemical control in conjunction with a program to competitively displace parthenium with a suite of beneficial plants appears suitable for majority of countries where parthenium is a major problem in urban areas (Table 12.5). However, a legislative framework to contain or eradiate parthenium within the urban areas either at local or state government level is lacking in many countries.

12.6 Conclusion

The ability of parthenium to grow in a wide range of areas (e.g. wastelands, disturbed lands, degraded pastures, crops, forests, along railway tracks and roadsides, and along streams and rivers), across a wide range of habitats (e.g. hot, arid, semi-arid, humid high-altitude areas), its persistent seed bank and its allelopathic potential all make the management of this weed more difficult. No single management option would be adequate to control this weed across all habitats, and there is a need to integrate various management options (e.g. grazing management, competitive displacement, cultural practices) with classical biological control as a core management option.

References

Adamson DC (1996) Introducing dynamic considerations when economically evaluating weeds. Masters Thesis, University of Queensland, Brisbane.

Adamson DC, Bray S (1999) The economic benefit from investing in insect biological control of parthenium weed (*Parthenium hysterophorus*). School of Natural and Rural Systems Management, University of Queensland, Brisbane, p. 44.

Adhikari B, Tiwari S (2004) *Parthenium hysterophorus* L.: highly allergic invasive alien plant growing tremendously in Nepal. Botanica Orientalis 4:36–37.

Adkins SW, Navie SC, Dhileepan K (2005) Parthenium weed in Australia: research progress and prospects. In: Prasad TVR, Nanjappa HV, Devendra R et al. (eds) Proceedings of the second international conference on parthenium management, University of Agricultural Sciences, Bangalore, India, pp. 11–27.

Akula B, Kondap SM (1997) Studies on control of *Parthenium hysterophorus* L. by allelopathy. In: Mahadevappa M, Patil VC (eds) First international conference on parthenium management, Volume II, University of Agricultural Sciences, Dharwad, India, pp. 44–49.

Angiras NN, Saini JP (1997) Distribution, menace and management of *Parthenium hysterophorus* L. in Himachal Pradesh. In: Mahadevappa M, Patil VC (eds) First international conference on parthenium management, Volume II, University of Agricultural Sciences, Dharwad, India, pp. 13–15.

Anjum T, Bajwa R (2005) Biocontrol potential of grasses against *Parthenium hysterophorus*. In: Prasad TVR, Nanjappa HV, Devendra R et al. (eds) Second international conference on parthenium management, University of Agricultural Sciences, Bangalore, India, pp. 143–146.

Anjum T, Bajwa R, Javaid A (2006) Biological control of Parthenium. I. Effect of *Imperata cylindrica* on distribution, germination and seedling growth of *Parthenium hysterophorus* L. In: Proceedings of the fourth world congress on allelopathy, Charles Sturt University, Wagga Wagga, New South Wales, Australia, pp. 297–300.

Anonymous (2003) Incursion of parthenium weed (*Parthenium hysterophorus*) in Papua New Guinea. Pest Alert No. 30, Plant Protection Service, Secretariat of the Pacific Community, p. 2.

Auld BA, Hosking J, McFadyen RE (1982–1983) Analysis of the spread of tiger pear and parthenium weed in Australia. Australian Weeds 2:56–60.

Bagyanarayana G, Manoharachary C (1997) Studies on *Puccinia abrupta* var. *partheniicola* a potential mycoherbicide. In: Mahadevappa M, Patil VC (eds) Proceedings of the first international conference on parthenium management, Volume II, University of Agricultural Sciences, Dharwad, India, pp. 95–96.

Basappa H (1997) Incidence of biocontrol agent *Zygogramma bicolorata* Pallister on *Parthenium hysterophorus* L. In: Mahadevappa M, Patil VC (eds) First international conference on parthenium management, Volume II, University of Agricultural Sciences, Dharwad, India, pp. 81–84.

Basappa H (2005) Parthenium an alternate host of sunflower necrosis disease and thrips. In: Ramachandra Prasad TV, Nanjappa HV, Devendra R et al. (eds) Second international conference on parthenium management, University of Agricultural Sciences, Bangalore, India, pp. 83–86.

Bennett FD (1976) A preliminary survey of the insects and diseases attacking *Parthenium hysterophorus* L. (Compositae) in Mexico and the USA to evaluate the possibilities of its biological control in Australia. Mimeographical Report, Commonwealth Institute of Biological Control, Trininad, 18 pp.

Bhatia S, Choudhary R, Singh M (2005) Current status of the invasive weed *Parthenium hysterophorus* (Asteraceae) and impact of defoliation by the biocontrol agent *Zygogramma bicolorata* (Coleoptera: Chrysomelidae) in Jammu (J&K), India. Eighth International Conference on the Ecology and Management of Alien Plant Invasions, 5–12 September 2005, Katowice, Poland (Abstract).

Bhumannavar BS, Balasubramanian C, Ramani S (1998) Life table of the Mexican beetle *Zygogramma bicolorata* Pallister on parthenium and sunflower. Journal of Biological Control 12:101–106.

Bowen (2001) A study of the extent of alien species invasion into the Lubombo Conservancy, Swaziland. http://www.questoverseas.com/activeafrica/documents /AstudyOfTheExtentSwaziland.doc.

Brooks SJ, Vitelli JS, Rainbow AG (2004) Developing best practice roadside *Parthenium hysterophorus* L. control. In: Sindel BM, Johnson SB (eds) Proceedings of the fourteenth Australian weeds conference, Weed Society of New South Wales, Australia, pp. 195–198.

CABI (2004) Parthenium fact sheet. In: Crop Protection Compendium CD-ROM. CAB International, UK.

Chakravarthy AK, Bhat NS (1994) The beetle (*Zygogramma conjuncta* Rogers), an agent for the biological control of weed, *Parthenium hysterophorus* L. in India feeds on sunflower (*Helianthus annus* L.). Journal of Oilseeds Research 11:122–125.

Chakravarthy AK, Bhat NS (1997) Ecology of the beetle *Zygogramma conjuncta* (Rogers) on *Parthenium hysterophorus* Linn. In: Mahadevappa M, Patil VC (eds) First international conference on parthenium management, Volume II, University of Agricultural Sciences, Dharwad, India, pp. 74–77.

Chakravarthy AK, Bhat NS, Sridhar S (1994) The beetle *Zygogramma conjuncta* (Rogers), a bioagent for the control of the weed, *Parthenium hysterophorus* L. is oligophagous. Science and Culture 60:61–62.

Chakravarthy AK, Cox ML, Bhat NS, Sridhar S, Thyagaraj NE (1996) Identification, host specificity and infestation of *Zygogramma conjuncta* Rogers on *Helianthus annus* L. In: Ambrose DP (ed) Biological and cultural control of insect pests, an Indian scenario, Adeline, Thirunelveli, Tamil Nadu, India, pp. 243–250.

Cheney M (1998) Determination of the prevalence of sensitivity to Parthenium in areas of Queensland affected by the weed. Master of Public Health thesis, Queensland University of Technology, Brisbane, Australia, pp. 118.

Chippendale JF, Panetta FD (1994) The cost of parthenium weed to the Queensland cattle industry. Plant Protection Quarterly 9:73–76.

Dale IJ (1981) Parthenium weed in the Americas: a report on the ecology of *Parthenium hysterophorus* in South, Central and North America. Australian Weeds 1:8–14.

Da Silva MC, Izidine S, Amuda AB (2004) A preliminary checklist of the vascular plants of Mozambique. South African Botanical Diversity Network Report No. 30. SABONET, Pretoria, 185 pp.

Dawson J, Sarkar PA (1997) Effect of pre-emergence herbicides in controlling *Parthenium hysterophorus* L. and associated weeds in greengram. In: Mahadevappa M, Patil VC (eds) First international conference on parthenium management, Volume II, University of Agricultural Sciences, Dharwad, India, p. 115.

Deshpande KS, Deshpande UK, Baig MMV (1997) Managing parthenium: Achilles' heel undamaged. In: Mahadevappa M, Patil VC (eds) First international conference on parthenium management, Volume II, University of Agricultural Sciences, Dharwad, India, pp. 97–98.

Dhawan DR, Dhawan P, Gupta SK (1997) Allelopathic potential of some leguminous plant species towards *Parthenium hysterophorus* L. In: Mahadevappa M, Patil VC (eds) First international conference on parthenium management, Volume II, University of Agricultural Sciences, Dharwad, India, pp. 53–55.

Dhileepan K (2001) Effectiveness of introduced biocontrol insects on the weed *Parthenium hysterophorus* (Asteraceae) in Australia. Bulletin of Entomological Research 91:167–176.

Dhileepan K (2003a) Current status of the stem-boring weevil *Listronotus setosipennis* (Coleoptera: Curculionidae) introduced against the weed *Parthenium hysterophorus* (Asteraceae) in Australia. Biocontrol Science and Technology 13:3–12.

Dhileepan K (2003b) Seasonal variation in the effectiveness of leaf-feeding beetle *Zygogramma bicolorata* (Coleoptera: Curculionidae) and stem-galling moth *Epiblema strenuana* (Lepidoptera: Totricidae) as biocontrol agents on the weed *Parthenium hysterophorus* (Asteraceae). Bulletin of Entomological Research 93:393–401.

Dhileepan K (2004) The applicability of the plant vigor and resource regulation hypotheses in explaining *Epiblema* gall moth-*Parthenium* weed interactions. Entomologia Experimentalis et Applicata 113:63–70.

Dhileepan K (2007) Biological control of parthenium (*Parthenium hysterophorus*) in Australian rangeland translates to improved grass production. Weed Science 55:497–501.

Dhileepan K, McFadyen RE (1997) Biological control of parthenium in Australia – progress and prospects. In: Mahadevappa M, Patil VC (eds) First international conference on parthenium management, Volume I, University of Agricultural Sciences, Dharwad, India, pp. 40–44.

Dhileepan K, McFadyen RE (2001) Effects of gall damage by the introduced biocontrol agent *Epiblema strenuana* (Lepidoptera: Tortricidae) on the weed *Parthenium hysterophorus* (Asteraceae). Journal of Applied Entomology 125:1–8.

Dhileepan K, Madigan B, Vitelli M et al. (1996) A new initiative in the biological control of parthenium. In: Shepherd RCH (ed) Proceedings of the eleventh Australian weeds conference, Weeds Society of Victoria, Australia, pp. 309–312.

Dhileepan K, Setter S, McFadyen RE (2000a) Impact of defoliation by the introduced biocontrol agent *Zygogramma bicolorata* (Coleoptera: Chrysomelidae) on parthenium weed in Australia. BioControl 45:501–512.

Dhileepan K, Florentine SK, Lockett CJ (2006) Establishment, initial impact and persistence of parthenium summer rust *Puccinia melampodii* in north Queensland. In: Preston C, Watts JH, Crossman ND (eds) Proceedings of the fifteenth Australian weeds conference, Weed Management Society of South Australia, Adelaide, pp. 577–580.

Dhiman SC, Bhargava ML (2005) Seasonal occurrence and bio-control efficacy of *Zygogramma bicolorata* Pallister (Coleoptera: Chrysomelidae) on *Parthenium hysterophorus*. Annals of Plant Protection Sciences 13:81–84.

Dixit A, Bhan VM (1997) Management of parthenium through herbicides. In: Mahadevappa M, Patil VC (eds) First international conference on parthenium management, Volume II, University of Agricultural Sciences, Dharwad, India, pp. 99–100.

Evans HC (1983) Parthenium project: report on a visit to Mexico to survey fungal pathogens of *Parthenium hysterophorus*. L. (Compositae). March–May 1983. Commonwealth Mycological Institute, Kew, Surrey, UK, p. 19.
Evans HC (1987) Life cycle of *Puccinia abrupta* var. *partheniicola*, a potential biological control agent of *Parthenium hysterophorus*. Transactions of the British Mycological Society 88:105–111.
Evans HC (1997a) *Parthenium hysterophorus*: a review of its weed status, and the possibilities for biological control. Biocontrol News and Information 18:89–98.
Evans HC (1997b) The potential of neotropical fungal pathogens as classical biological control agents for management of *Parthenium hysterophorus* L. In: Mahadevappa M, Patil VC (eds) First international conference on parthenium management, University of Agricultural Sciences, Dharwad, India, pp. 55–62.
Fauzi MT, Tomley AJ, Dart PJ, Ogle HJ, Adkins SW (1999) The rust *Puccinia abrupta* var. *partheniicola*, a potential biocontrol agent of Parthenium weed: environmental requirements for disease progress. Biological Control 14:141–145.
Fessehaie R, Chichayibelu M, Giorgis MH (2005) Spread and ecological consequences of *Parthenium hysterophorus* in Ethiopia. Arem 6:11–21.
Firehun Y, Tamado T (2006) Weed flora in the Rift Valley sugarcane plantations of Ethiopia as influenced by soil types and agronomic practises. Weed Biology and Management 6:139–150.
Florentine SK, Raman A, Dhileepan K (2002) Response of the weed *Parthenium hysterophorus* (Asteraceae) to the stem gall-inducing weevil *Conotrachelus albocinereus* (Coleoptera: Curculionidae). Entomologia Generalis 26:195–206.
Frew M, Solomon K, Mashilla D (1996) Prevalence and distribution of *Parthenium hysterophorus* L in eastern Ethiopia. In: Fessehaie R (ed) Proceedings of the first annual conference, Ethiopian Weed Science Society, 24–25 Nov. 1993, Addis Ababa, Ethiopia. Arem 1:19–26.
Gautam RD, Aslam Khan M, Samyal A et al. (2005a) Survey of plants suppressing *Parthenium hysterophorus* Linnaeus in Delhi. In: Prasad TVR, Nanjappa HV, Devendra R et al. (eds) Second international conference on parthenium management, University of Agricultural Sciences, Bangalore, India, pp. 94–97.
Gautam RD, Aslam Khan M, Samyal A (2005b) Release, recovery and establishment of Mexican beetle, *Zygogramma bicolorata* Pallister (Chrysomelidae: Coleoptera) on *Parthenium hysterophorus* Linnaeus proliferating in and around Delhi. Journal of Entomological Research 29:167–172.
Gautam RD, Khan MA, Garg AK (2006) Ecological adaptability and variations among population of Mexican beetle, *Zygogramma bicolorata* Pallister (Chrysomelidae: Coleoptera). Journal of Entomological Research 30:21–23.
Govindappa MR, Chowda Reddy RV, Devaraja et al. (2005) *Parthenium hysterophorus*: a natural reservoir of tomato leaf curl begomovirus. In: Prasad TVR, Nanjappa HV, Devendra R et al. (eds) Second international conference on parthenium management, University of Agricultural Sciences, Bangalore, India, pp. 80–82.
Graham GC, Lang CL (1998) Genetic analysis of relationship of parthenium occurrences in Australia and indications of its origins. Cooperative Research Centre for Tropical Pest Management, University of Queensland, Brisbane, Australia (Internal Report).
Grice AC, Brown JR (1996) Repeated fire in the management of invasive tropical shrubs. In: Hunt LP, Sinclair R (eds) Proceedings of 9th biennial Australian rangeland conference, Australian Rangeland Society, Port Augusta, South Australia, pp. 171–172.
Griffiths MW, McFadyen RE (1993) Biology and host-specificity of *Platphalonidia mystica* (Lep, Cochylidae) introduced into Queensland to biologically control *Parthenium hysterophorus* (Asteraceae). Entomophaga 38:131–137.
Gupta PR, Sood A (2002) Spread of *Zygogramma bicolorata* Pallister in Himachal Pradesh. Insect Environment 8:101–102.
Gupta PR, Sood A (2005) Biological observations on *Zygogramma bicolorata* Pallister on congress grass (*Parthenium hysterophorus* L.) and its activity in mid-hills of Himachal Pradesh. Pest Management and Economic Zoology 13:21–27.

Gupta RK, Khan MS, Bali K et al. (2004) Predatory bugs of *Zygogramma bicolorata* Pallister: an exotic beetle for biological suppression of *Parthenium hysterophorus* L. Current Science 87:1005–1010.

Handa S, Sahoo B, Sharma VK (2001) Oral hyposensitization in patients with contact dermatitis from *Parthenium hysterophorus*. Contact Dermatitis 44:299–282.

Haseler WH (1976) *Parthenium hysterophorus* L. in Australia. PANS 22:515–580.

Hilliard OM (1977) *Compositae in Natal*. University of Natal Press, Pietermaritzburg, South Africa, 659 pp.

Holden ANG, Parker A, Tomley AJ (1995) Host range screening of *Puccinia abrupta var. partheniicola* for the biological control of *Parthenium hysterophorus* in Queensland. In: Delfosse ES, Scott RR (eds) Proceedings of the eight international symposium on biological control of weeds, DSIR/CSIRO, Melbourne, pp. 555–560.

Holman DJ (1981) Parthenium weed threatens Bowen Shire. Queensland Agricultural Journal 107:57–60.

Jadhav RB, Varma A (2001) Results of field introduction and establishment of *Zygogramma bicolorata* Pallister against *Parthenium hysterophorus* L., weed around sugarcane growing areas of Pravaranagar (Maharashtra). Indian Journal of Sugarcane Technology 16:66–69.

Jaipal S (2007) Potential of Mexican beetle, *Zygogramma bicolorata* in suppression of *Parthenium hysterophorus* in the Indian subtropics (abstract). XII International Symposium on Biological Control of Weeds, La Grande Motte, France, 22–27 April 2007.

Javaid A, Anjum T (2005) *Parthenium hysterophorus* L. – a noxious alien weed. Pakistan Journal of Weed Science Research 11:1–6.

Javaid A, Shabbir A (2007) First report of biological control of *Parthenium hysterophorus* by *Zygogramma bicolorata* in Pakistan. Pakistan Journal of Phytopathology 18:199–200.

Javaid A, Anjum T, Bajwa R (2005) Biological control of *Parthenium*. II. Allelopathic effect of *Desmostachya bipinnata* on distribution and early seedling growth of *Parthenium hysterophorus* L. International Journal of Biology and Biotechnology 2:459–463.

Jayanth KP (1987a) Introduction and establishment of *Zygogramma bicolorata* on *Parthenium hysterophorus* in Bangalore, India. Current Science 56:310–311.

Jayanth KP (1987b) Investigations on the host-specificity of *Epiblema strenuana* (Walker) (Lepidoptera: Tortricidae), introduced for biological control trials against *Parthenium hysterophorus* in India. Journal of Biological Control 1:133–137.

Jayanth KP, Bali G (1993a) Temperature tolerance of *Zygogramma bicolorata* (Coleoptera: Chrysomelidae) introduced for biological control of *Parthenium hysterophorus* (Asteraceae) in India. Journal of Entomological Research 17:27–34.

Jayanth KP, Bali G (1994a) Biological control of parthenium by the beetle *Zygogramma bicolorata* in India. FAO Plant Protection Bulletin 42:207–213.

Jayanth KP, Nagarkatti S (1987) Investigations on the host-specificity and damage potential of *Zygogramma bicolorata* Pallister (Coleoptera: Chrysomelidae) introduced into India for biological control of *Parthenium hysterophorus*. Entomon 12:141–145.

Jayanth KP, Visalakshy PNG (1994a) Dispersal of the parthenium beetle *Zygogramma bicolorata* (Chrysomelidae) in India. Biocontrol Science and Technology 4:363–365.

Jayanth KP, Visalakshy PNG (1994b) Field evaluation of sunflower varieties for susceptibility to the parthenium beetle *Zygogramma bicolorata*. Journal of Biological Control 8:48–52.

Jayanth KP, Visalakshy PNG (1996) Succession of vegetation after suppression of parthenium weed by *Zygogramma bicolorata* in Bangalore, India. Biological Agriculture and Horticulture 12:303–309.

Jayanth KP, Mohandas S, Ashokan R et al. (1993) Parthenium pollen induced feeding by *Zygogramma bicolorata* (Coleoptera: Chrysomelidae) on sunflower. Bulletin of Entomological Research 83:595–598.

Jayanth KP, Visalakshy PNG, Ghosh SK et al. (1997) Feasibility of biological control of *Parthenium hysterophorus* by *Zygogramma bicolorata* in the light of the controversy due to its feeding on sunflower. In: Mahadevappa M, Patil VC (eds) First international conference on parthenium management, University of Agricultural Sciences, Dharwad, India, pp. 45–51.

Jayanth KP, Visalakshy PNG, Chaudhary M et al. (1998) Age-related feeding by the Parthenium beetle *Zygogramma bicolorata* on sunflower and its effect on survival and reproduction. Biocontrol Science and Technology 8:117–123.
Jayasurya AHM (2005) Parthenium weed – status and management in Sri Lanka. In: Prasad TVR, Nanjappa HV, Devendra R et al. (eds) Second international conference on parthenium management, University of Agricultural Sciences, Bangalore, India, pp. 36–43.
Jeyalakshmi C, Doraisamy S, Paridasan VV (2005) Biodiversity of fungal pathogens affecting *Parthenium hysterophorus* L. in Tamil Nadu, India. In: Prasad TVR, Nanjappa HV, Devendra R et al. (eds) Second international conference on parthenium management, University of Agricultural Sciences, Bangalore, India, pp. 98–101.
Joel DM, Liston A (1986) New adventive weeds in Israel. Israel Journal of Botany 35:215–223.
Joshi S (1991a) Biological control of *Parthenium hysterophorus* L. (Asteraceae) by *Cassia uniflora* Mill (Leguminosae), in Bangalore, India. Tropical Pest Management 37:182–184.
Joshi S (1991b) Interference effects of *Cassia uniflora* Mill on *Parthenium hysterophorus* L. Plant and Soil 132:213–218.
Kadhane DL, Jangde CR, Sadekar RD et al. (1992) *Parthenium* toxicity in buffalo calves. Journal of Soils and Crops 21:69–71.
Kandasamy OS, Sankaran S (1997) Management of parthenium using competitive crops and plants. In: Mahadevappa M, Patil VC (eds) First international conference on parthenium management, Volume I, University of Agricultural Sciences, Dharwad, India, pp. 33–36.
Kauraw LP, Chile A, Bhan VM (1997a) Effect of marigold (*Tegetes papula* Linn.) population on the growth and survival of *Parthenium hysterophorus* L. In: Mahadevappa M, Patil VC (eds) First international conference on parthenium management, Volume II, University of Agricultural Sciences, Dharwad, India, pp. 39–40.
Kauraw LP, Chile A, Bhan VM (1997b) Evaluation of *Fusarium pallidorseum* (Cooke) Sacc. for the biocontrol of *Parthenium hysterophorus* L. In: Mahadevappa M, Patil VC (eds) First international conference on parthenium management, Volume I, University of Agricultural Sciences, Dharwad, India, pp. 70–74.
Khosla SN, Sobti SN (1979) Parthenium – a national health hazard, its control and utility – a review. Pesticides 13:121–127.
Kohli RK, Batish DR, Singh HP (1997) Management of *Parthenium hysterophorus* L. through an integrated approach. In: Mahadevappa M, Patil VC (eds) First international conference on parthenium management, Volume II, University of Agricultural Sciences, Dharwad, India, pp. 6–62.
Kololgi PD, Kololgi SD, Kololgi NP (1997) Dermatologic hazards of parthenium in human beings. In: Mahadevappa M, Patil VC (eds) First international conference on parthenium management, Volume I, University of Agricultural Sciences, Dharwad, India, pp. 18–21.
Kulkarani KA, Kulkarani NS, Santoshkumar GH (2000) Loss estimation in sunflower due to *Zygogramma bicolorata* Pallister. Insect Environment 6:10–11.
Kumar ARV (1992) Is the Mexican beetle *Zygogramma bicolorata* (Coleoptera: Chrysomelidae) expanding its host range? Current Science 63:729–730.
Kumar PS, Evans HC (2005) The mycobiota of *Parthenium hysterophorus* in its native and exotic ranges: opportunities for biological control in India. In: Prasad TVR, Nanjappa HV, Devendra R et al. (eds) Proceedings of the second international conference on parthenium management, University of Agricultural Sciences, Bangalore, India, pp. 107–113.
Macdonald IAW, Reaser JK, Bright C et al. (eds) (2003) Invasive alien species in southern Africa: national reports and directory of resources. Global Invasive Species Programme, Cape Town, South Africa.
Mahadevappa M (1997) Ecology, distribution, menace and management of parthenium. In: Mahadevappa M, Patil VC (eds) First international conference on parthenium management, Volume I, University of Agricultural Sciences, Dharwad, India, pp. 1–12.
Mahadevappa M, Patil VC (eds) (1997) First international conference on parthenium management, Volumes I & II, University of Agricultural Sciences, Dharwad, India.
Manickam K, Doraisamy S, Sankaran S (1997a) Epidemiology and host range studies on powdery mildew (*Oidium parthenii* S & U) of *Parthenium hysterophorus* L. In: Mahadevappa M, Patil

VC (eds) First international conference on parthenium management, Volume I, University of Agricultural Sciences, Dharwad, India, pp. 75–76.

Manickam K, Doraisamy S, Sankaran S (1997b) Inhibitory effect of *Fusarium moniliforme* on seed germination of *Parthenium hysterophorus* L. In: Mahadevappa M, Patil VC (eds) First international conference on parthenium management, Volume II, University of Agricultural Sciences, Dharwad, India, pp. 90–91.

Maninder S, Singh J, Brar KS, Bakhetia DRC (1998) Spread of *Zygogramma bicolorata* Pallister on *Parthenium* in Punjab and adjoining states. Insect Environment 4:83.

McClay AS (1979) Preliminary report on the biology and host-specificity of *Smicronyx lutulentus* Dietz (Col.: Curculionidae), a potential biocontrol agent for *Parthenium hysterophorus* L. Unpublished report, Commonwealth Institute of Biological Control, Mexican sub-station, Monterrey, Mexico, 10 pp.

McClay AS (1980) Studies of some potential biocontrol agents for *Parthenium hysterophorus* in Mexico. In: Delfosse ES (ed) Proceedings of the fifth international symposium on biocontrol of weeds, CSIRO, Australia, pp. 471–482.

McClay AS (1985) Biocontrol agents for *Parthenium hysterophorus* from Mexico. In: Delfose ES (ed) Proceedings of the sixth international symposium on biological control of weeds, Agriculture Canada, Vancouver, Canada, pp. 771–778.

McClay AS (1987) Observations on the biology and host specificity of *Epiblema strenuana* (Lepidoptera, Tortricidae), a potential biocontrol agent for *Parthenium hysterophorus* (Compositae). Entomophaga 32:23–34.

McClay AS, McFadyen RE, Bradley JD (1990) Biology of *Bucculatrix parthenica* Bradley sp. n. (Lepidoptera: Bucculatricidae) and its establishment in Australia as a biological control agent for *Parthenium hysterophorus* (Asteraceace). Bulletin of Entomological Research 80:427–432.

McClay AS, Palmer WA, Bennett FD et al. (1995) Phytophagous arthropods associated with *Parthenium hysterophorus* (Asteraceae) in North America. Environmental Entomology 24:796–809.

McFadyen PJ (1976) A survey of insects attacking *Parthenium hysterophorus* L. (F. Compositae) in South America. Internal report, Department of Lands, State Government of Queensland, Australia, p. 8

McFadyen PJ (1979) A survey of insects attacking *Parthenium hysterophorus* L. (Compositae) in Argentina and Brazil. Dusenia 11:42–45.

McFadyen RE (1985) The biological control programme against *Parthenium hysterophorus* in Queensland. In: Delfosse ES (ed) Proceedings of the sixth international symposium on the biological control of weeds, Agriculture Canada, Ottawa, Canada, pp. 789–796.

McFadyen RE (1987) The effect of climate on the stem-galling moth *Epiblema strenuana*. In: Lemerle D, Leys AR (eds) Proceedings of the Seventh Australian Weeds Conference, Weed Society of New South Wales, Sydney, Australia, pp. 97–99.

McFadyen RE (1992) Biological control against parthenium weed in Australia. Crop Protection 24:400–407.

McFadyen RE (1995) Parthenium weed and human health in Queensland. Australian Family Physician 24:1455–1459.

McFadyen RE (2000) Biology and host specificity of the stem galling weevil *Conotrachelus albocinereus* Fielder (Col: Curculionidae), a potential biocontrol agent for parthenium weed *Parthenium hysterophorus* L. (Asteraceae) in Queensland, Australia. Biocontrol Science and Technology 10:195–200.

McFadyen RE, McClay AR (1981) Two new insects for the biological control of parthenium weed in Queensland. In: Wilson BJ, Swarbrick JD (eds) Sixth Australian Weeds Conference, Weed Science Society of Queensland, Australia, pp. 145–149.

McFadyen RE, Withers TM (1997) Report on biology and host-specificity of *Carmenta ithacae* (Lep: Sesiidae) for the biological control of parthenium weed (*Parthenium hysterophorus*). Internal Report, Queensland Department of Natural Resources, Australia.

Narasimhan TR, Ananth M, Narayana Swamy M et al. (1980) Toxicity of *Parthenium hysterophorus* L. in cattle and buffaloes. Indian Journal of Animal Science 50:173–178.

Nath R (1981) Note on the effect of *Parthenium* extract on seed germination and seedling growth in crops. Indian Journal of Agricultural Science 51:601–603.

Nath R (1988) *Parthenium hysterophorus* L. – a general account. Agricultural Review 9:171–179.

Navie SC, McFadyen RE, Panetta FD et al. (1996) The biology of Australian weeds 27. *Parthenium hysterophorus* L. Plant Protection Quarterly 11:76–88.

Navie SC, Panetta FD, McFadyen RE et al. (1998a) Behaviour of buried and surface-lying seeds of parthenium weed (*Parthenium hysterophorus* L.). Weed Research 38:338–341.

Navie SC, Priest TE, McFadyen RE et al. (1998b) Efficacy of the stem-galling moth *Epiblema strenuana* Walk. (Lepidoptera: Tortricidae) as a biological control agent for the ragweed parthenium (*Parthenium hysterophorus* L.). Biological Control 13:1–8.

Navie SC, Panetta FD, McFadyen RE et al. (2004) Germinable soil seedbanks of Central Queensland rangelands invaded by the exotic weed *Parthenium hysterophorus* L. Weed Biology and Management 4:154–167.

Njoroge JM (1986) New weeds in Kenya coffee. A short communication. Kenya Coffee 51:333–335.

Njoroge JM (1989) Glyphosate (round-up 36% a.i.) low rate on annual weeds in Kenya coffee. Kenya Coffee 54:713–716.

Njoroge JM (1991) Tolerance of *Bidens pilosa* and *Parthenium hysterophorus* L. to paraquat (Gramaxone) in Kenya coffee. Kenya Coffee 56:999–1001.

Ntushelo K, Wood AR (2007) Supplementary host specificity testing of *Puccinia melampodii*, a biocontrol agent of *Parthenium hysterophorus* (abstract). XII International Symposium on Biological Control of Weeds, 22–27 April 2007, La Grande Motte, France.

O'Donnell C, Adkins SW (2005) Management of parthenium weed through competitive displacement with beneficial plants. Weed Biology and Management 5:77–79.

Page AR, Lacey KL (2006) Economic impact assessment of Australian weed biological control. Technical Series no. 10, CRC for Australian Weed Management, Adelaide, Australia, p. 150.

Pandey S, Joshi BD, Tiwari LD (2001) The incidence of Mexican beetle *Zygogramma bicolorata* Pallister (Coleoptera: Chrysomelidae) on *Parthenium hysterophorus* L. (Asteraceae) from Haridwar and surrounding areas. Journal of Entomological Research 25:145–149.

Parker A, Holden ANG, Tomley AJ (1994) Host specificity testing and assessment of the pathogenicity of the rust, *Puccinia abrupta* var. *partheniicola*, as a biological control agent of parthenium weed (*Parthenium hysterophorus*). Plant Pathology 43:1–16.

Parsons WT, Cuthbertson EG (2001) Noxious weeds of Australia, 2nd Edition. CSIRO Publishing, Melbourne, Australia, p. 698.

Patel VN, Viraktamath CA (2005) Dispersal of parthenium biocontrol agent *Zygogramma bicolorata* Pallister (Coleoptera: Chrysomelidae) in sunflower and *Parthenium hysterophorus* L. In: Prasad TVR, Nanjappa HV, Devendra R et al. (eds) Second international conference on parthenium management, University of Agricultural Sciences, Bangalore, India, pp. 120–122.

Patil SA, Jatti PD, Chetti MB, Hiremath SM (1997) A survey of parthenium in Bijapur district – a case study. In: Mahadevappa M, Patil VC (eds) First international conference on parthenium management, Volume II, University of Agricultural Sciences, Dharwad, India, pp. 20–23.

Peng CI, Hu LA, Kao MT (1988) Unwelcome naturalisation of *Parthenium hysterophorus* (Asteraceae) in Taiwan. Journal of Taiwan Museum 41:95–101.

Picman AK, Towers GHN (1982) Sesquiterpene lactones in various populations of *Parthenium hysterophorus*. Biochemical Systematics and Ecology 10:145–153.

Rao PVS, Mangala A, Rao BSS et al. (1977) Clinical and immunological studies on persons exposed to *Parthenium hysterophorus* L. Experientia 33:1387–1388.

Rao RDVJP, Govindappa MR, Devaraja et al. (2005) Role of parthenium in perpetuation and spread of plant pathogens. In: Ramachandra Prasad TV, Nanjappa HV, Devendra R et al. (eds) Second international conference on parthenium management, University of Agricultural Sciences, Bangalore, India, pp. 65–72.

Rao RS (1956) Parthenium – a new record for India. Journal of Bombay Natural History Society 54:218–220.

Robertson LN, Kettle BA (1994) Biology of *Pseudoheteronyx* sp. (Coleoptera, Scarabaeidae) on the central highlands at Queensland. Journal of the Australian Entomological Society 33:181–184.

Sarkar PA (1997) Effect of isoproturon and 2,4-D on *Parthenium hysterophorus* L. and other weeds in wheat. In: Mahadevappa M, Patil VC (eds) First International conference on parthenium management, Volume II, University of Agricultural Sciences, Dharwad, India, pp. 111–113.

Sarkate MB, Pawar VM (2006) Establishment of Mexican beetle (*Zygogramma bicolorata*) against *Parthenium hysterophorus* in Marathwada. Indian Journal of Agricultural Sciences 76:270–271.

Sashidhar KC, Hiremath NV, Bhat ARS, Patil VC (1997) Studies on aeropalynological survey of pollen of parthenium and other species in Bangalore city. In: Mahadevappa M, Patil VC (eds) First international conference on parthenium management, Volume II, University of Agricultural Sciences, Dharwad, India, pp. 28–31.

Seier MK (1999) Studies on the rust *Puccinia melampodii* Diet. and Holw. – a potential biological control agent for parthenium weed (*Parthenium hysterophorus* L.) in Australia. CABI Bioscience UK Centre, Silwood Park, UK (Report submitted to the Queensland Department of Natural Resources & Mines, February 1999).

Seier MK, Tomley AJ (2000) Host range of *Puccinia melampodii*: implications for its use as a biocontrol agent of parthenium weed in Australia (abstract). In: Spencer NR (ed) Proceedings of the X international symposium on biological control of weeds, Bozeman, MT, p. 686.

Seier MK, Harvey JL, Romero A et al. (1997) Safety testing of the rust *Puccinia melampodii* as a potential biocontrol agent of *Parthenium hysterophorus* L. In: Mahadevappa M, Patil VC (eds) First international conference on parthenium management, Volume I, University of Agricultural Sciences, Dharwad, India, pp. 63–69.

Shabbir A, Bajwa R (2006) Distribution of parthenium weed (*Parthenium hysterophorus* L.), an alien invasive weed species threatening the biodiversity of Islamabad. Weed Biology and Management 6:89–95.

Sharma TK, Shujauddin (2006) Incidence and biology of Mexican beetle, *Zygogramma bicolorata* on *Parthenium hysterophorous*. Bionotes 8:51.

Sharma VK, Sethuraman G, Radhakrishna B (2005) Evolution of clinical pattern of parthenium dermatitis: a case study of 74 cases. Contact Dermatitis 53:84.

Singh SP (1997) Perspectives in biological control of parthenium in India. In: Mahadevappa M, Patil VC (eds) First international conference on parthenium management, University of Agricultural Sciences, Dharwad, India, pp. 22–32.

Sridhara S, Basavaraja BK, Ganeshaiah KN (2005) Temporal variation in relative dominance of *Parthenium hysterophorus* and its effect on native biodiversity. In: Prasad TVR, Nanjappa HV, Devendra R et al. (eds) Proceedings of the second international conference on parthenium management, University of Agricultural Sciences, Bangalore, India, pp. 240–242.

Strathie LW (2007) Managing the impact of parthenium invasions in Africa. Biocontrol News and Information 28:54N–55N.

Strathie LW, Wood AR, van Rooi C et al. (2005) *Parthenium hysterophorus* (Asteraceae) in southern Africa, and initiation of biological control against it South Africa. In: Prasad TVR, Nanjappa HV, Devendra R et al. (eds) Second international conference on parthenium management, University of Agricultural Sciences, Bangalore, India, pp. 127–133.

Susilkumar (2000) Impact of the introduced Mexican beetle *Zygogramma bicolorata* on the suppression of *Parthenium hysterophorus*: a case study (Abstract). Third International Weed Science Congress, Foz do Iguassu, Brazil.

Susilkumar, Bhan VM (1997) Natural population replacement by *Cassia tora* at Jabalpur and adjoining areas of Madhya Pradesh. In: Mahadevappa M, Patil VC (eds) First international conference on parthenium management, Volume II, University of Agricultural Sciences, Dharwad, India, pp. 41–43.

Susilkumar, Bhan VM (1998) Establishment and dispersal of introduced exotic *Parthenium* controlling bioagent *Zygogramma bicolorata* in relation to ecological factors at Vindhyanagar. Indian Journal of Ecology 25:8–13.

Swamiappan M, Sethupitchai U, Geetha B (1997a) Feeding potential of freshly emerged *Z. bicolorata* adults on sunflower and parthenium. In: Mahadevappa M, Patil VC (eds) First international conference on parthenium management, Volume I, University of Agricultural Sciences, Dharwad, India, pp. 52–54.

Swamiappan M, Sethupitchai U, Geetha B (1997b) Evaluation of the susceptibility of sunflower cultivars to the Mexican beetle, *Zygogramma bicolorata* Pallister. In: Mahadevappa M, Patil VC (eds) First international conference on parthenium management, Volume II, University of Agricultural Sciences, Dharwad, India, pp. 177–179.

Tamado T, Milberg P (2000) Weed flora in arable fields of eastern Ethiopia with emphasis on the occurrence of *Parthenium hysterophorus*. Weed Research 40:507–521.

Tamado T, Ohlander L, Milberg P (2002a) Interference by the weed *Parthenium hysterophorus* L. with grain sorghum: influence of weed density and duration of competition. International Journal of Pest Management 48:183–188.

Taye T (2005) Investigation of pathogens for biological control of parthenium (*Parthenium hysterophorus* L.) in Ethiopia. Programme and Abstracts of the seventh annual conference of the Ethiopian Weed Science Society, 24–25 November 2005, Addis Ababa, Ethiopia, pp. 21–22.

Taye T (2006) Biological control research on parthenium weed in Ethiopia. In: Fessehaie R, Teklemariam A, Ali K et al. (eds) Proceedings of the second national workshop on invasive alien weeds and insect pests, 6–8 June, 2005, Bahir Dar, Ethiopia.

Taye T, Gossmann M, Einhorn G et al. (2002) The potential of pathogens as biological control of parthenium weed (*Parthenium hysterophorus* L.) in Ethiopia. Mededelingen – Faculteit Landbouwkundige en Toegepaste Biologische Wetenschappen, Universiteit Gent 67:409–420.

Taye T, Einhorn G, Gossmann M et al. (2004a) The potential of parthenium rust as biological control of parthenium weed in Ethiopia. Pest Management Journal of Ethiopia 8:39–50.

Taye T, Einhorn G, Metz R (2004b) *Parthenium hysterophorus*, an invasive species in Ethiopia – investigations on the occurrence and on its pathogens. Journal of Plant Diseases and Protection 19:271–278.

Taye T, Obermeier C, Einhorn G et al. (2004c) Phyllody disease of parthenium weed in Ethiopia. Pest Management Journal of Ethiopia 8:83–95.

Tomley AJ (1990) Parthenium weed rust, *Puccinia abrupta* var. *partheniicola*. In: Heap JW (ed) Proceedings of the ninth Australian weeds conference, Crop Science Society of South Australia, Adelaide, pp. 511–512.

Tomley AJ (2000) *Puccinia melampodii* (summer rust) a new biocontrol agent for parthenium weed. In: Swarbrick JT (ed) Proceedings of the sixth Queensland weed symposium, Weed Society of Queensland, Brisbane, pp. 126–129.

Towers GHN (1981) Allergic exzematous contact dermatitis from parthenium weed (*Parthenium hysterophorus*). In: Wilson BJ, Swarbrick JT (eds) Proceedings of the sixth Australian weeds conference, Gold Coast, Queensland, Australia, pp. 565–569.

Towers GHN, Rao PVS (1992) Impact of the pan-tropical weed, *Parthenium hysterophorus* L. on human affairs. In: Richardson RG (ed) Proceedings of the first international weed control congress, Volume I, Weed Science Society, Victoria, Melbourne, Australia, pp. 134–138.

Tudor GD, Ford AL, Armstrong TR et al. (1982) Taints in meat from sheep grazing *Parthenium hysterophorus*. Australian Journal of Experimental Agriculture and Animal Husbandry 22:43–46.

Uniyal VP, Mukherjee SK, Goyal CP et al. (2001) Defoliation of parthenium by Mexican beetle (*Zygogramma bicolorata*) in Rajaji National Park. Annals of Forestry 9:327–330.

van der Laan M (2006) Allelopathic interference potential of the alien invader plant *Parthenium hysterophorus*. MSc Thesis, University of Pretoria, Pretoria, 95 pp.

Vikrant P, Verma KK, Rajak RC et al. (2007) Characterization of a phytotoxin from *Phoma herbarum* for management of *Parthenium hysterophorus* L. Journal of Phytopathology 154:461–468.

Viraktamath CA, Bhumannavar BS, Patel VN (2004) Biology and ecology of *Zygogramma bicolorata* Pallister, 1953. In: Joliver P, Santiago-Blay JA, Schmitt M (eds) New developments in the biology of Chrysomelidae, SPB Academic, The Hague, The Netherlands, pp. 767–777.

Vogler W, Navie S, Adkins S et al. (2002) Use of fire to control parthenium weed. A report for the Rural Industries Research and Development Corporation, Australia, p. 41.

Wedner HJ, Zenger V, Lewis W (1986) Identification of American feverfew (*Parthenium hysterophorus*) as an allergen in the United-States Gulf-Coast. Journal of Allergy and Clinical Immunology 77:198.

Wild CH (1980) Preliminary report on the biology and host-specificity of *Hyperodes* sp. (Coleoptera: Curculionidae), a potential biological control agent for *Parthenium hysterophorus* L. (Compositae). Unpublished report, Alan Fletcher Research Station, South American Field Office, Londrina, Brazil, 31 pp.

Wild CH, McFadyen RE, Tomley AJ et al. (1992) The biology and host specificity of the stem-boring weevil *Listronotus setosipennis* (Coleoptera: Curculionidae) – a potential biocontrol agent for *Parthenium hysterophorus* (Asteraceae). Entomophaga 37:591–598.

Withers TM (1998) Influence of plant species on host acceptance behaviour of the biocontrol agent *Zygogramma bicolorata* (Col.: Chrysomelidae). Biological Control 13:55–62.

Withers TM (1999) Examining the hierarchy threshold model in a no-choice feeding assay. Entomologia Experimentalis et Applicata 91:89–95.

Withers TM, McFadyen RE, Marohasy J (1999) Importation protocols and risk assessment of weed biological control agents in Australia: the example of *Carmenta nr. ithacae*. In: Follett PA, Duan JJ (eds) Nontarget effects of biological control, Kluwer, Boston, pp. 195–214.

Wood AR, Scholler M (2002) *Puccinia abrupta* var. *partheniicola* on *Parthenium hysterophorus* in Southern Africa. Plant Disease 86:327.

Yadav A, Malik RK, Balyan RS (1997) Evaluation of herbicides against carrot weed (*Parthenium hysterophorus* L.). In: Mahadevappa M, Patil VC (eds) First international conference on parthenium management, Volume II, University of Agricultural Sciences, Dharwad, India, pp. 103–105.

Yaduraju NT, Sushilkumar, Prasad Babu MBB et al. (2005) *Parthenium hysterophorus* – distribution, problems and management strategies in India. In: Prasad TVR, Nanjappa HV, Devendra R et al. (eds) Second international conference on parthenium management, University of Agricultural Sciences, Bangalore, India, pp. 6–10.

Chapter 13
Black and Pale Swallow-Wort (*Vincetoxicum nigrum* and *V. rossicum*): The Biology and Ecology of Two Perennial, Exotic and Invasive Vines

C.H. Douglass, L.A. Weston, and A. DiTommaso

Abstract Black and pale swallow-worts are invasive perennial vines that were introduced 100 years ago into North America. Their invasion has been centralized in New York State, with neighboring regions of southern Canada and New England also affected. The two species have typically been more problematic in natural areas, but are increasingly impacting agronomic systems such as horticultural nurseries, perennial field crops, and pasturelands. While much of the literature reviewed herein is focused on the biology and management of the swallow-worts, conclusions are also presented from research assessing the ecological interactions that occur within communities invaded by the swallow-wort species. In particular, we posit that the role of allelopathy and the relationship between genetic diversity levels and environmental characteristics could be significant in explaining the aggressive nature of swallow-wort invasion in New York. Findings from the literature suggest that the alteration of community-level interactions by invasive species, in this case the swallow-worts, could play a significant role in the invasion process.

Keywords Allelopathy • Genetic diversity • Invasive plants • Swallow-wort spp. • *Vincetoxicum* spp.

Abbreviations AMF: Arbuscular mycorrhizal fungi; BSW: Black swallow-wort; PSW: Pale swallow-wort

---

C.H. Douglass (✉)
Department of Bioagricultural Sciences and Pest Management, Colorado State University, Fort Collins, CO 80523, USA
Cameron.Douglass@colostate.edu

A. DiTommaso
Department of Crop and Soil Sciences, Cornell University, Ithaca NY 14853, USA

L. A. Weston
Charles Sturt University, E.H. Graham Centre for Agricultural Innovation, Wagga Wagga NSW 2678, Australia

Inderjit (ed.), *Management of Invasive Weeds*,

13.1 Introduction

Black swallow-worts (BSW) and pale swallow-worts (PSW) are invasive, herbaceous perennial vines that were introduced over a century ago into North America. Currently both species are on banned or prohibited plant species lists in Connecticut, Massachusetts, and New Hampshire, and BSW is classified as a "noxious" weed in Vermont (USDA Plants Database 2008). Like many invasive species, the two swallow-worts exhibit numerous attributes of ideal weeds (Baker 1974). That is, they are strong competitors for available and sometimes scarce resources, are prolific reproducers, and can significantly alter invaded habitats (Ernst and Cappuccino 2005; Greipsson and DiTommaso 2006; Smith 2006). The focus of our research on the swallow-wort species has been to better understand the specific similarities and differences among these two congeneric species and evaluate if there is a physiological or genetic basis for their rapid invasion in regions of North America. This chapter is a summary of the literature that has been presented with respect to both species, and in addition, an overview of our recent work related to their spread, allelopathic potential, and genetic diversity among populations in New York. By developing a broader understanding of a plant species' biology and ecology in particular locations, we can try to develop more effective strategies for the management of these species and other problematic nonnative invaders, which have thus far evaded effective control.

Unlike some more infamous plant invaders, swallow-worts produce small flowers and their often prostrate growth habit allows them to easily blend in with resident vegetation. Swallow-worts often persist largely unnoticed by landowners or managers until they are well established and have displaced prior vegetation (Lawlor 2003; West and Fowler 2008). Mature vines can grow to several meters in length, and infestations can often become dense impenetrable thickets of intertwined vines (hence the common name synonym dog-strangling vine) (DiTommaso et al. 2005b; Sheeley and Raynal 1996). Most significantly, cultural control methods that effectively reduce mature infestations are not available at present. Current recommendations for control are limited to the use of broad-spectrum herbicides and mechanical controls that must be repeated both during the growing season and annually for several years. Unfortunately, even these laborious and expensive strategies only provide reliable and sometimes temporary control of smaller satellite populations (Averill et al. 2008; Lawlor and Raynal 2002; Weston et al. 2005).

13.2 Taxonomy

PSW [*Vincetoxicum rossicum* (Kleopow) Babar. = *Cynanchum rossicum* (Kleopow) Borhidi] and BSW [*Vincetoxicum nigrum* (L.) Moench = *Cynanchum nigrum* (L.) Pers. = *Cynanchum louiseae* (L.) Kartesz & Gandhi] are generally placed in the Asclepiadaceae (Gleason and Cronquist 1991; USDA Plants Database 2008). Recent work, however, suggests that the species should more accurately be placed in the

Apocynaceae (S.J. Darbyshire, personal communication). In Europe there is evidence of successful hybridization between PSW and another relative, white swallow-wort (*V. hirundinaria* Medik.) (Lauvanger and Borgen 1998). This evidence not only contributes to taxonomic confusion regarding identification of swallow-worts, but also points to the potential for pale and black wallow-worts to hybridize (DiTommaso et al. 2005b).

There remains a great deal of confusion as to the correct taxonomic placement of the invasive swallow-wort species, with some taxonomists placing them under the genus *Cynanchum* and others in the genus *Vincetoxicum*. Given that there are a number of native North American plants in the genus *Cynanchum* [22 according to the USDA Plant Database (2008)], DiTommaso et al. (2005b) proposed that *Vincetoxicum* should be used solely for the alien species in order to denote their old world origins. This view is shared by Liede (1996), who, primarily on the basis of the presence of unique alkaloids and glycosides, separates *Vincetoxicum* from *Cynanchum* and places PSWs and BSWs within the genus *Vincetoxicum* (also Liede and Tauber 2002).

From a review of the literature, no studies have been performed to determine the ploidy level or chromosome numbers for swallow-wort species or populations in the USA, and other evaluations that have been performed were limited in their geographic focus. A chromosome count of $2n = 22$ was reported for PSW plants in Ottawa, ON, Canada (Moore 1959), while chromosome counts for BSW plants vary from $n = 11$ in Spain (Diosdado et al. 1993) to $2n = 22$ and $2n = 44$ for two populations in Italy (Aparicio and Silvestre 1985; Moore 1959). At this time, we have no information on chromosome numbers of swallow-wort populations in the USA, which could provide valuable information to more fully describe the genotypic relationships among and between these two species.

PSW is native to eastern regions of the Ukraine and southwestern portions of Russia north of the Black Sea and Caucasus; BSW (*V. nigrum*) is endemic to southwestern Europe, particularly regions of the Iberian Peninsula, southern France, and northern Italy (DiTommaso et al. 2005b). In their native ranges, the two species are relatively rare and do not overlap, with scattered patches of 3–15 stems of BSWs found to be typical of native populations in southern France (DiTommaso et al. 2005b; L.R. Milbrath, personal communication; Tewksbury et al. 2002). While PSW has been reported as invasive in one case in Norway, the third *Vincetoxicum* species, *V. hirundinaria*, is actually much more widespread and has greater invasive potential in Europe than the two species that are problematic in North America (Lauvanger and Borgen 1998).

13.3 Reproductive Biology and Phenology

Flowering can begin as early as mid to late May for PSW populations in Central New York and will peak several weeks later in early to mid June (Sheeley 1992). Floral development can be delayed by 10 days in populations farther north in New York, and by up to 4 weeks for populations in Ontario and other regions of southern Canada (DiTommaso et al. 2005b; Lawlor 2000). BSW flowering tends to peak in mid to late June, though in shadier sites it can be delayed by up to a month (DiTommaso et al. 2005b). The swallow-wort species are self-compatible, but are also insect pollinated by a variety of fly, ant, bee, wasp, and beetle species (DiTommaso et al. 2005b; Lumer and Yost 1995).

Fruit development for PSW typically begins in early June in Central New York, with maturity occurring 4–5 weeks after flowering, and finally dehiscence peaking at the end of July; development in BSW is normally 2–4 weeks slower (DiTommaso et al. 2005b; Lawlor 2000). There appears to be a physiological dormancy requirement for PSW seeds, and this likely applies as well to the black species. Though minimal germination is often found experimentally with PSW without subjecting seeds to a cold stratification period, greater germination occurs in seeds subjected to a stratification treatment (Cappuccino et al. 2002; DiTommaso et al. 2005a).

While some studies have found a significant positive correlation between seed size and the probability of germination in PSW (DiTommaso et al. 2005a), other studies have reported no correlation between seed size and germinability (Cappuccino et al. 2002; C.H. Douglass, unpublished data). Cappuccino et al. (2002), however, did find that seed size in PSW was positively correlated with final dispersal distance, especially for seeds that had been subjected to a stratification period of 3 months. This effect was weaker in a later study (Ladd and Cappuccino 2005) though the positive trend did generally hold true. While these authors also found that larger seeds tended to produce taller seedlings during the first growing season, they concluded that there was a nominal advantage in survivorship associated with a greater initial seed weight during three growing seasons.

Swallow-wort species produce polyembryonic seed, and estimates suggest that 45–75% of PSW seeds are polyembryonic (Sheeley 1992; St. Denis and Cappuccino 2004). Our own work indicates that the occurrence of polyembrony is much lower in BSW in comparison to PSW, with the probability of a PSW seed being polyembryonic roughly ten times greater than that for a BSW seed (C.H. Douglass, unpublished data). Research suggests that polyembryonic seeds are more successful than monoembryonic seeds in undisturbed habitats and in the absence of strong competitors (which often occurs in disturbed areas) (Cappuccino et al. 2002; Ladd and Cappuccino 2005). However, in a recent 3-year field study in central New York State, Hotchkiss et al. (2008) reported that polyembryonically derived PSW plants were not afforded a survival or growth advantage over plants derived from single embryo seeds under both high and low light environments within a forest site.

PSW has a stout and often large root crown that produces perennating buds and extensive, fleshly, fibrous roots (DiTommaso et al. 2005b). Many plants also possess a horizontal, woody rhizome, though this structure does not appear to substantially facilitate dispersion of the plants (Cappuccino 2004; Weston et al. 2005). The root-to-shoot biomass ratio of PSW can be substantial, up to 6.7 in New York soils that contained beneficial arbuscular mycorrhizal fungal (AMF) species (Smith 2006). Root structures in BSW are similar but tend to be thicker and more fibrous, and rhizomes in this species are reported to contribute more significantly to population expansion (DiTommaso et al. 2005b; Lumer and Yost 1995). For example,

Lumer and Yost (1995) often found adjoining plants that were connected by "horizontal underground stems" growing at a depth of nearly 50 cm.

Both swallow-wort species, and PSW in particular, have high seed output potentials. At a heavily infested site in northern New York State, Smith (2006) reported a potential seedling output of 63,439 seedlings/m^2 when polyembryonic offsprings were taken into account. However, it is not clear whether newly emerged or older seedlings contribute more relatively to the expansion of swallow-wort patches. Ladd and Cappuccino (2005) found that when they planted (buried 1 cm) overwintered PSW seeds in an experimental old field, 71% of the seeds germinated in the first year. Swallow-wort seeds generally mature dormant, and while experimentally germination can be doubled in seeds provided with a cold treatment, the nature and extent of this dormancy is unknown (DiTommaso et al. 2005a, b; Lumer and Yost 1995).

13.4 Introduction and Current Distributions

The earliest North American collection of PSWs was made in 1885 from Victoria, British Columbia (DiTommaso et al. 2005b; Sheeley and Raynal 1996). The earliest collection in the USA was 6 years later (1891), when it was simultaneously recorded in both Monroe and Nassau Counties in New York state. The first specimen of BSW in North America was collected in Ipswich, Massachusetts (MA) in 1854. In Gray's 1867 *Manual of Botany*, BSW was cited as a garden escape in Cambridge, MA. By 1871 there was a report of the plant "running wild" along a road in what is now likely Flatlands, in modern day Brooklyn, New York (Anonymous 1871). Eleven years later, it was described as naturalized in West Point and New Rochelle, NY (Bailey 1882; Day 1881).

The most likely source of introduction of both species was importation as specimens for botanical or estate gardens, though this remains uncertain (DiTommaso et al. 2005b; Sheeley 1992). For many years the two swallow-wort species were cultivated and sold as ornamental plants, though this is no longer common (DiTommaso et al. 2005b; Monachino 1957).

PSW invasion in North America is centralized in upstate New York, specifically Central New York, the Finger Lakes Region, and the region surrounding Lake Ontario in both the USA and southern Canada. There are additional extensive populations throughout Long Island, NY and other states in the Northeast, and there have been isolated reports of plant sightings in Indiana, Michigan, Missouri, and Wisconsin (DiTommaso et al. 2005b; Weston et al. 2005). BSW has a wider distribution longitudinally, with populations reported as far west as Kansas, Nebraska, Minnesota, and even California. However, its invasion is also centered in New York, with the heaviest infestations found in the Hudson River Valley, but also in Massachusetts and Connecticut (DiTommaso et al. 2005b). The wider distribution of BSW has been attributed to its apparent ability to adapt to more severe climatic conditions than encountered in its native Mediterranean range, unlike PSW that has largely remained within its predicted climatic boundaries (DiTommaso et al. 2005b).

Both swallow-wort species are typically found in habitats with temperature ranges in the winter of −11 to 0.7°C and in the summer of 20.7–26.4°C, while mean annual precipitation levels in these areas range from 776–1,206 mm (DiTommaso et al. 2005b).

In contrast, annual temperatures in Ukraine (PSW's native range) vary from −8 to 24°C with a mean precipitation of 629 mm. Temperatures in southwestern France (part of BSW's native range) vary from 2 to 28°C with 668 mm of precipitation and, in northeastern Spain, the climate varies from 2 to 31°C with only 317 mm of precipitation (World Meteorological Organization 2008).

13.5 Impacts

PSW in particular has invaded sensitive and rare alvar communities both in eastern Ontario, Canada, and in Jefferson County, NY, and has displaced endemic flora and fauna (DiTommaso et al. 2005b). A survey in the affected areas revealed a significant negative correlation between PSW cover and the number and diversity of previously common grassland bird species (DiTommaso et al. 2005b). Ernst and Cappuccino (2005) found fewer arthropods both dwelling on PSW plants and ground-dwelling insects adjacent to sampled plants. The authors concluded that the decline in old-field arthropod populations because of the invasion of swallow-worts could negatively impact bird and small mammals that also depend on insects for food.

Lawlor (2000) reported that habitats of the Hart's tongue fern [*Asplenium scolopendrium* L. var. *americanum* (Fern.) Kartesz & Gandhi], a rare plant species native to regions of New York, have been invaded by PSW. Similarly, PSW has invaded sites at The Nature Conservancy's Mashomack Preserve on Shelter Island, NY where the federally listed endangered species sandplain gerardia (*Agalinis acuta* Pennell) is found (M. Scheibel, personal communication). BSW is a relatively less well-studied species and thus few targeted studies have been carried out to assess the ecological impacts of its invasion. One study found that the species threatens the survival of the endemic Jessop's milkvetch [*Astragalus robbinsii* (Oakes) A. Gray] along the banks of the Connecticut River in Windsor, VT, one of only three locations in which the plant is known to remain (DiTommaso et al. 2005b).

While the swallow-worts have had a substantial negative impact in a variety of natural areas, the species pose a substantial and looming threat to New York states' important agricultural industry. The detection of PSW plants in no-till corn and soybean fields is problematic given the relative difficulty of controlling either of the swallow-wort species effectively with commonly used herbicides in crop systems (DiTommaso et al. 2005b; Lawlor 2003; Weston et al. 2005). There have been numerous reports of landowners abandoning horse pastures due to unmanageable infestations of PSW, possibly due to the physical obstruction posed by dense swallow-wort stands or the suspected toxicity to mammals of plant tissues (Lawlor 2003; Weston et al. 2005). A feeding trial with fresh PSW plant material resulted in the death of a goat from suspected cardiac arrest 4 days after the last tissue treatment, which seems to support evidence from Scandinavia that sheep avoid grazing on PSW plants (DiTommaso et al 2005b; Haeggstrom 1990).

The New York State Forest Owner's Association and many foresters have claimed that swallow-wort infestations in understories are also compromising forest regeneration (Lawlor 2003). Horticultural nursery owners and Christmas tree producers affected by swallow-wort infestations reported that due to lack of effective control methods and regeneration impacts, land abandonment was often the only reasonable option. Indeed, several orchard owners east of Rochester, NY cited PSW as their most problematic weed species (A. Fowler, personal communication; Lawlor 2003).

The potential for both swallow-wort species to serve as fatal hosts for Monarch butterflies (*Danaus plexippus* L.), a condition in which adults lay eggs on the plants but the larvae do not survive, has been well reported (Casagrande and Dacey 2001; DiTommaso and Losey 2003). Casagrande and Dacey (2007) found that in fields with little or no common milkweed (*Asclepias syriaca* L. – the butterflies' normal host species), the density of eggs found on BSW stems was five times greater than that found in a more diverse old-field site with abundant common milkweed. Although there have been studies that questioned whether swallow-worts play a significant role as fatal hosts for Monarch butterflies (Mattila and Otis 2003), it is likely that through the competitive displacement of common milkweed populations, the two swallow-wort species could ultimately pose a serious threat to Monarch butterfly populations in infested areas (DiTommaso et al. 2005b; Tewksbury et al. 2002).

13.6 Management

13.6.1 Manual Control

Manual methods can often be effective at controlling established patches of perennial weeds (Radosevich et al. 1997; Ross and Lembi 1999). However, both PSWs and BSWs can rapidly regrow from buds on the root crown, rendering mowing, tillage, clipping, and other frequently used control strategies less effective against these perennials (Averill et al. 2008; Lawlor 2002; Lawlor and Raynal 2002; Weston et al. 2005). Mowing can contain invasive populations of the swallow-worts when timed to suppress seed production, but must be repeated for the duration of the growing season as plants tend to regrow more rapidly than nonmowed plants and produce seed at a more immature stage of growth than is typical (C.H. Douglass, personal observations). Averill et al. (2008) found that clipping of PSW stems once annually at the beginning of summer (June) led to a 44% reduction in cover at an infested site in northern New York over a 2-year period. Because of their tall, brittle stems, swallow-worts are also particularly sensitive to trampling, which has resulted in a substantial reduction of PSWs in some localized fields (DiTommaso et al. 2005b).

13.6.2 Chemical Control

There are several herbicides that provide relatively effective control of black or PSWs when applied postemergence (Averill et al. 2008; Lawlor 2002; Weston et al. 2005). Foliar applications are generally more difficult to apply than cut stem applications because of the intertwining growth habit of the swallow-worts and high patch densities at maturity, but are generally more effective (Lawlor and Raynal 2002). Furthermore, Lawlor and Raynal (2002) found that foliar applications were significantly more effective at controlling plants in shaded plots than drier, full sun plots. In particular, the most effective chemical treatments were glyphosate (10.4 kg ai ha^{-1}) applied at an early stage of flowering and triclopyr (2.6 kg ai ha^{-1}) applied at early fruit formation, both of which resulted in a 73% reduction in cover, decreased densities, and a loss of apical dominance (Lawlor and Raynal 2002).

Recent work has demonstrated that glyphosate applied at a much lower rate (1.79 kg ai ha^{-1}) was equally as effective (77% reduction in cover when applied in late June) as a higher rate, and more effective overall than triclopyr alone or combinations of triclopyr and 2,4-D or dicamba and 2,4-D (F. Lawlor, unpublished data in Weston et al. 2005). Similarly, Averill et al. (2008) found that triclopyr applied at a lower rate (1.9 kg ae ha^{-1}) reduced PSW stem densities by 80% 2 years after a single June application. In any case, an adequate surfactant must be included in postemergent foliar applications so that uptake is maximized, particularly because of the waxy cuticle present on the leaf surfaces and stems of both swallow-wort species (Radosevich et al. 1997).

13.6.3 Biological Control

To date, research on the biological control of black and PSW species with insects has been limited (Weston et al. 2005). Since most larval stages of insects do not thrive on foliage of the alkaloid-containing leaves of the two swallow-wort species, effective biocontrol with insects presents a strong challenge to researchers in finding an herbivorous insect for specific long-term control (Christensen 1998; Tewksbury et al. 2002). Potential pathogens of the two swallow-wort species have not been found, although several pathogenic organisms do infect members of the milkweed family (Weston et al. 2005).

Recently, the USDA's Agricultural Research Service (ARS) Laboratory located in Ithaca, NY and headed by L.R. Milbrath initiated a biocontrol program targeting both of these invasive species. The search for potential biocontrol agents has focused primarily in Europe and Eurasia. The criteria for the biocontrol program specifies that the candidate organism is able to be propagated in culture, can be successfully released and established in affected regions, and remains specific to the swallow-worts (Milbrath and Gibson 2006). Given the lack of success from chemical and manual tactics to date, biocontrol of the two swallow-wort species might offer the greatest potential for successful long-term control in North America (Tewksbury et al. 2002).

13.7 Characteristics and Patterns of Invasion

13.7.1 Dispersal and Establishment

Cappuccino et al. (2002) demonstrated that 50% of PSW seeds landed within 2.5 m of their release points, but that seed weight was inversely correlated with dispersal distance so that lighter seeds dispersed further. However, in a more recent study, 83% of seeds produced by a PSW plant landed directly beneath the parent plant (Ladd and Cappuccino 2005). Moreover, 51% of seeds placed on the soil surface germinated and resulted in emergent seedlings while 71% of seeds buried at a soil depth of 1 cm resulted in emergent seedlings. These high rates of emergence could contribute to the ability of the two swallow-wort species to rapidly and successfully establish satellite populations, with maximum dispersal of seeds found to be up to 60 m from the parent plant (Ladd and Cappuccino 2005). First-year PSW seedlings also have unusually high survivorship (71–100%) when compared with many other herbaceous plant species. This particular study was performed in an undisturbed old-field community, suggesting that while the swallow-worts can be invasive in disturbed habitats, they can also become established in intact natural plant communities (Ladd and Cappuccino 2005). The ability of swallow-worts to invade apparently stable plant communities is remarkable given the conventional wisdom in the field of invasion ecology that intact native communities will be "ecologically resistant" to invasive species (Elton 1958).

The production of polyembryonic seedlings by swallow-worts has not been found to be a significant competitive advantage in the presence of neighbors, but was beneficial in the absence of plant competition (Cappuccino et al. 2002). Polyembryony is likely to be most beneficial in disturbed habitats because of the enhanced ability of multiple seedlings to successfully establish in the absence of native vegetation and with increased light availability (Cappuccino 2004; DiTommaso et al. 2005a; Hotchkiss et al. 2008).

13.7.2 Plant–Plant Competition

Direct competition with monocots was found to significantly decrease the average size of PSW seedlings, especially for seedlings produced from large-sized seeds (Cappuccino et al. 2002). All but two of the seedlings grown in the presence of grasses were smaller than expected, while 90% of those grown in the absence of competition were of above average size. PSW can have strong drought tolerance as shown by the relatively low water tension levels recorded ($\Psi = -0.062$ mPa at midday and −0.079 at predawn) (DiTommaso et al. 2005b). This may be due, in part, to its extensive root system and waxy leaf cuticle, suggesting that this species can effectively tolerate environmental stresses that may reduce the vigor and performance of associated plant species (DiTommaso et al. 2005b).

Soils at sites invaded by PSW have been found to have greater AMF inoculum potentials than adjacent, uninvaded sites (DiTommaso et al. 2005b; Greipsson and DiTommaso 2006; Smith 2006; Smith et al. 2008). Swallow-wort plants also showed significantly greater growth in the presence of locally associated microbial communities than nonlocal communities. The authors proposed that by altering the species of mycorrhizal fungi at sites, swallow-wort could facilitate its establishment and expansion by displacing resident flora dependent on native fungal species.

13.7.3 Habitat and Environmental Variability

In Central New York, PSW plants are normally found on shallower soils over limestone bedrock or deep, well-drained silt-loam soils in wooded ravines, calcareous cliffs, talus slopes, alluvial woods, pastures, and grasslands (DiTommaso et al. 2005b; Weston et al. 2005). PSW exhibits a wide tolerance to light and moisture conditions, but appears to be particularly successful and aggressive on shallow, droughty soils or deeper silt loams with partial to full sun (Lawlor 2002; Lawlor and Raynal 2002; DiTommaso et al. 2005a; Smith 2006). BSW seems to share comparable habitat preferences, but is often limited to sunny, open field sites rather than shaded forest sites (Lumer and Yost 1995).

Shaded sites are characterized by greater densities of PSW, and taller plants with longer internodes (Lawlor 2000; Sheeley 1992; Smith 2006). Smith et al. (2006) reported a seasonal variation in seedling stem densities of PSW in a northern New York State site, with an almost fivefold decrease between late July and August. Moreover, growth and fecundity of both swallow-wort species is substantially greater in open, sunny sites or gaps in the forest understory (DiTommaso et al. 2005a; Hotchkiss et al. 2008; Sheeley 1992). Seeds produced by plants at shaded sites are significantly more likely to posses dormant or nonviable embryos (Smith 2006; Smith et al. 2006).

Our recent findings with regard to impact of soil type upon establishment of both swallow-wort species suggest that soil pH and precipitation levels play significant roles in influencing their success of establishment (Douglass 2008). Soil pH levels in particular were significantly negatively correlated ($P < 0.05$) with overall plant height of both species. Also, it appears that sites invaded by PSW were characterized by significantly ($P < 0.05$) high pH, calcium and magnesium levels than those invaded by the black species. Further studies are needed to assess these factors and their relationship to successful invasion with respect to both the swallow-worts and other nonnatives in New York.

13.8 Invasiveness

Cappuccino (2004) found that smaller-sized patches (1 and 9 plants) of PSW had higher reproductive success (measured as maturation of follicles), and higher root-to-shoot biomass ratios than larger-sized patches (81 plants). However, larger patches produced three times more follicles and thus had a greater net production of seeds. When considered in the context of her earlier work (Cappuccino et al. 2002), the latest results suggest that in general larger swallow-wort patches will produce large quantities of seed and that at least a portion of these seeds will lead to the successful establishment of pioneering satellite populations.

These smaller satellite patches appear to invest proportionally greater resources into root biomass, presumably to ensure establishment, before allocating resources to vegetative or reproductive structures (Cappuccino et al. 2002; Cappuccino 2004; Smith 2006). Indeed, Smith (2006) found that PSW had a significantly greater root-to-shoot biomass ratio than its close relative common milkweed. Likewise, when in competition with common milkweed, young PSW plants had greater overall reproductive output. While diminishing its competitive ability relative to common milkweed, this allocation of resources to reproduction could ensure the presence of a large seed bank from which satellite populations could be produced (Cappuccino 2004, Lockwood et al. 2007; Myers and Bazely 2003). Given that the seed output of a single PSW population in central New York was 35,244–62,439 seeds/m^2 (the higher figure takes into account an average proportion of polyembryonic seeds), the seed bank of this species can be quite large, and its impact over the course of several years significant (Smith 2006).

Once established, both swallow-wort species grow profusely and aggressively. PSW and BSW can rapidly alter the abiotic and biotic features of their understory and surrounding areas: decreasing sunlight penetration, increasing nutrient acquisition through large root biomasses, and altering rhizosphere dynamics both through shifts in the AMF community and the exudation of allelopathic chemicals (Douglass 2008; Greipsson and DiTommaso 2006; Lawlor 2002; Sheeley and Raynal 1996; Weston et al. 2005). Despite the increasing number of studies on the two invasive swallow-wort species, there has been little focus to date on three potentially significant factors that may influence the invasiveness of the species, namely, allelopathy activity by tissue leachates or root exudates; adaptive morphological plasticity; and the genetic diversity of introduced populations.

13.8.1 Allelopathy

Some exotic plant species competitively exclude and eliminate their neighbors in invaded "recipient" communities, but generally are found to coexist with neighbors in species-diverse systems in their native habitat. Allelopathy has been suggested as one of the mechanisms responsible for this success (Callaway et al. 2005,

Inderjit et al. 2006a). Although allelopathy has not yet been clearly implicated in association with the field dominance of any invasive species, evidence suggests that many invasive species produce an array of secondary plant products or allelochemicals that are released through decomposition of plant material or directly exuded from roots or shoots into the soil rhizosphere (Inderjit et al. 2006b).

Cappuccino (2004) first reported that PSW root extracts inhibited the germination of radish seeds and showed broad antifungal activity. DiTommaso et al. (2005b) suggested that the purported allelopathic activity of root exudates could indirectly affect competitive interactions of plants through the alteration of the structure of the rhizosphere community. Interactions between the allelopathic activity of invasive plant species and mycorrhizal associations of affected plant communities can have marked effects on the population dynamics of invaded sites (Roberts and Andersen 2001; Stinson et al. 2006).

Both swallow-wort species have been found to have high concentrations of cytotoxic secondary products in their roots, stems, and leaves (Capo and Saa 1989; Lee et al. 2003; Nowak and Kiesel 2000; Staerk et al. 2000, 2002). In unpublished laboratory work, N. Cappuccino demonstrated that the foliage of PSW produces phytotoxins when these tissues are ground and their chemical constituents extracted using water (N. Cappuccino, personal communication). Other investigators have postulated that the dense monocultural stands created by swallow-worts following establishment may have resulted from the exudation of root-released allelochemicals that limit growth of neighboring species. The decomposition of swallow-wort foliage and stems underneath a dense stand may contribute to seedling suppression from the effects of both allelopathy and limited light reaching the soil surface due to a mulch effect from swallow-wort plant material (Weston et al. 2005).

Our own findings in laboratory simulations of allelopathic activity in agar gel box assays and Parker bioassays suggest that the role of allelopathy may be limited in contributing to the establishment and interference of either swallow-wort species and their respective effects on nearby competitors (Douglass 2008). Indeed, PSW root exudates in particular resulted in substantial (up to 40%) reductions in the root length of a number of indicator species, and leachates of leaf tissues of both swallow-wort species caused similar reductions in both root and shoot length of indicators. However, both stimulation of indicator species growth and autotoxicity were observed during the experiments, and a comparison of inhibitory effects with common milkweed (generally not considered to be invasive) found that the swallow-worts did not exhibit significantly greater negative allelopathic abilities than the related nonnative species.

13.8.2 Phenotypic and Genetic Diversity

Ellstrand and Schierenbeck (2000) proposed that hybridization (both inter- and intraspecific) could play an important role in enhancing the invasiveness of introduced species. In particular, they suggested that hybridization between populations of the same taxa could lead to adaptive evolution in cases where the species was intentionally introduced multiple times (resulting in a diverse gene pool), and that this process would occur only after a lag period. Given the history of swallow-wort introduction and invasion, and evidence of hybridization occurring between two *Vincetoxicum* species, we suggest that there is a high potential for occurrence of rapid evolutionary changes among introduced swallow-worts (DiTommaso et al. 2005b; Lauvanger and Borgen 1998).

There is also evidence that the rapid evolution of plasticity for ecologically advantageous traits is relatively common among invasive species, and that this may partially explain the success of many invasive species only after an initial lag time during which necessary evolutionary adjustments have occurred (Pigliucci 2005; Richards et al. 2006). The potential roles of trait plasticity in the invasiveness of the two swallow-wort species are of great interest given reports of wide variation in reproductive and phenological traits among invasive swallow-wort populations (DiTommaso et al. 2005b; Lawlor 2000; Sheeley 1992; St. Denis and Cappuccino 2004).

Nevo (1988) and Nevo et al. (1984) found that plant species tend to be more genetically polymorphic if they occur in broader climatic, ecological, or biotic spectra, both at the macro and microgeographic scale. Agrawal (2001) predicted that species whose phenotypic traits are exposed to (and thus respond to) larger ranges of environmental stimuli will be more likely to influence both the ecology and perhaps evolution of that species' interactions in novel habitats. Our preliminary work characterizing genetic diversity levels amongst PSW and BSW populations throughout New York state has suggested that while the two species are genetically distinct, intraspecies genetic diversity is actually relatively low (Douglass 2008). The further determination of molecular patterns and degrees of adaptive morphological variation in introduced swallow-wort populations should be a research priority as the species clearly display strong abilities to acclimatize to diverse environments.

13.9 Conclusions

It has been proposed that it is possible to predict the invasiveness of particular plant species from some life history traits, including the capacity to reproduce vegetatively, seed size and volume of production, and the persistence of the seed bank, among others (Myers and Bazely 2003; Rejmanek and Richardson 1996). Pale and BSWs are characterized by many of these traits, arguably making them archetypal invasive plants. While we do not currently fully understand the role of vegetative reproduction in the rapid spread of the swallow-worts, we are gaining information related to their physiology, ecology, and invasiveness (DiTommaso et al. 2005b; Weston et al. 2005). The apparent ability of these species to become invasive in many locations in New York state indicates that swallow-worts have an enhanced propensity toward establishing in and becoming invasive in novel habitats that are outside of their currently described range (Lockwood et al. 2007).

Myers and Bazely (2003) proposed two key areas of study that are important in determining the invasiveness of a given species: (1) interactions between invasive species specifically (as well as the impact this has on existing community structure and function), and (2) the interactions between and evolution of plants and soil organisms. Given the high reproductive outputs of swallow-worts, the unknown role of vegetative expansion, and the clear impacts that swallow-wort invasions have on both belowground and aboveground communities, it is vital to prioritize the investigation of interactions between the swallow-worts and the biotic communities in invaded habitats (Greipsson and DiTommaso 2006; Mitchell et al. 2006; Smith 2006; Smith et al. 2008). This information is necessary to further our understanding of the invasion process in these congeneric perennials and to develop strategies that may prove effective for eventual control and management of the species. Given their rapid spread across the Northeastern USA and our inability to successfully control these invasions, we must rapidly devote additional resources to study the ecology and management of these interesting and unusual vines. Considering the lack of success in management of these species using traditional means of control, possibilities for successful biocontrol options are now being rapidly explored. We hope that the information we have generated regarding the ability of these species to spread and reproduce, and exhibit allelopathic interference may lend itself to the successful development of biocontrol strategies before further uncontrolled spread occurs.

References

Agrawal A (2001) Phenotypic plasticity in the interactions and evolution of species. Science 294: 321–326

Anonymous (1871) B Torrey Bot Club 2: 43

Aparicio A, Silvestre S (1985) Numeros cromosomicos para la flora Espanola 422–434. Lagascalia 13: 318–323

Averill KM, DiTommaso A, Morris SH (2008) Response of pale swallow-wort (*Vincetoxicum rossicum*) to triclopyr application and clipping. Invas Plant Sci Manage 1(2): 196–206

Bailey WW (1882) B Torrey Bot Club 9: 106

Baker HG (1974) The evolution of weeds. Annu Rev Eco Syst 5: 1–24

Callaway RM, Ridenour WM, Laboski T, Weir T, Vivanco JM (2005) Natural selection for resistance to the allelopathic effects of invasive plants. J Ecology 93: 576–583

Capo M, Saa JM (1989) (–)-antofine: a phenanthroindolizidine from *Vincetoxicum nigrum*. J Nat Prod 52(2): 389–390

Cappuccino N (2004) Allee effect in an invasive alien plant, pale swallow-wort *Vincetoxicum rossicum* (Asclepiadaceae). Oikos 106: 3–8

Cappuccino N, Mackay R, Eisner C (2002) Spread of the invasive alien vine *Vincetoxicumrossicum*: tradeoffs between seed dispersability and seed quality. Am Midl Nat 148: 263–270

Casagrande RA, Dacey J (2001) Monarch butterfly (*Danaus plexippus*) oviposition on black swallow-wort (*Vincetoxicum nigrum*). RI Nat Hist Surv 8: 2–3

Casagrande RA, Dacey J (2007) Monarch butterfly oviposition on swallow-worts (*Vincetoxicum* spp.). Environ Entomol 36(3): 631–636

Christensen T (1998) Swallow-worts. Wildflower, Summer issue 21–25

Day EH (1882) Proceedings of the Torrey Botanical Club. B Torrey Bot Club 9: 120

Diosdado JC, Ojeda F, Pastor J (1993) In: Stace CA (ed) IOPB chromosome data 5. Newsletter Int Org Plant Biosyst 20: 6–7

DiTommaso A, Losey JE (2003) Oviposition preference and larval performance of monarch butterflies (*Danaus plexippus*) on two invasive swallow-wort species. Entomol Exp Appl 108: 205–209

DiTommaso A, Brainard DC, Webster BR (2005a) Seed characteristics of the invasive alien vine *Vincetoxicum rossicum* are affected by site, harvest date, and storage duration. Can J Bot 83: 102–110

DiTommaso A, Lawlor FM, Darbyshire SJ (2005b) The biology of invasive alien plants in Canada. II. *Cynanchum rossicum* (Kleopow) Borhidi [= *Vincetoxicum rossicum* (Kleopow) Barbar.] and *Cynanchum louiseae* (L.) Kartesz & Gandhi [=*Vincetoxicum nigrum* (L.) Moench]. Can J Plant Sci 85: 243–263

Douglass CH (2008) The role of allelopathy, morphology and genetic diversity in the invasion of swallow-wort species in New York state. Thesis, Cornell University, Ithaca, NY

Ellstrand NC, Schierenbeck KA (2000) Hybridization as a stimulus for the evolution of invasiveness in plants? Proc Natl Acad Sci USA 97(13): 7043–7050

Elton CS (1958) The ecology of invasions by animals and plants. University of Chicago Press, Chicago, IL

Ernst CM, Cappuccino N (2005) The effect of an invasive alien vine, *Vincetoxicum rossicum* (Asclepiadaceae), on arthropod populations in Ontario old fields. Biol Invas 7: 417–425

Gleason HA, Cronquist A (1991) Manual of vascular plants of northeastern United States and adjacent Canada, 2nd edn. New York Botanical Gardens Press, Bronx, NY

Gray E (1867) Manual of botany of the Northern United States. Including the district east of the Mississipi and north of North Carolina and Tennessee, 5th edn. Ivison, Blakeman, Taylor & Co., New York, NY

Greipsson S, DiTommaso A (2006) Invasive non-native plants alter the occurrence of arbuscular mycorrhizal (AMF) and benefit from that association. Ecol Restor 24: 236–241

Haeggstrom CA (1990) The influence of sheep and cattle grazing on wooded meadows in Åland, SW Finland. Acta Bot Fenn 141: 1–28

Hotchkiss EE, DiTommaso A, Brainard DC, Mohler CL (2008) Survival and performance of the invasive vine *Vincetoxicum rossicum* (Apocynaceae) from seeds of different embryony under two light environments. Am J Bot 95(4): 447–453

Inderjit, Callaway RM, Vivanco JM (2006a) Can plant biochemistry contribute to an understanding of invasion ecology? Trends Plant Sci 11(12): 574–580

Inderjit, Weston LA, Duke SO (2006b) Challenges, achievements and opportunities in allelopathy research. J Plant Int 1(2): 69–81

Ladd D, Cappuccino N (2005) A field study of seed dispersal and seedling performance in the invasive exotic vine *Vincetoxicum rossicum*. Can J Bot 83: 1181–1188

Lauvanger EG, Borgen L (1998) The identity of Vincetoxicum in Norway. Nordic J Bot 18(3): 353–364

Lawlor FM (2000) Herbicidal treatment of the invasive plant *Cynanchum rossicum* and experimental post control restoration of infested sites. Thesis, State University of New York College of Environmental Science and Forestry, Syracuse, NY

Lawlor FM (2002) Element Stewardship Abstract for *Vincetoxicum nigrum* (L.) Moench and *Vincetoxicum rossicum* (Kleopow) Barbarich. http://tncweeds.ucdavis.edu/esadocs/documnts/vinc_sp.pdf. Accessed 27 January 2008

Lawlor FM (2003) The swallow-worts. N Y Forest Owner 41(4): 14–15

Lawlor FM, Raynal DJ (2002) Response of swallow-wort to herbicides. Weed Sci 50: 179–185

Lee SK, Nam KA, Heo YH (2003) Cytotoxic activity and G2/M cell cycle arrest mediated by antofine, a phenanthroindolizidine alkaloid isolated from *Cynanchum paniculatum*. Planta Med 69: 21–25

Liede S (1996) *Cynanchum – Rhodostegiella – Vincetoxicum – Tylophora (Asclepiadaceae):* new considerations on an old problem. Taxon 45: 192–211

Liede S, Tauber A (2002) Circumscription of the genus *Cynanchum* (Apocynaceae – Asclepiadoideae). Syst Bot 27(4): 789–800

Lockwood JL, Hoopes MF, Marchetti MP (2007) Invasion ecology. Blackwell, Malden, MA
Lumer C, Yost SE (1995) The reproductive biology of *Vincetoxicum nigrum* (L.) Moench (Asclepiadaceae), a Mediterranean weed in New York state. B Torrey Bot Club 122(1): 15–23
Mattila HR, Otis GW (2003) A comparison of the host preference of monarch butterflies (*Danaus plexippus*) for milkweed (*Asclepias syriaca*) over dog-strangling vine (*Vincetoxicum rossicum*). Entomol Exp Appl 107: 193–199.
Milbrath L, Gibson D (2006) Biological control of swallow-wort and other invasive weeds of the Northeastern United States, 2006 Annual Report. http://www.ars.usda.gov/research /projects/projects.htm?ACCN_NO = 406908&showpars = true&fy = 2006. Accessed January 27 2008
Mitchell CE, Agrawal AA, Bever JD, Gilbert GS, Hufbaeur RA, Klironomos JN, Maron JL, Morris WF, Parker IM, Power AG, Seabloom EW, Torchin ME, Vasquez DP (2006) Biotic interaction and plant invasions. Ecol Lett 9: 726–740
Monachino J (1957) Cynanchum in the New York area. B Torrey Bot Club 84(1): 47–48
Moore RJ (1959) The dog-strangling vine *Cynanchum medium*, its chromosome number and its occurrence in Canada. Can Field Nat 73: 144–147
Myers JH, Bazely DR (2003) Ecology and control of introduced plants. Cambridge University Press, Cambridge, UK
Nevo E (1988) Genetic diversity in nature: patterns and theory. Evol Biol 23: 217–246
Nevo E, Beiles A, Ben-Shlomo R (1984) The evolutionary significance of genetic diversity: ecological, demographic and life history correlates. In: Mani GS (ed) Evolutionary dynamics of genetic diversity, Vol 53. Springer, Berlin
Nowak R, Kisiel W (2000) Hancokinol from *Vincetoxicum officinale*. Fitoterapia 71: 584–586
Pigliucci M (2005) Evolution of phenotypic plasticity: where are we going now? Trends Ecol Evol 20(9). doi: 10.1016/j.tree.2005.06.001
Radosevich S, Holt J, Ghersa C (1997) Weed ecology: implications for management, 2nd edn. Wiley, New York
Rejmanek M, Richardson RM (1996) What attributes make some plant species more invasive? Ecology 77(6): 1655–1661
Richards CL, Bossdorf O, Muth N, Gurevitch J, Pigliucci M (2006) Jack of all trades, master of some? On the role of phenotypic plasticity in plant invasions. Ecol Lett 9: 981–993
Roberts KJ, Anderson RC (2001) Effect of garlic mustard [*Alliaria petiolata* (Beib. Cavara & Grande)] extracts on plants and arbuscular mycorrhizal (AM) fungi. Am Midl Nat 146: 146–152
Ross MA, Lembi CA (1999) Applied weed science, 2nd edn. Prentice Hall, Upper Saddle River, NJ
Sheeley SE (1992) Life history and distribution of *Vincetoxicum rossicum* (Asclepiadaceae): an exotic plant in North America. Thesis, State University of New York College of Environmental Science and Forestry, Syracuse, NY
Sheeley SE, Raynal DJ (1996) The distribution and status of species of *Vincetoxicum* in eastern North America. B Torrey Bot Club 123(2): 148–156
Smith LL (2006) Arbuscular mycorrhizal fungi and the competitiveness of the invasive vine pale swallow-wort (*Vincetoxicum rossicum*). Thesis, Cornell University, Ithaca, NY
Smith LL, DiTommaso A, Lehmann J, Greipsson S (2006) Growth and reproductive potential of the invasive exotic vine *Vincetoxicum rossicum* in northern New York State. Can J Bot 84(12): 1771–1780
Smith LL, DiTommaso A, Greipsson SL (2008) Effects of arbuscular mycorrhizal fungi on the exotic invasive vine pale swallow-wort (*Vincetoxicum rossicum*). Invas Plant Sci Manage 1(2): 142–152
Staerk D, Christensen J, Lemmich E, Duus JO, Olsen CE, Jaroszewski JW (2000) Cytotoxic activity of some phenanthroindolizidine *N*-oxide alkaloids from *Cynanchum vincetoxicum*. J Nat Prod 63: 1584–1586
Staerk D, Lykkeberg AK, Christensen J, Budnik BA, Abe F, Jaroszewski JW (2002) In vitro cytotoxic activity of phenanthroindolizidine alkaloids from *Cynanchum vincetoxicum* and

Tylophora tanakae against drug-sensitive and multidrug-resistant cancer cells. J Nat Prod 65(9): 1299–1302

St. Denis M, Cappuccino N (2004) Reproductive biology of *Vincetoxicum rossicum* (Kleo.) Barb. (Asclepiadaceae), an invasive alien in Ontario. J Torrey Bot Soc 131: 8–15

Stinson KA, Campbell SA, Powell JR, Wolfe BE, Callaway RM, Thelen GC, Hallett SG, Prati D, Klironomos JN (2006) Invasive plant suppresses the growth of native tree seedlings by disrupting belowground mutualisms. PLoS Biol 4(5): e140. doi: 10.1371/journal.pbio. 0040140

Tewksbury L, Casagrande R, Gassmann A (2002) Swallow-worts. In: Van Driesche R, Lyon RS, Blossey B, Hoddle M, Reardon R (eds) Biological control of invasive plants in the eastern United States. US Department of Agriculture (USDA) Forestry Service Publication FHTET-2002–04. USDA, Washington, DC

USDA Plants Database (2008) http://plants.usda.gov. Accessed 27 January 2008

West J, Fowler A (2008) Testimonials. http://www.swallow-wort.com. Accessed 27 January 2008

Weston LA, Barney JN, DiTommaso A (2005) A review of the biology, ecology and potential management of three important invasive perennials in New York state: Japanese knotweed (*Polygonum cuspidatum*), mugwort (*Artemisia vulgaris*) and pale swallow-wort (*Vincetoxicum rossicum*). Plant Soil 277(1–2): 53–69

World Meteorological Organization (2008) World weather information service. http://www.worldweather.org. Accessed 27 January 2008

Chapter 14
Management of *Phalaris minor*, an Exotic Weed of Cropland

Inderjit and Shalini Kaushik

Abstract *Phalaris minor* is a troublesome nonnative weed, particularly in wheat fields of northwestern India. In spite of protracted efforts to manage this weed with herbicides, it is still a significant challenge. Here, we discuss some agroecological practices that could influence establishment and survival of *P. minor*. Although this chapter deals with a specific example in purely agricultural settings, it illustrates the magnitude of the problem created by a nonnative weed of cropland. This weed is largely restricted to wheat fields. Future research should include examination of the ecological factors for the restricted distribution of *P. minor* in wheat fields.

Keywords *Phalaris minor* • Isoproturon • Allelopathy • Rice straw • Sulfosulfuron

14.1 Introduction

Phalaris minor Retz. (littleseed canarygrass, Poaceae) is an annual exotic cropland weed, which is particularly common in wheat (*Triticum aestivum* L.) fields (Fig. 14.1a). It is a native of much of the Mediterranean region, extending through the Middle East to the Persian Gulf (Singh et al. 1999; Kaushik et al. 2005; Kaushik and Inderjit 2007). In part of its native range, e.g., Turkey, it commonly occurs in cultivated fields (Fig. 14.1b). *P. minor* was reported by Hooker (1896) and Stewart (1945) in India as mentioned in the Flora of British India and Brittonia, respectively. Consequently, *P. minor* had certainly reached India well before the Green Revolution during 1960s, when dwarf Mexican wheat varieties were imported from CYMMIT (International Maize and Wheat Improvement Center). In northwestern India, *P. minor* is largely restricted to wheat fields and rarely escapes these cultivated

Inderjit(✉) and S. Kaushik
Centre for Environmental Management of Degraded Ecosystems (CEMDE), University of Delhi, Delhi 110007, India
inderjit@cemde.du.ac.in

Fig. 14.1 **(a)** *Phalaris minor* (indicated by an *arrow*) growing mainly in wheat fields; **(b)** A cultivated field heavily infested with *Phalaris spp*. at the sea coast in Yulova, Turkey. In its native range, Turkey, *P. minor* grows in mixed crops

fields (Fig. 14.1a). During wheat harvest, *P. minor* seeds are incorporated into the soil and serve as a seed bank for the subsequent year.

Phalaris minor is responsible for significant economic losses in wheat production due to substantial declines in yield and quality of wheat (Singh et al. 1999). Exponential reduction in wheat yield with increasing density of *P. minor* has been reported repeatedly (see Khera et al. 1995; Dhaliwal et al. 1997). An increase in

P. minor density from 40 to 80 plants/m^2 resulted in a corresponding loss of 530 (11%) to 568 kg/ha (13.5%) of wheat yield (Khera et al. 1995).

Much attention is currently focused on exotic invasive weeds that may damage native biodiversity (Inderjit and Drake 2006; Mack et al. 2000). Noninvasive exotics such as *P. minor*, however, could also cause severe damage in terms of economic losses. We discuss here the control of *P. minor* with an emphasis on research topics that deserve more attention.

14.2 Agricultural Practices that Influence the Ecology and Management of *P. minor*

Agricultural practices, such as incorporation of rice straw or application of herbicide mixtures, or both, can play an important role in the successful establishment and growth of *P. minor* (Kaushik et al. 2005). Farmers in northwestern India often incorporate unburned and burned stubble/straw into the soil; this practice is followed by irrigation and sowing of the next crop. Straw (particularly rice and wheat straw) is reported to cause allelopathic suppression of crops (Inderjit et al. 2004). Incorporation of rice straw into the soil followed by irrigation, for example, exerts a negative effect on the seedling performance of mustard (*Brassica napus* var. *toria* L.) (Inderjit et al. 2004). After rice harvest, wheat is usually sown next, particularly in the States of Punjab and Haryana. We studied the influence of rice straw incorporation on the seedling growth of *P. minor* (Kaushik and Inderjit 2007). Incorporation of rice straw into soil suppressed the growth of *P. minor*.

Allelochemicals may influence plant growth directly or indirectly by influencing abiotic and/or biotic factors (Inderjit and Weiner 2001). Although allelopathic potential of Indian rice cultivars was not examined, several US and Chinese varieties are reported to exude chemicals with potential allelopathic activities (Olofsdotter 1998; Olofsdotter et al. 1999). It would be interesting to explore the allelopathic potential of Indian rice cultivars and determine if the straw itself (compared with root exudates) of any rice cultivar is phytotoxic to *P. minor*. We tried to examine whether inhibition of *P. minor* seedling growth in rice straw-incorporated soil could be explained by direct effect of allelochemicals or their indirect effect through altered soil abiotic and/or biotic factors, by using washed, unwashed rice straw or by amending the unwashed rice straw-incorporated soil with activated charcoal (Kaushik and Inderjit 2007). Washed rice straw as well as activated carbon did not ameliorate the phytotoxic effects of rice straw, which ruled out the direct involvement of rice straw allelopathy in growth inhibition of *P. minor*.

Microbes can act as a sink for mineral nutrients (Schmidt et al. 1997). The suppression of seed germination of *P. minor* in straw-incorporated soil, however, could also be due to immobilization of N by soil microbes, which would be fostered by carbon-rich straw. We observed higher levels of exchangeable phosphate in soil amended with rice straw compared with unamended straw, which might result in higher microbial activity (Kaushik and Inderjit 2007). Om et al. (2002) suggested that the allelopathic potential of certain crop species of wheat-rice cropping system could be used to manage *P. minor*. They found that the use of sunflower (*Helianthus annuus* L.) and dhaincha (*Sesbania aculeate* L.) as green manures could decrease the seed germination of *P. minor*. In wildland settings, a carbon source has been added to soil to reduce growth and seed set of fast-growing weeds (see chapter "USA: Applying Ecological Concepts to the Management of Widespread Grass Invasions" by D'Antonio et al., this volume). Needed now are field studies to determine the impact of rice straw on the emergence and establishment of *P. minor* seedlings.

14.3 Management

Several herbicides are employed to combat *P. minor* in wheat fields (Table 14.1). Isoproturon has been used widely to control *P. minor* in wheat fields for the last 35 years by farmers of Haryana and Punjab (Singh et al. 1999). Since the first observation by Malik and Singh (1995), reports of *P. minor* biotypes with isoproturon resistance have been increasing across northwestern India, likely as the products of continuous and excessive use of isoproturon (Singh et al. 1999; Kaushik et al. 2005). Isoproturon, when applied as a preemergence herbicide at 0.5–1.5 kg ha^{-1}, inhibited seedling emergence of Delhi biotypes but not a Haryana biotype of *P. minor* (Sharma and Pandey 1997). Doubling the recommended dose of isoproturon to 1.9 kg ha^{-1} did not control the resistant *P. minor* biotypes from Punjab (Walia et al. 1997). Continued use of isoproturon kills susceptible *P. minor* individuals year after year. Increased herbicide selection pressure causes resistant biotypes to outcompete susceptible biotypes. This practice results in a soil seed bank of resistant biotypes and slow elimination of susceptible individuals from the population.

New herbicides with different modes of action have been introduced periodically to manage resistant biotypes of *P. minor* (Table 14.1). Tralkoxydim (acetyl-*co*-A carboxylase (ACCase) inhibitor, 0.25 and 0.35 kg/ha) and diclofop-methyl (ACCase inhibitor, 0.75 and 1.0 kg/ha) were preferred over chlortoluron for their efficiency in reducing the population and biomass allocation of *P. minor* and increasing wheat production (Walia and Brar 1996). Similarly, the application of sulfosulfuron (acetolactate synthase (ALS) inhibitor, 25 g ai/ha), fenoxaprop-*p*-ethyl (ACCase inhibitor, 100–120 g ai/ha), or clodinafop (ACCase inhibitor, 35 g ai/ha) successfully controlled isoproturon-resistant biotypes of *P. minor* and enhanced wheat yield by 200% compared to that of weedy plots (Chhokar and Malik 2002). Isoxaflutole (5-cyclopropyl-4-isoxazolyl)[2-(methylsulfonyl)-4-(trifluoro methyl)phenyl]methanone) is a preemergent, systemic, soil applied and nontoxic herbicide (Mitra et al. 2001). It undergoes rapid conversion to the toxic by-product diketonitrile, which inhibits the enzyme 4-hydroxyphenylpyruvate dioxygenase of carotenoid biosynthesis. Isoxaflutole is an effective herbicide against some broadleaf and grass weeds (e.g., redroot pigweed, velvetleaf, and barnyard grass) (Bhowmik et al. 1999). Kaur et al.

Table 14.1 Herbicides recommended for controlling *Phalaris minor* in India

Herbicide	Target site	Chemical class	Time of application	Application rate (Kg/ha)	References
Sulfosulfuron (Leader)	ALS	Sulfonylurea	Pre- and postemergence	0.025–0.03	Chhokar et al. (2006)
Tribenuron (Express)	ALS	Sulfonylurea	Postemergence	0.005–0.03	Bhattacherjee and Dureja (1999)
Isoproturon (Techical)	PS II	Phenylurea	Pre- and postemergence	1	Gill et al. (1979), Walia and Gill (1985)
Chlortoluron (Dicuran)	PS II	Phenylurea	Pre- and early postemergence	0.5–1	Gill and Brar (1977), Walia and Brar (1996)
Metribuzin (Sencor)	PS II	Triazine	Pre- and postemergence	0.21	Chhokar et al. (2006)
Terbutryn (Prebane)	PS II	Triazine	Pre- and postemergence	1	Joshi and Singh (1981), Kataria and Kumar (1981)
Metaxuron (Dosanex)	PS II	Subsituted urea	Pre- and postemergence	1.2–1.75	Joshi and Singh (1981)
Methabenzthiazuron	PS II	Subsituted urea	Postemergence	0.7–1.3	Joshi and Singh (1981), Rathi and Tiwari (1981)
Clodinafop (Topik)	ACCase	Aryloxyphenoxy propionate	Postemergence	0.06	Chahal et al. (2003)
Diclofop-methyl (Hoegrass)	ACCase	Aryloxyphenoxy propionate	Postemergence	0.88	Singh and Kundra (2003), Walia and Brar (1996)
Fenoxaprop-P (Cheetah super)	ACCase	Aryloxyphenoxy propionate	Postemergence	0.1	Singh and Kundra (2003)
Tralkoxydim (Grasp)	ACCase	Cyclohexane dinone	Postemergence	0.35	Singh and Kundra (2003); Walia and Brar (1996)
Pendimethalin (Stomp)	Microtubules	Dinitroaniline	Preemergence	1.5	Yadav et al. (1984), Singh and Kundra (2003)
Nitrofen (TokWP-50[R])	PPO	Diphenylether	Pre or PPI emergence	1.2–2.5	Gill et al. (1979)
Flufenacet (Define)	FAE	Oxyacetamide	Pre or early post emergence	0.24	Chhokar et al. (2006)

ALS acetolactate synthase, *PS* photosystem, *ACCase* acetyl-CoA carboxylase, *PPO* protoporphyrinogen oxidase, *FAE* fattyacid elongase

(2004) carried out greenhouse experiments to examine the phytotoxicity of isoxaflutole against *P. minor*. They found that 0.5 mg/L of isoxaflutole reduced shoot height of *P. minor* but had no adverse effect on seedling growth of wheat. Although isoxaflutole did not influence soil properties, the observed response may be restricted to the soil type and dose of herbicide used in the study. Further field trials are needed to verify the findings of the greenhouse study.

Herbicide mixtures are preferred due to their ability to delay the evolution of resistant biotypes of weeds, cost effectiveness, and their ability to control a broad spectrum of weeds (Kudsk and Mathiassen 2004). Mixtures of two or more herbicides with different molecular targets can produce synergistic effects (Kudsk and Mathiassen 2004). Herbicide mixtures could be useful in optimizing the herbicide dose so that negative environmental effects can be minimized (Kudsk 2008). Herbicide mixtures may also provide a better strategy to control *P. minor* or other weeds in wheat-rice cropping systems. Singh et al. (1993) found that tank mixtures of tralkoxydim (0.2 kg/ha) and isoproturon (0.5 kg/ha) were more effective in controlling *P. minor* compared with the effect of these herbicides applied singly; the joint action of herbicides was not, however, examined. Kaushik et al. (2006) found that mixtures of pretilachlor (very long chain fatty acid inhibitor) and sulfonylureas (ALS inhibitor) showed synergistic interactions, whereas mixtures of pendimethalin (microtubule assembly inhibitor) and sulfonylureas showed either antagonistic or additive activities on rice. These authors stress that the specificity of herbicides could be enhanced by using herbicide mixtures, which could lower the total dose of herbicides required for weed control. The use of herbicide mixtures with different molecular targets could possibly delay the development of resistance to a particular herbicide. Needed are studies on the joint action of herbicides, mixed according to their relative potency to effectively design better management strategies for *P. minor*.

14.4 Conclusion

Some parts of the States of Haryana and Punjab are badly infested with *P. minor*. A survey conducted in Haryana by Franke et al. (2003) revealed that agroecological conditions and socioeconomical factors were responsible for the current status of *P. minor* infestation. Farmers with >4-ha land under cultivation could easily combat the increasing resistance of *P. minor* to isoproturon by switching to alternate herbicides, such as sulfosulfuron (leader), fenoxaprop-*p*-ethyl (Cheetah super), and clodinafop-propagyl (Topik). Farmers with <2-ha land under cultivation could remove *P. minor* manually. However, the farmers with intermediate sized fields (2–4 ha) were most affected by the selection for resistance biotypes to isoproturon because alternate herbicides could be too expensive. The use of low quality or inadequate amounts, or both, of isoproturon continues in some areas; these practices likely accelerate the selection of resistant *P. minor* biotypes. India is undergoing economic reforms. To ensure that economic reforms go together with agricultural advancement, benefits must reach less privileged farmers. Weed infestations and their sustainable management could be better handled by joint efforts of agricultural scientists, social scientists, and economists, working in close cooperation with farmers.

Acknowledgment We thank K. Neil Harker, Carla D'Antonio, and Richard Mack for their comments on the earlier version of the manuscript. We appreciate Ms. Rajwant Kaur for help to improve the quality of the manuscript.

References

Bhattacherjee AK, Dureja P (1999) Light-induced transformations of tribenuron-methyl in aqueous solution. Pestic Sci 55:183–188

Bhowmik PC, Kushwaha S, Mitra S (1999) Response of various weed species and corn (*Zea mays*) to RPA 201772. Weed Technol 13:504–509

Chahal PS, Brar HS, Walia US (2003) Management of Phalaris minor in wheat through integrated approach. Indian J Weed Sci 35:1–5

Chhokar RS, Malik RK (2002) Isoproturon-resistant littleseed canarygrass (*Phalaris minor*) and its response to alternate herbicides. Weed Technol 16:116–123

Chhokar RS, Sharma RK, Chauhan DS et al (2006) Evaluation of herbicides against *Phalaris minor* in wheat in north-western plains. Weed Res 46:40–49

Dhaliwal BK, Walia US, Brar LS (1997) Response of wheat to *Phalaris minor* Retz. population density. Proc Brighton Crop Prot Conf Weeds 1021–1024

Franke AC, McRoberts N, Marshell G et al (2003) A survey of *Phalaris minor* in the Indian rice–wheat system. Exp Agric 39:253–265

Gill HS, Brar LS (1977) Chemical control of *Phalaris minor* and *Avena ludoviciana* in wheat. Pest Art News Summ 23:293–296

Gill HS, Walia US, Brar LS (1979) Chemical weed control in wheat (*Triticum aestivum*) with particular reference to *Phalaris minor* Retz. and wild oats (*Avena ludoviciana* Dur.). Pesticides 13:15–20

Hooker JD (1896) Flora of British India, Vol VII. Reeve L and Co, London

Inderjit, Drake JA (2006) The ecology of nonnative invasive plant species: are there consistent patterns? CAB Rev: Perspect Agric Veter Sci Nutr Natl Resour 1:36

Inderjit, Rawat D, Foy CL (2004) Multifaceted approach to determine rice straw phytotoxicity. Can J Bot 82:168–176

Inderjit, Weiner J (2001) Plant allelochemical interference or soil chemical ecology? Perspect Plant Ecol Evol Syst 4:3–12

Joshi NL, Singh HG (1981) Chemical control of grassy weeds in wheat. Indian J Agron 26:302–306

Kataria OP, Kumar V (1981) Response of dwarf wheat (*Triticum aestivum*) and four weed species to herbicides. Weed Sci 29:521–524

Kaur H, Inderjit, Bhowmik PC (2004) Phytotoxicity of isoxaflutole to *Phalaris minor* Retz. Plant Soil 256:161–168

Kaushik S, Blackshaw RE, Inderjit (2005) Ecology and management of an exotic weed *Phalaris minor*. In: Inderjit (ed) Invasive plants: ecological and agricultural aspects. Birkhauser-Verlag AG, Basel, pp 181–194

Kaushik S, Inderjit (2007) Oryza sativa restricts *Phalaris minor* growth: allelochemicals or soil resource manipulation? Biol Fert Soils 43:557–563

Kaushik S, Inderjit, Streibig JC et al (2006) Activities of mixtures of soil-applied herbicides with different molecular targets. Pest Manage Sci 62:1092–1097

Khera KL, Sandhu BS, Aujla TS (1995) Performance of wheat (*Triticum aestivum*) in relation to small canarygrass (*Phalaris minor*) under different levels of irrigation, nitrogen and weed population. Indian J Agric Sci 65:717–722
Kudsk P (2008) Optimising herbicide dose: a straightforward approach to reduce the risk of side effects of herbicides. Environmentalist 28:49–55
Kudsk P, Mathiassen SK (2004) Joint action of amino acid biosynthesis inhibitors. Weed Res 44:313–322
Mack RN, Simberloff D, Lonsdale WM et al (2000) Biotic invasion: causes, epidemiology, global consequences and control. Ecol Appl 10:689–710
Malik RK, Singh S (1995) Littleseed canarygrass (*Phalaris minor*) resistance to isoproturon in India. Weed Technol 9:419–425
Mitra S, Bhowmik PC, Xing B (2001) Physical and chemical properties of soil influence the sorption of diketonitrile metabolite of RPA 201772. Weed Sci 49:423–430
Olofsdotter M (1998) Allelopathy in Rice. International Rice Research Institute (IRRI), Manila
Olofsdotter MD, Navarez D, Rebulanan M et al (1999) Weed-suppressing rice cultivars: does allelopathy play a role? Weed Res 39:441–454
Om H, Dhiman SD, Kumar S et al (2002) Allelopathic response of *Phalaris minor* to crop and weed plants in rice–wheat system. Crop Prot 21:699–705
Rathi KS, Tiwari AN (1981) Chemical weed control in wheat with special reference to *Phalaris minor* Retz. Indian J Weed Sci 13:125–128
Schmidt IK, Mischelsen A, Jonasson S (1997) Effects of labile soil carbon on nutrient partitioning between an arctic graminoid and microbes. Oecologia 112:557–565
Sharma R, Pandey J (1997) Studies on isoproturon resistance in different biotypes of *P. minor* Retz. Ann Agric Res 18:344–347
Singh K, Kundra HC (2003) Bio-efficacy of herbicide against isoproturon resistant biotypes of *Phalaris minor* in wheat. Indian J Weed Sci 35:15–17
Singh S, Kirkwood RC, Marshall G (1999) Biology and control of *Phalaris minor* Retz. (littleseed canarygrass) in wheat. Crop Prot 18:1–16
Singh S, Malik RK, Bishnoi LK et al (1993) Effect of tank mixture of isoproturon and tralkoxydim on the control of *Phalaris minor* in wheat. Indian J Weed Sci 25:11–13
Stewart RR (1945) The grasses of northwest India. Brittonia 5:404–468
Valvarde BE, Itoh K (2001) World rice and herbicide resistance. In: Powles SB, Shaner DL (eds) CRC Press, Boca Raton, FL, pp 195–249
Walia US, Brar LS (1996) Performance of new herbicides for controlling wild canary grass (*Phalaris minor*) in wheat. Indian J Weed Sci 28:70–73
Walia US, Brar LS, Dhaliwal BK (1997) Resistance to isoproturon in *Phalaris minor* Retz. in Punjab. Plant Prot Q 12:138–140
Walia US, Gill HS (1985) Interactions between herbicides and nitrogen in the control of *Phalaris minor* in wheat. Trop Pest Manage 31:226–231
Yadav SK, Bhan VM, Singh SP (1984) Post-emergence herbicides for control of *Phalaris minor* in wheat. Trop Pest Manage 30:467–469

ALS acetolactate synthase, *PS* photosystem, *ACCase* acetyl-CoA carboxylase, *PPO* protoporphyrinogen oxidase, *FAE* fattyacid elongase

Chapter 15
Ecology and Management of the Invasive Marine Macroalga *Caulerpa taxifolia*

Linda Walters

Abstract In coastal waters of Australia, the USA, and Europe, aquarium strains of the green macroalga *Caulerpa taxifolia* have invaded and caused ecological and economic disasters. As a result, this alga was placed on the International Union for the Conservation of Nature's list of 100 worst invasive species. Two things have promoted the invasions. First, *C. taxifolia* asexually reproduces by vegetative fragmentation. Fragments as small as 4 mm can survive and attach within 2 days. Second, this species has been and continues to be very popular with the aquarium industry, prized by both home hobbyists and public aquaria. Although regulations are now in place in many countries, retail shops and e-commerce continue to sell many species of feather *Caulerpa*, including *C. taxifolia*. "Aquarium dumping" is thought to be the reason for most, if not all, of the major invasions. Field eradication efforts have included manual and vacuum pump harvesting, covering colonies with opaque tarpaulins, subjecting *C. taxifolia* to a range of noxious chemicals, temperature, and salinity shocks, while outreach, monitoring, and modeling are promoted as ways to prevent future incursions. To date, only the USA and the West Lakes area of South Australia have eradicated *C. taxifolia*. Further research and outreach are needed to prevent future invasions of this noxious alga.

Keywords Marine Macroalga • *Caulerpa taxifolia* • Salinity tolerances • Competition • *Posidonia oceanica* • Aquarium Industry • Chlorophyta • vegetative fragmentation • secondary chemicals

15.1 Introduction

Caulerpa taxifolia (Vahl) C. Agardh (Caulerpales, Chlorophyta) is a brilliant green marine macroalga with multiple upright, feathery blades and a basal rhizome (stolon)

L. Walters
Department of Biology, University of Central Florida, 4000 Central Florida Blvd., Orlando, FL 32816, USA
ljwalter@pegasus.cc.ucf.edu

that runs across the sandy or muddy bottom and is anchored to the substrate by bundles of colorless, filamentous, root-like rhizoids (Figs. 15.1 and 15.2). Growth is indeterminate (Collado-Vides and Robledo 1999), and *C. taxifolia* is native in subtidal waters in tropical and subtropical areas, including the Caribbean, Indonesia, Southeast Asia, Australia, and Hawaii (Phillips and Price 2002; Guiry and Dhonncha 2004). *Caulerpa taxifolia* can be found as isolated individuals on reef flats, sandy areas, or the undersides of floating docks (e.g., Coconut Island, Hawaii) or in a

Fig. 15.1 Two morphologies of invasive *Caulerpa taxifolia* from New South Wales, Australia. The more spiral morphology (individual on left) occurs in areas with higher levels of water motion

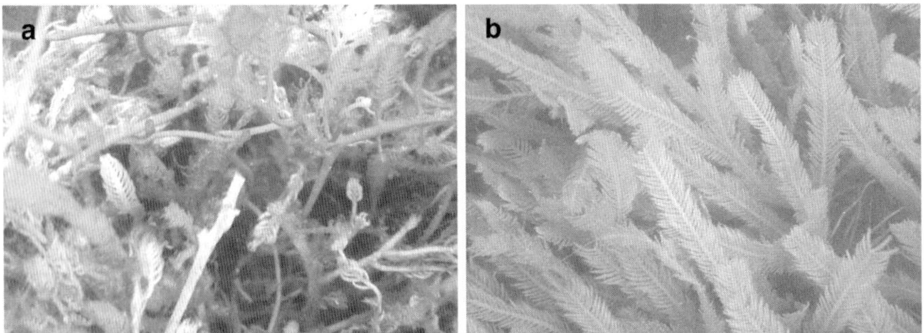

Fig. 15.2 *Caulerpa taxifolia* meadows. (**a**) A meadow of invasive *C. taxifolia* in early spring in Lake Conjola, NSW, Australia in which you can see new growth emerging from where the colonies had died back in the winter. (**b**) A lush meadow of native *C. taxifolia* from Moreton Bay, QLD, Australia

distinct, narrow band in soft sediments immediately seaward the low tide line (e.g., Moreton Bay, Australia) (personal observation). *Caulerpa taxifolia* is known by the aquarium industry as "feather *Caulerpa*" or "feather algae" and is lumped with a number of other species of *Caulerpa* with feathery blades (e.g., *C. sertularioides*, *C. ashmeadii*, *C. mexicana*), and this group has dominated the flora in both personal and public saltwater aquaria for many decades (Walters et al. 2006).

15.2 Invasion History of *Caulerpa taxifolia*

The aquarium strain of *C. taxifolia* has the distinction of being one of the world's 100 worst invasive species listed by the International Union for the Conservation of Nature (ICUN) and Europe's second worst macroalgal invasion on record (www.issg.org/database). The invasion history of the aquarium strain of *Caulerpa taxifolia* (*aka the killer alga) started with an accidental introduction into the Mediterranean Sea while cleaning tanks at the Monaco Oceanographic Museum in 1984 (Fig.15.3) (Meinesz and Hesse 1991, Meinesz 1999). Reports documented expansion from a small patch adjacent to the Museum to many areas in the Mediterranean at a rate of approximately 50 km per year, with boating activity, fishing nets, and water currents largely responsible for the spread (Meinesz et al. 1993, 2001; Sant et al. 1996). Dense populations of *C. taxifolia* are concentrated in zones with extensive development (Madl and Yip 2005). Monocultures of the aquarium strain of *C. taxifolia* now can be found at over 100 locations in Mediterranean waters extending for hundreds of kilometers (Madl and Yip 2005 for chronology). In the Mediterranean, *C. taxifolia* is found on steep slopes as well as flat bottom areas. Dense meadows are found at depths ranging from a few meters to over 40 m, while sparse meadows extend to 55 m and isolated individuals have been observed at 100 m (Meinesz and Hesse 1991; Belsher and Meinesz 1995). Through DNA forensics, the global origin of Mediterranean *C. taxifolia* was found to be Moreton Bay in Queensland, Australia (Wiedenmann et al. 2001). *Caulerpa taxifolia* was imported in the early 1970s by the Wilhelmina Zoologischbotanischer Garden in Stuttgart, Germany, which displayed the alga in its tropical aquarium (Jousson et al. 1998). Between 1980 and 1983, clones originating from Moreton Bay were given to the tropical aquarium of Nancy in northern France and subsequently to the Monaco Oceanographic Museum (Jousson et al. 1998). Clones were cultivated in various aquaria for 14 years prior to being found in Monaco waters.

In Japan, Komatsu et al. (2003) surveyed 65 public aquaria for *Caulerpa*. Sixteen of the 51 aquaria that responded to the survey cultured or exhibited *C. taxifolia*. Six purchased the alga from aquarium shops, one obtained *C. taxifolia* from another public aquarium (originally bought at a retail shop), and the rest were uncertain about origin but stated that the alga was from somewhere in Japan. The Notojima Aquarium staff reported temporary establishment of *C. taxifolia* in the Sea of Japan due to culturing practices. The aquarium initially received *C. taxifolia* when purchasing "foreign shrimp" of unknown taxonomy from an aquarium shop in Osaka. The aquarium staff grew the fragment in a tank. The fragment did well and was transferred to an open pool of 1,000 tons of seawater with an open-circuit

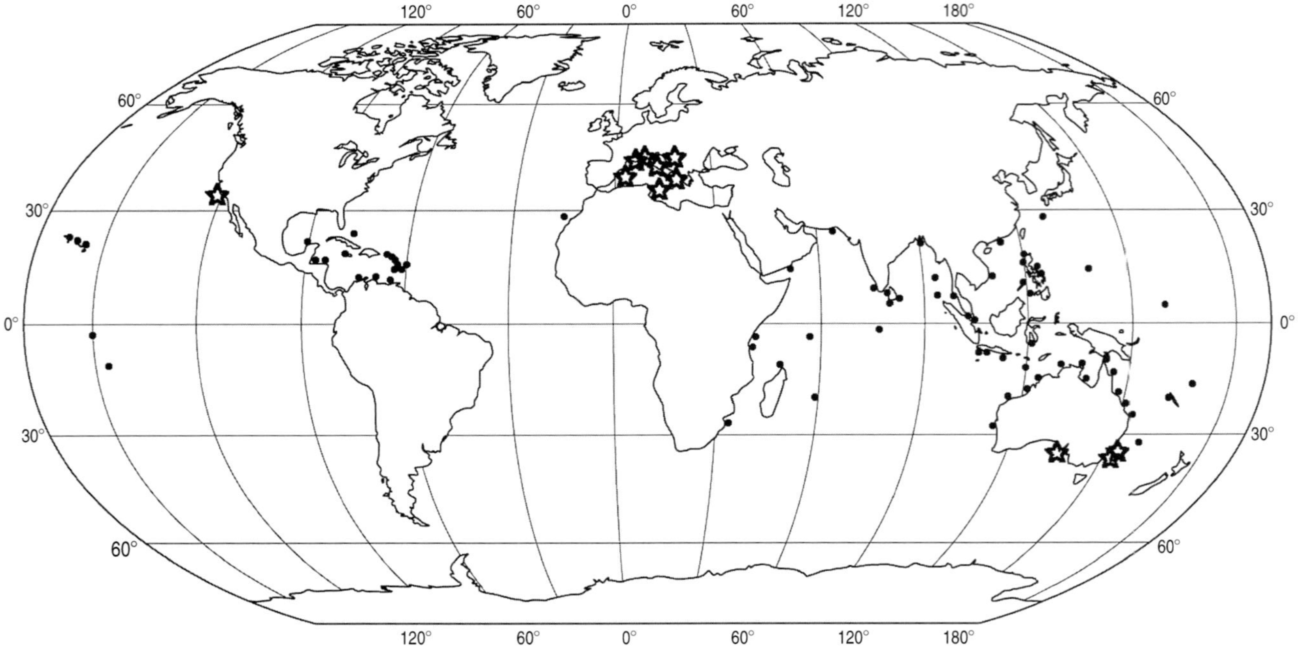

Fig. 15.3 Global distribution of *Caulerpa taxifolia*. Circles indicate native ranges and stars indicate known invasions (modified from Zaleski and Murray 2006)

water flow system. In August 1992, the aquarium staff observed two colonies of *C. taxifolia* (0.2 and 1.0 m diameter) in a coarse sand bed about 5 m from the pool's discharge pipe in the Sea of Japan (Komatsu et al. 2003). Both colonies disappeared in the winter of 1993. Two new colonies were observed in 1994; both disappeared in the winter (Komatsu et al. 2003). After that, the Notojima Aquarium began keeping their *C. taxifolia* cultures in a closed system. This Japanese aquarium strain genetically was identical to the Mediterranean aquarium strain (Komatsu et al. 2003). *C. taxifolia* did not become established in Japanese waters most likely because the water temperatures in the winter dropped below *C. taxifolia's* lower lethal limit (Komatsu et al. 2003). No *C. taxifolia* has been documented in Japanese waters since that time.

Populations of *C. taxifolia* were next discovered in April 2000 in Fisherman's Bay, Port Hacking along the southern outskirts of Sydney, Australia, more than 800 km south of its closest native distribution in the state of Queensland, where observational records date back to 1860 (Schaffelke et al. 2002). Also in April 2000, *C. taxifolia* was documented 200 km south of Sydney in Lake Conjola (Fig. 15.2). Researchers predict that the Port Hacking infestation started 2 years prior and the Lake Conjola infestation began 5–13 years earlier (Creese et al. 2004). In Australia, the number of invaded locations continues to increase with ten documented coastal lakes/estuary infestations in New South Wales (NSW), both north and south of Sydney, and two waterways in South Australia (Schaffelke et al. 2002; Millar 2004). Presently, Lake Conjola has the distinction of being the most severely infested location in NSW, and possibly Australia (West and West 2007). Here, *Caulerpa taxifolia* has replaced seagrass as the dominant macrophtye, covering approximately 30% of the lake floor (Creese et al. 2004). All invaded locations in Australia are relatively sheltered areas, are less than 10-m deep, and are soft sediment habitats that were either previously uncolonized or occupied by seagrasses (Creese et al. 2004; Davis et al. 2005).

Schaffelke et al. (2002) and Murphy and Schaffelke (2003) used molecular markers and determined that invasive *C. taxifolia* in Australia was not identical to the aquarium strain. Nor were NSW populations the result of a natural, southward range expansion along the eastern coastline (Phillips and Price 2002; Millar 2002). Most likely, the new infestations were the result of multiple domestic translocation(s), aided by boating, fishing, and the domestic aquarium trade (Schaffelke et al. 2002, 2006). Some areas, such as Port Hacking, have aquarium shops near shorelines, which stocked and sold *C. taxifolia* at the time the infestations were thought to have occurred (Creese et al. 2004). Other infested areas were much more remote, suggesting that boating and fishing activities were important for the transport of fragments (Relini et al. 2000). Indeed, many remote NSW infestations were popular fishing spots (Creese et al. 2004). Additionally, some areas where *C. taxifolia* currently is found were important commercial grounds prior to the lakes being closed to commercial fish netting in May 2002.

The year 2000 was a critical year for recognition of global infestations of *C. taxifolia*. In addition to the Australian invasions, populations of the aquarium strain of *C. taxifolia* were found in two lagoons in southern California: Agua

Hediona Lagoon and Huntington Harbor (Jousson et al. 2000; Anderson 2005). The source of both infestations was hypothesized to be personal aquaria, and the DNA forensics found that Californian *C. taxifolia* virtually was identical to the Mediterranean aquarium strain (Jousson et al. 2000). Already at high density, the Huntington Harbor invasion is thought to have started at least 2 years earlier, while the timing of the invasion in Agua Hediona Lagoon remains unknown (Williams and Grosholz 2002). Eradication efforts began immediately in both California locations and, to date, the USA is the only country with complete *C. taxifolia* eradication. A large celebration was held on12 July 2006 for this victory in spite of the price tag of over 7 million dollars and the necessary, but seemingly temporary, associated environmental harm (Anderson 2005; R. Woodfield, personal communication).

15.3 Why Is *Caulerpa* Such a Potent Invasive Species?

15.3.1 Reproductive Capacity

"Invasive success" refers to traits of a species that promote establishment, spread, and proliferation in the new range (Inderjit 2005). One hallmark trait of an invasive plant or macrophyte is its ability to rapidly spread to both close and distant locations. Short-range expansion occurs regularly via the basal rhizome in *C. taxifolia*. Long-range dispersal can occur via sexual or asexual reproduction. Sexual reproduction occurs in *C. taxifolia* with separate male and female individuals releasing gametes that are fertilized externally (Zuljevic and Antolic 2000). However, only male gametes, probably from a single clone that initially entered Monaco waters, have been documented in the Mediterranean (Zuljevic and Antolic 2000). No sexual reproduction has been described for any other invasive population either because only single sex clones entered the waterways or from the lack of site-specific research.

Many diverse genera of marine macroalgae excel at asexual reproduction via vegetative fragmentation (e.g., Fig. 15.4, Walters and Smith 1994; Herren et al. 2006). For this type of reproduction to be important, fragments must be generated regularly, have the ability to disperse widely, land safely, and then rapidly attach and grow under the new suite of biotic and abiotic conditions (Smith and Walters 1999; Walters 2003). With the genus *Caulerpa*, the capability is especially impressive as all members of the genus are siphonous, with no internal cell walls to reduce loss of cytoplasm when damage occurs that leads to fragmentation. Long before the genus *Caulerpa* became well known because of invasive characteristics, research was focused on understanding the underlying chemical properties that healed wounds in less than 1 min and kept fragments and the parent thallus from losing all the nucleus-rich cytoplasm (Dreher et al. 1978; Goddard and Dawes 1983).

Laboratory bioassays have confirmed that very small fragments of *C. taxifolia* can survive wounding and continue growing by producing new attachment rhizoids

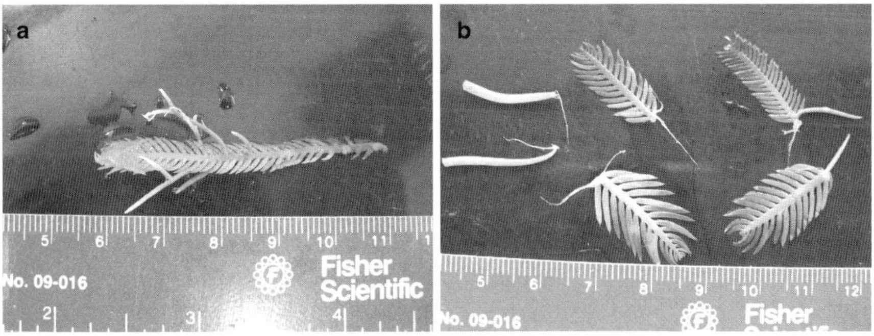

Fig. 15.4 Fragments of *Caulerpa taxifolia*. (**a**) New growth from a field-collected fragment. (**b**) New growth from laboratory trials

and new rhizomes from which new blades emerge. Much more robust than *C. verticillata* or *C. prolifera*, Smith and Walters (1999) found that native Hawaiian *C. taxifolia* fragments as small as 1 cm survived and produced new attachment rhizoids and rhizomes. To compare this to the success of native and invasive *C. taxifolia* from Australia, similar laboratory bioassays were run with invasive Lake Conjola and native Moreton Bay, Australia populations (L. Walters, P. Sacks, A. Davis and D. Burfiend, unpublished data). For both Australian populations, we found that blade fragments as small as 4 mm were successful and new growth was visible within 2–3 days. At 20-mm length, 100% of the blade fragments from Hawaii and both Australian populations survived and grew (Smith and Walters 1999; L. Walters, P. Sacks, A. Davis and D. Burfiend, unpublished data). Stolon-only fragments from Australia were only successful if they were at least 15 (Lake Conjola) or 20 mm in length (Moreton Bay) (L. Walters, P. Sacks, A. Davis and D. Burfiend, unpublished data). A small percentage of Hawaiian *C. taxifolia* stolons (10, 15, and 20-mm length) also survived (20%), but none attached (Smith and Walters 1999). For both Hawaiian and Australian *C. taxifolia*, new growth occurred at the wound sites, undamaged growing edges, and along branchlets. Unsuccessful fragments were obvious because all cytoplasm oozed out within 1 min and only the colorless cell walls remained. Finally, Smith and Walters (1999) determined the forces required to create fragments of *C. taxifolia* using a puncturometer (Pennings and Paul 1992). Forces needed were significantly less for fronds than that for stolons (Smith and Walters 1999).

15.3.2 Survival of Fragments in Field

Both storms and herbivory naturally create fragments, and these fragments add to fragmentation associated with recreational and commercial use of the waterways. Although normally negatively buoyant, fragments may float if covered with mucus or filamentous epiphytes (Chisholm et al. 2000). Lake Conjola (NSW, Australia) is very popular with recreational boaters, water skiers, jet skiers, and fishermen, especially during the summer holiday season. On ten dates between 17 December and 25 January 2007, 13.9 ± 2.3 boats per hour (mean ± S.E.) in Lake Conjola passed adjacent to *Caulerpa* meadows (L. Walters, P. Sacks, A. Davis, D. Burfiend, unpublished data). On the same dates and at the same locations, a mean of 34.0 ± 19.8 fragments ($n = 10$) were found in a 100 × 2 m band transect along the shoreline where *C. taxifolia* meadows were less than 2 m offshore. Boating traffic in Moreton Bay waterways in February 2007 was lower and consisted primarily of large sailboats, small fishing boats, and ferries (5.9 ± 1.0 boats/h), and we found about half as many fragments (19.7 ± 4.4) in similar transects in Moreton Bay waters ($n = 14$) (L. Walters, P. Sacks, A. Davis, D. Burfiend, unpublished data). Using Kruskal–Wallis nonparametric ANOVAs, there was significantly more boating activity in Lake Conjola ($F = 19.17$; $p = 0.001$) but no significant difference in the number of fragments ($F = 0.19$; $p = 0.669$). Ninety-eight percent and 99.9% of the fragments from Lake Conjola and Moreton Bay, respectively, were greater than 10 mm in length, allowing us to predict that the fragments will be successful. On the contrary, Wright (2005) found more asexual reproduction via fragmentation in Moreton Bay populations than at the invasive locations.

Fragments can be spread by anthropogenic activities related to boating and fishing (Sant et al. 1996; West 2003). In the Mediterranean, more fragments of *C. taxifolia* were found in locations where boats moored (Relini et al. 2000) and most new outbreaks appeared in areas of high boating activity (Meinesz et al. 1993). West et al. (2007) determined that fragments of Australian *C. taxifolia* created by all anchor types tested were of similar sizes. Eighty-two percent of anchors lowered into *C. taxifolia* beds fragmented individuals, and the biomass removed on a single anchor was as large as 49 g dry weight (West et al. 2007). Chains and ropes also generated fragments when lowered onto a meadow of *C. taxifolia* (96% chains, 4% ropes). Once removed from the water, fragment survivorship increased with clump size, protection from desiccation, and decreased exposure time (West et al. 2007). Likewise, Sant et al. (1996) found that fragments of aquarium *C. taxifolia* survived out of water in dark, humid conditions. Combined, the results suggest that multiple anchorings within a limited area can greatly promote dispersal of *C. taxifolia* while much longer dispersal (greater than hundreds of meters) is also possible when vessels travel long distances with fragments harbored inside anchor lockers, attached to ropes, or entangled on boat trailers (West et al. 2007). West et al. (2007) suggest that boaters may be more likely to discard larger, obvious clumps of *C. taxifolia* from anchors, rope, and chains while smaller fragments may be overlooked.

Fragment retention and attachment in *C. taxifolia* depends on season, hydrodynamics, and water depth. In Italy, aquarium-strain fragments (15-cm stolon with five fronds) survived best in summer months (Ceccherelli and Cinelli 1999a). Thibaut (2001) found that fragment survival on the French coast in the summer in 10-m water on sandy bottoms was as great as 98%, whereas only 50% establishment was found in shallower waters with more water motion. Many studies also have documented a positive correlation between fragment retention and structural complexity (e.g., A.R. Davis unpublished data). Wright and Davis (2006) determined experimentally that stolon growth and fragment success were linked in Australian *C. taxifolia*, and fragment recruitment was enhanced when stolons were present. Ceccherelli et al. (2002) found that dense, low-growing (turf) species promoted spread of aquarium *C. taxifolia*, while habitats with encrusting and erect macrophytes were less likely to be invaded.

15.3.3 Temperature and Salinity Tolerances of C. Taxifolia

Tolerance of a wide range of temperatures is another attribute of a successful invasive species. *Caulerpa* is endemic in tropical and subtropical regions around the world (Fig. 15.3), and latitude is a significant predictor of native occurrences (Creese et al. 2004; Glardon et al. 2008). Silva (2003) stated that members of the genus *Caulerpa* also can grow in locations up to 34°N along the southeastern US coastline, where the Gulf Stream typically warms the water. From known historical temperatures, Keppner (2002) suggested that the potential thermal distribution for *C. taxifolia* in the USA is (1) just south of Virginia Beach, VA on Atlantic coast, (2) Stonewall Bank, OR on the Pacific Coast, (3) throughout the Gulf of Mexico, (4) Hawaii, (5) Puerto Rico, (6) the US Virgin islands, (7) American Samoa, (8) Guam, and (9) the Northern Mariana Islands. In Europe, Ivesa et al. (2006) reported that invasive *C. taxifolia* was first recorded in Malinska (Island of Krk), Croatia in 1994, the highest northern latitude (45°7′30′) documented for *C. taxifolia*. However, only a few thalli were present in 2004 surveys, and low temperatures (9.5–10.5°C) in the previous winter were the suspected cause of the decline.

Many lab and field studies have determined experimentally the thermal tolerance range for *C. taxifolia*. Komatsu et al. (1997) determined that the upper temperature limit was 31.5–32.5°C, allowing *C. taxifolia* to thrive in tropical waters around the globe. The lower lethal temperature is much more important for understanding the potential range of *C. taxifolia*, and Komatsu et al. (1997) found the lower threshold to be between 9 and 10°C for fragments of the Mediterranean *C. taxifolia*. Additionally, Komatsu et al. (1997) found that minimum temperatures of 15 and 17.5°C were required for new growth of blades and stolons, respectively, for aquarium *C. taxifolia*. At the Monaco Oceanographic Museum, winter daily seawater temperatures from 1978 to 1991 were only below 11°C for 3 days (Meinesz and Hesse 1991). Minimum water temperatures in Moreton Bay, Australia are similar to minimum temperatures in the Mediterranean Sea (12°C) (Wright and Davis 2006). Chisholm et al. (2000) reported that *C. taxifolia* from Moreton Bay had a lower lethal temperature between 9 and 11°C. The lower limit for Lake Conjola was 11°C while it was 14°C for Port Hacking (Phillips and Price 2002; Wright and Davis 2006). Fragments did not grow at temperatures less than 20°C and had 100% mortality at 15°C or less in NSW trials (West 2003). When introduced into the Sea of Japan, *C. taxifolia* failed to establish where the mean surface water temperature averaged 9°C for 2 months (Komatsu et al. 2003). In Croatia, the biomass of *C. taxifolia* was reduced after cold winters (9.5–10.5°C). In addition to suggesting a sharp lower limit for survival, the results suggest that *C. taxifolia* naturally has a wide temperature tolerance. These results, in turn, refute early suggestions that the Mediterranean clone had become increasingly cold-adapted while propagated and dispersed through the aquarium industry.

In the Mediterranean and NSW, *C. taxifolia* is considered pseudoperennial because individuals dieback in winter months (Fig. 15.2). In both locations, researchers have documented that *C. taxifolia* reached maximal size at the end of the summer months and then decreased in dimensions, especially blade height, during cold weather (Meinesz et al. 1995; Ceccherelli and Cinelli 1999b; Glasby et al. 2005a). Temporary diebacks in shallow areas with only knots of stolons and a few bleached blades were often observed during winter when temperatures reached as low as 11°C (Wright and Davis 2006). In Italian waters, percent cover went from 2.7% in April to 100% in October (Balata et al. 2004). Williams and Schroeder (2004) found that chloroplasts were translocated to buried portions of tissue when fragments were either heat or cold-shocked.

West and West (2007) simultaneously compared the impact of six salinities (range: 15–30 ppt) and four temperatures (15–30°C) on *C. taxifolia* from Lake Conjola, NSW. Blades, rhizomes, and thalli (= rhizome plus one blade) had similar responses in the lab trial. Some fragments in all morphological categories doubled in size over the week-long trial by producing new stolons and fronds; the maximum growth rate was 174 mm/week (West and West 2007). Fragments grew well at salinities ≥ 22.5 ppt and temperatures ≥ 20°C, while mortality approached 100% at lower salinities and temperatures. The result was especially interesting in light of the changes to some NSW waterways. In 2001, the entrance to Lake Conjola was manipulated to keep the lake permanently open to the sea (West and West 2007). The current salinity in Lake Conjola is always above 30 ppt. Prior to 2001, Lake Conjola often was less than 17 ppt for extended periods of time. Entrance manipulation may have improved the success of invasive *C. taxifolia*.

15.3.4 Blade Lengths, Depths, and Densities

The ability to vary photosynthetic capacity (e.g., blade length, pigment concentration) when spatial competition occurs and along a depth gradient promotes invasiveness of marine macrophytes. Dumay et al. (2002) found an increase in blade length in response to competition between *C. taxifolia* and the seagrass *Posidonia oceanica*. In Mediterranean waters, blades of aquarium strain *C. taxifolia* as long as 60 cm were recorded in water up to 100-m deep (Meinesz et al. 1995; Belsher and Meinesz 1995), while the average range in shallower waters was 4–20 cm (Ceccherelli and Cinelli 1997, 1998). In Croatia, frond lengths ranged from 10 to 18 cm (Ivesa et al. 2006). In California, the maximum frond length was 24.6 cm, with a mean of 10.4 cm (Williams and Grosholz 2002). In Australia, Wright and

Davis (2006) found that the fronds of invasive populations of *C. taxifolia* were significantly shorter than native individuals when populations from 1 to 3 m of water were compared. In Moreton Bay, fronds ranged from 7.8 to 11 cm, while at Lake Conjola the range was 3.5 to 6.5 cm and the Port Hacking range was 1.6 to 5.5 cm (Wright and Davis 2006). Combined, the data suggest that large fronds are possible but only occur if required for survival in deeper waters or when competing for space.

Huge densities of *C. taxifolia* have been documented in invaded habitats. Average densities at invaded sites in Australia are currently the highest recorded globally, with 4,700 stolons and 9,000 blades/m^2 (Wright and Davis 2006). Mediterranean populations follow closely behind with 8,000 blades/m^2 and fresh weights of 11.5 kg/m^2 (de Villele and Verlaque 1995). Thibaut et al. (2004) found colonies to be denser at 5-m depths than 20-m depths in French waters; the former ranged from 203 to 518 g dry wt/m^2 and the latter was 62–466 g dry wt/m^2. Williams and Grosholz (2002) examined *C. taxifolia* that invaded Huntington Harbor in CA and found that the mean number of stolon meristems (horizontal shoots/runners) was 555 ± 182/m^2 (±SE) and the mean number of blades/m^2 = 1,478.

15.3.5 Competition

Seagrass habitat is an excellent predictor for finding species in *Caulerpa* (Glardon et al. 2008). In Florida, Glardon et al. (2008) surveyed 132 sites and of the 31 where *Caulerpa* was found, 24 were in seagrass beds. Native *Caulerpa* species are known to grow adjacent to seagrass beds or to be the first colonizers of areas that later are colonized by seagrasses (Williams 1984, 1990; Magalhaes et al. 2003). The interaction differs from invasive *C. taxifolia* and seagrasses in the Mediterranean. In the Mediterranean, *C. taxifolia* is especially adept at establishing on the edges of seagrass beds in the warmer months (Ceccherelli and Cinelli 1999b) and then outcompeting the seagrass *Posidonia oceanica* (Chisholm and Jaubert 1997; de Villele and Verlaque 1994; Montefalcone et al. 2007), but not always *Cymodocea nodosa* (Ceccherelli and Sechi 2002). Both live and dead rhizomes of *P. oceanica* proved to be suitable substrate for retention of fragments of invasive *C. taxifolia* (Cuny et al. 1995). Researchers additionally have found that *Caulerpa* fragments were more likely to recruit in *Posidonia* beds in deeper waters because the seagrass blades moved more by wave motion in shallower waters and fragments were dislodged or abraded (Cecherelli and Cinelli 1999b).

Posidonia oceanica meadows cover large areas of the Mediterranean seabed and are an important primary producer, providing food, shelter, spawning ground, and nursery for a huge variety of fishes and invertebrates (e.g., Madl and Yip 2005). Phenolic compounds in *Posidonia* that are known to increase if damaged were not allelopathic to *C. taxfolia* (Agostini et al. 1998), while competition with aquarium *C. taxifolia* caused *Posidonia* to decrease leaf longevity, possibly due to *Caulerpa's* allelopathic compounds (Dumay et al. 2002). Other authors have suggested that

Posidonia may be able to effectively compete with *C. taxifolia* in areas with limited urban pollution, but *C. taxifolia* will win in polluted areas (Jaubert et al. 1999). Williams and Grosholz (2002) found that in California, biomass of the seagrass *Ruppia maritima* was 20 times lower if mixed with *C. taxifolia* than alone.

15.3.6 Secondary Chemistry and Predation

Many marine algae are defended from herbivory by a wide diversity of secondary metabolites (e.g., Paul et al. 2001). Some green algae, including *C. taxifolia*, use an activated defense system whereby damage from feeding or abrasion results in the conversion of a stored secondary metabolite with minimal to moderate biological activity into a product with greater bioactivity (Paul and Van Alstyne 1992). If damaged, caulerpenyne, the dominant toxin in *C. taxifolia*, is transformed into a more toxic and deterrent cytotoxic sesquiterpene, which has greater antifeedant, antibiotic, and antifouling properties (Paul and Fenical 1986; Paul et al. 1987; Jung and Pohnert 2001). Reactive chemicals that are present within seconds after tissue damage act locally as defensive metabolites during the relatively slow feeding process by urchins or slugs (Sureda et al. 2006). *Caulerpa taxifolia* also has 10, 11-epoxy-caulerpenyne and caulerpenynol, two minor sesquiterpenoids, taxifolials, and other terpenes (Raffaelli et al. 1997; Paul 2002).

Caulerpenyne helps with wound response (Adolph et al. 2005) and is very unstable in seawater (Amade and Lemee 1998). Samples degraded by 50% in 4h and 95% in 24h (Amade and Lemee 1998). More caulerpenyne was found in blades than stolons in *C. taxifolia* (Dumay et al. 2002) and content varied seasonally in the Mediterranean (Amade and Lemee 1998), with the lowest concentrations in the winter followed by a sharp increase in summer (Dumay et al. 2002). Additionally, caulerpenyne content decreased with depth (Amade and Lemee 1998). Caulerpenyne levels were greater in the aquarium strain than native strains (Guerriero et al. 1992) and accounted for up to 1.3% of algal fresh weight or 2% + of algal dry mass (Paul 2002). Caulerpenyne was lethal in tropical waters to the sea urchin *Lytechinus pictus* (fertilized eggs, sperm, larvae) and toxic to the damselfish *Pomacentrus coruleus* and *Dascyllus aruanus* (Paul and Fenical 1986). Lemee et al. (1993) found that aquarium *C. taxifolia* whole extracts were toxic to mammals and eggs of the sea urchin *Paracentrotus lividus*. However, *Oxynoe olivacea*, a Mediterranean sacoglossan opisthobranch, expands its diet to consume *C. taxifolia* after the alga invades an area and uses caulerpenyne for self-protection by transforming caulerpenyne into oxytoxin-2, the mollusc's main defensive metabolite (Cutignano et al. 2004; Gianguzza et al. 2007).

Foraging behaviors and population structures of vertebrates and invertebrates were negatively impacted by *C. taxifolia* in numerous studies (e.g., Boudouresque et al. 1996; Relini et al. 1998; Davis et al. 2005). Densities and biomasses of fish assemblages were significantly lower in *Caulerpa*-invaded Mediterranean *Posidonia* beds (Francour et al. 1995; Harmelin-Vivien et al. 1999; Levi and Francour 2004;

Galil 2007). In southeastern Australia, total numbers of fishes were similar when *Caulerpa* and native seagrass beds were observed (York et al. 2006). However, species richness was significantly reduced in *Caulerpa* patches with high proportions of gobiid fishes and limited numbers of syngnathid and monacanthid fish species (York et al. 2006). In NSW, Gollan and Wright (2006) found that there were only four herbivores that co-occurred with *C. taxifolia*. The fish *Girella tricuspidata*, the sea hare *Aplysia dactylomela,* and two mesograzers, the amphipod *Cymadusa setosa* and the polychaete *Platynereis dumerilii antipoda,* all preferentially fed on other food sources in lab and field trials. Pinnegar and Polunin (2000) examined *C. taxifolia* impacts on the labrid fish *Coris julis*, which has limited migration and dispersal. Oxidative stress was examined for foraging *C. julis* in three habitat types: meadows of *C. taxifolia*, *C. prolifera*, and *P. oceanica* in waters surrounding Mallorca Island, Spain. Increased activity of liver antioxidant enzymes in *Caulerpa* meadows suggested ongoing detoxification of caulerpenyne by *C. julis* if algal blades or organisms that previously have consumed *Caulerpa* blades were ingested (Pinnegar and Polunin 2000). Even humans have suffered ill effects from caulerpenyne. Patients have been diagnosed with food poisoning after consuming *Sarpa salpa*, a fish that consumes *C. taxifolia* in the Mediterranean; doctors also documented neurological disorders such as amnesia, vertigo, and hallucinations associated with caulerpenyne consumption (DeHaro et al. 1993).

Predators have been documented to be negatively impacted by *C. taxifolia* in ways not related directly to caulerpeyne. For example, *Caulerpa's* dense clumps of rhizomes and stolons can form obstructions to fish trying to feed on benthic invertebrates (Fig. 15.2). Levi and Francour (2004) documented obstructions with the striped red mullet *Mullus surmuletus*. Longpierre et al. (2005) additionally found that the mullet's foraging effort increased with increased *density of C. taxifolia* with significantly fewer large individuals in *Caulerpa* meadows (1.2%) vs. 27.8% in *Posidonia* seagrass beds. The number of individuals of the bivalve *Anadara trapezia* increased in areas of *C. taxifolia* relative to unvegetated controls in Australian waters (Gribben and Wright 2006), possibly the result of increased structural complexity. However, delayed reproductive development, changes in timing of spawning, and fewer oocytes and sperm were all associated with *Caulerpa* beds relative to controls (Gribben and Wright 2006).

15.3.7 Other Characteristics that Promote Invasiveness

Other aspects of the life history of the genus *Caulerpa* that promote "invasiveness" include (1) ability to survive burial in sediment, and (2) ability to extract nutrients from multiple sources. Typically, unicellular species such as *Caulerpa* can translocate chloroplasts away from portions of the thallus if buried or held in darkness. Glasby et al. (2005b) documented that partial burial of *Caulerpa* in sediment had very limited impacts on individuals, while total burial for 17 days resulted in only 35% survival. While many seagrasses obtain a large fraction of nutrients from the sediment via roots and leaf uptake is considered of secondary importance (Pedersen and Borum 1993; Ceccherelli and Cinelli 1997), *Caulerpa* utilizes both sediment and water column nutrients (Williams 1984; Chisholm and Jaubert 1997). *Caulerpa* also can modify organic and inorganic components of the sediment (Chisholm and Moulin 2003). Thus, *Caulerpa* may receive a selective advantage in nutrient-limited environments when competing with seagrasses (Williams 1984; Ceccherelli and Cinelli 1997, 1999b; Ceccherelli and Sechi 2002).

15.4 Popularity of *Caulerpa* in the Aquarium Industry

While many of the economic and ecological impacts of the aquarium hobbyist industry can be enumerated (e.g., Padilla and Williams 2004; Walters et al. 2006; Zaleski and Murray 2006), one critical aspect that cannot be quantified is the number of accidental and purposeful releases of organisms from aquaria into coastal waterways. In spite of missing information, some of the most harmful invasive species that have become established in global waters are presumed to be the result of aquarium releases (e.g., Whitfield et al. 2002; Semmens et al. 2004; Ruiz-Carus et al. 2006). The source of the US and Australian invasions of *C. taxifolia* will never be known, but the similarity to the Mediterranean invasion lends support to aquarium releases (Stam et al. 2006). Currently in the USA there are over 11 million aquarium hobbyists spending billions of dollars annually to have colorful marine communities in their homes or businesses (Kay and Hoyle 2001).

In spite of the invasive reputation, many members of the genus *Caulerpa* remain extremely popular with aquarium hobbyists (Walters et al. 2006; Zaleski and Murray 2006). *Caulerpa* sp. are quintessential-looking marine aquarium plants, difficult to kill, propagate easily, remove nutrients, and some are also fish food. In three 2006 publications, the popularity and ease with which *Caulerpa* is dispersed within US boundaries via the aquarium industry was documented. Zaleski and Murray (2006) focused on availability of the genus *Caulerpa* in retail shops in southern California immediately after the first Californian invasion was reported. Zaleski and Murray (2006) found no seaweeds for sale in large corporate/franchise pet stores, so focused on independent, nonfranchise stores that specialized in ornamental organisms for hobbyists. In total, ten species of *Caulerpa* were for sale by at least one shop for 52% of 50 stores visited between November 2000 and August 2001. Fourteen percent of the shops visited had *C. taxifolia* (Zaleski and Murray 2006). None was the invasive strain (Stam et al. 2006). In 2006–2007, S. Diaz and S. Murray (unpublished data) resurveyed the same southern California shops. Forty-four were still in business. In spite of the California code banning nine species of *Caulerpa*, including *C. taxifolia*, and all the publicity associated with the two Californian invasions, 52% of the 44 remaining shops still sold at least one species of *Caulerpa* and four had *C. taxifolia* (not identified to strain, S. Diaz and S. Murray unpublished data). Two other banned species, *C. racemosa* and feathery *C. sertularioides*, also were for sale (S. Diaz and S. Murray unpublished data).

Walters et al. (2006) surveyed 47 salt water aquarium retailers in central Florida. Fifty-three percent sold *Caulerpa*, but none had *C. taxifolia*. A total of 9 species of *Caulerpa* were available for sale in central Florida.

Electronic commerce (e-commerce) increasingly is the primary, if not only, shopping venue for aquarium hobbyists. From 30 internet commercial retailers and 60 internet auction sites (eBay), Walters et al. (2006) made online purchases of 12 species of *Caulerpa* from 25 US states and Great Britain (Walters et al. 2006; Stam et al. 2006). Fifty-two percent of the states were landlocked, suggesting prior transport by hobbyists or the US postal/private shipping services. Walters et al. (2006) purchased *Caulerpa taxifolia* only once from a commercial online retailer when listed as "green feather *Caulerpa*" on the seller's web site. The purchased *Caulerpa* was shipped from a southern California retailer to Florida in November 2004 raising concern that this purchase might be the first documented case of human transport of the invasive strain of *C. taxifolia*. It, however, turned out to be a specimen of Caribbean origin based on the DNA sequence analysis (Stam et al. 2006). Hence, this clone of *C. taxifolia* was somehow transported from the Caribbean basin to California and then to Florida. On eBay you can buy entire aquarium setups that need to be collected in person. Walters et al. (2006) noted 13 auctions representing 10 states, but did not acquire *C. taxifolia* via this dispersal mechanism. They did, however, purchase *C. racemosa* and *C. mexicana* with one aquarium purchase.

Live rock is another way *Caulerpa* can be globally distributed (Walters et al. 2006). Live rock is coral (or other substrate) that is either directly quarried from reefs or kept in waters under aquaculture conditions to allow a diversity of organisms to attach. Live rock is extremely popular with hobbyists because the rock can be inexpensive and with each purchase there is the possibility of receiving a diversity of novel species. Zaleski and Murray (2006) found that 94% of southern California shops sold live rock and 18% of these had visible growth of *Caulerpa* on them. Walters et al. (2006) purchased live rock from ten retailers in central Florida and had small quantities of rock (≤10 kg) shipped to a Florida address from 11 internet retailers and 9 auction sites. After a minimum of 1 month in quarantine culture, *C. racemosa* was visible on three purchases of live rock, and *C. sertularioides*, *C. mexicana*, and *C. verticillata* were each found on one purchase. No *C. taxifolia* was transported via live rock in this study (Walters et al. 2006).

15.5 Control Methods Used to Try Dealing with Recent Invasions

There is significant pressure on marine managers to immediately remove or control any invasive marine species when an incursion occurs using the best science available (Bax et al. 2001). Thresher and Kuris (2004) list issues with the marine environment that make managing marine invasions very difficult. These include the following: (1) the ocean is perceived as an open system, (2) the public perceive oceans and coastlines as pristine, (3) a defeatist attitude by coastal managers, (4) limited knowledge about

Table 15.1 Control methods tested to eradicate invasive *Caulerpa taxifolia*

Method	Location	Success?
Manual harvest	Mediterranean, Australia	Not successful due to fragmentation, costs
Suction pumps	Mediterranean, Australia	Not successful due to fragmentation, residual attached biomass
Opaque tarpaulins	USA	Successful when combined with liquid chlorine (tarpaulins not tested alone in USA)
	Mediterranean	Not successful due to damage to tarps
	Australia	Not successful due to damage to tarps, cost for labor, nontarget mortality
Altering salinity	Australia	Successful
Liquid chlorine	USA	Successful
Copper	Mediterranean, USA	Not successful with short exposures as 100% mortality not achieved
Hydrogen peroxide	Mediterranean	Not successful with short exposures as 100% mortality not achieved
Aquatic herbicides	USA	Not successful with short exposures as 100% mortality not achieved
Coarse sea salt	Australia	Successful
Dry ice	Mediterranean	Not successful as 100% mortality not achieved
Heated water	Mediterranean	Not successful as 100% mortality not achieved
Ultrasound	Mediterranean	Not successful as 100% mortality not achieved
Biological control	Mediterranean	Not successful due to limited numbers of native herbivores and government restrictions on nonnative herbivores

the biology of most invasive species, and (5) uncertainty about control outcomes. With any marine invasive, it obviously is best to attempt eradication with a small infestation using methods that have limited ecosystem impacts. The ability of *C. taxifolia* to grow successfully from very small fragments has hampered many control efforts, but there have been some success stories. Some were based on well-planned experimental manipulations, while others were shotgun approaches. A variety of tested control methods and the reported effectiveness of each are described later and in Table 15.1. Mangers must remember that all relevant biotic and abiotic variables at the infestation site need to be considered when deciding on an eradication plan.

15.5.1 Manual Harvesting and Suction Pumps

In Mediterranean waters, harvesting by hand or with suction pumps was one of the first eradication methods attempted. Unfortunately, all attempts were unsuccessful because of *C. taxifolia's* ability to propagate clonally from small fragments missed by divers and because any residual attached biomass regrew (Meinesz et al. 1993; Rierra et al. 1994). Zuljevic and Meinesz (2002) described later efforts to use suction pumps in Croatian waters. Eradication with suction pumps was tested in four locations in 1996 and 1997 after *C. taxifolia* invaded Croatia in 1994. There was extensive regrowth in three of the four locations. To be effective, Zuljevic and Meinesz (2002) noted that suction pumping needed to be repeated after short while and eradication was more effective if the patch was small. Initial eradication attempts in NSW, Australia also included physical removal by hand and suction pump (Glasby et al. 2005a). Again, hand removal was abandoned because of the intense labor involved (<1–3 m^2 per diver per hour) and fragmentation caused by the removal process (Glasby et al. 2005a). In only one place is hand removal continuing. At the French National Park of Port Cros, annual hand removal plus applying cloths soaked in copper salts has occurred since 1994 to remain local biodiversity for SCUBA divers (Rierra et al. 1994; Thibaut 2001; Madl and Yip 2005).

15.5.2 Smothering Colonies

Black tarpaulins (20–30-mil PVC), surrounded by PVC frames and weighted down by gravel-filled bags, were used in the successful eradication campaign in California (Anderson 2005). Chlorine was injected into the confined space under all tarps, so it was not possible to determine if both treatments were required for *C. taxifolia* eradication. In other locations, tarps to smother individuals or inhibit photosynthesis were not successful. Zuljevic and Meinesz (2002) tested 0.15-mm thick black plastic tarps that were 4-m wide and suggested that this technique would be useful for all types of substrata. However, success was limited because of damage to tarps from anchoring, fishing, and storm events (Zuljevic and Meinesz 2002). In NSW, heavy rubber conveyor belts and jute matting (hessian) were tested and soon abandoned because the method was excessively labor intensive, tears in the tarps allowed for survival of some *C. taxifolia*, and high mortality of many species in the impacted area (Glasby et al. 2005a).

15.5.3 Changes to Local Salinity

Caulerpa taxifolia was found in West Lakes and the upper reaches of the Port River in South Australia in 2002 (Cheshire et al. 2002; Westphalen and Rowling 2005), prompting a significant eradication program. The West Lakes system is an artificial marine water body constructed in the 1970s and filled from Gulf St. Vincent at the south end by the tide. With an average salinity of 35 ppt, the system extends 7 km from north to south, is less than 500-m wide, and ranges in depth from 3 to 7 m (Collings et al. 2004). After extensive literature review, understanding the local flow regime, and mesocosm salinity trials, managers closed West Lakes to the sea and replaced the saltwater with freshwater. Freshwater was pumped from the nearby Torrens River. Any negative impacts were deemed acceptable by managers because the river historically drained through the area that included West Lakes and infrastructure for transfer of water was mostly in place. Engineers designed a pumping station on the bank of the Torrens River, which transferred freshwater to the southern end of the lake by gravity. Pumping began on 23 July 2003. Two additional barge-mounted pumps were deployed on the lake to pump high salinity water out and into the Port River. The project was completed on 1 December 2003. Although limited mixing of fresh and saltwater slowed progress, there was a salinity reduction at the depth where *C. taxifolia* grew (Collings et al. 2004). Most locations in West Lakes went below 17 ppt for 0–30 days, although one area never dropped below 24.8 ppt. Fragments of *C. taxifolia* from West Lakes tested after 3 months failed to grow, and there was no evidence of any regrowth after salinity was increased. Within 2 weeks of refilling from the sea, normal salinity returned to West Lakes. No *C. taxifolia* has been found in West Lakes since 2003 (Collings et al. 2004).

15.5.4 Chemical Controls, Including Chlorine and Sea Salt

Many chemicals have been tested in the laboratory and in the field with *C. taxifolia*. These include chlorine, copper (electrodes and cloths soaked in copper salts), hydrogen peroxide, and domestic herbicides known to kill nuisance freshwater algae and angiosperms (e.g., Uchimura et al. 2000; Thibaut 2001; Madl and Yip 2005). While some chemicals produced the desired results at high doses, toxicity extended to all organisms in the locality. Possible exceptions included liquid chlorine (California) and sea salt (Australia). Liquid chlorine (sodium hypochlorite) at a 12% stock solution was injected under black tarpaulins in California shortly after the infestation was reported (Anderson 2005). Over time, liquid sodium hypochlorite was replaced with 2.5-cm diameter, solid, chlorine-releasing tablets used for swimming pools (Anderson 2005). Monitoring of sediments under the tarps occurred in December 2001 and August 2002 to determine if the treated sediments continued to preclude growth of fragments of *C. taxifolia* (Anderson 2005). Cores from untreated areas promoted fragment growth while none emerged in treated sediments. Anderson (2005) did find, however, that seagrass and living invertebrates were present in all cores. Williams and Schroeder (2004) examined the role of chlorine for eradication of *C. taxifolia* in finer detail in the laboratory. They tested apical fragments at 10, 15, 50, and 125 ppm doses at three temperature regimes (7–10°C shocks, 10–11°C, 20–23°C). At the highest temperatures, chlorine at 50 ppm killed all but one fragment of *C. taxifolia* and at 125 ppm killed all fragments. Williams and Schroeder (2004) concluded that field eradication would require a chlorine concentration of 125 ppm for at least 30 min in the water column directly surrounding the *C. taxifolia* blades. Chlorine must also penetrate a minimum of 15 cm into sediments to reach rhizoids and buried stolons (Williams and Schroeder 2004).

Although chlorine was successful in the USA, chlorine has not been investigated in Australia. The most popular chemical treatment in Australia has been coarse sea salt (99.5% NaCl, mean particle diameter: 2.7 mm) and the most effective dosage was determined experimentally to be 50 kg/m^2 (Glasby et al. 2005a). The salt dissolved within 10h and killed the alga via osmotic shock and cell lysis while having only minor effects on native biota. Additionally, no fronds were present in salted areas 6 months later (Glasby et al. 2005a; O'Neill et al. 2007). Researchers determined that the salting process worked best in cooler months when *C. taxifolia* died back naturally and that a thick, continuous covering of salt worked significantly better than salting discrete patches. For deploying salt, researchers compared hand dispersal vs. using a hopper on a flat-bottomed boat. Although waters deeper than 5m allowed for too much horizontal dispersal in the water column for salting to be successful, in shallower waters the hopper method cost $A7 per square meter (2005 values) (Glasby et al. 2005a). Salt deployment by hand cost an average of $30 per square meter at the same time. Glasby et al. (2005a) suggested that colonies should be mapped during the warm season, followed by repeated salting of the infestations during colder months. Glasby et al. (2005a) additionally calculated a cost of over $A60 million to cover all *C. taxifolia* in NSW with one application of salt, using the hopper method.

15.5.5 Other Treatments: Temperature Shock, Ultrasound, and Genetic Control

A variety of additional methods to kill *C. taxifolia* under field conditions have been tried unsuccessfully. While cold shock killed fragments in the laboratory (Williams and Schroeder 2004), dry ice applications were not successful in the field. With dry ice, only sublethal necrosis was obtained (Thibaut 2001). Likewise, when fragments of *C. taxifolia* in the laboratory were heat-shocked at 72°C for 1 or 2h, the fragments died (Williams and Schroeder 2004). However, underwater applications of hot water at or above 40°C appeared to work initially, but recovery was observed after 3 weeks (Thibaut 2001). Underwater welding devices to boil the plants also were not successful (Madl and Yip 2005). In situ application of ultrasound did not destroy plant tissue (Boudouresque et al. 1996). However, in a feasibility study for genetic control of *C. taxifolia*, Thresher and Grewe (2004) stated that a species-specific biocide based on an enzyme critical for photosynthesis or osmoregulation could be developed and then delivered in pellet form.

15.5.6 Biological Control

Many scientists and managers hypothesize that the only hope for control of *C. taxifolia* will involve biological control. Thresher and Kuris (2004) suggested that most current efforts to eradicate or control high-impact marine invasive species that are deemed acceptable to stakeholders are low risk and publically acceptable, while biological control remains more contentious for both social and political reasons. Thresher and Kuris (2004) continue by stating that contentious possibilities will not occur on a large-scale until scientists and managers learn more about biological control agents.

Two biological control agents initially deemed most likely to reduce *C. taxifolia* biomass in Mediterranean waters were the native sea slugs *Oxynoe olivacea* and *Lobiger serradifalci*. Both species perforated cell walls with uniserial radula and sucked up the algal contents. Both species have been tested in aquariums and in open ocean waters as potential biological control agents of *C. taxifolia* (Thibaut and Meinesz 2000). In later laboratory studies, *L. serradifalci* created viable fragments of *C. taxifolia* when feeding; *Oxynoe olivacea* did not produce fragments (Zuljevic et al. 2001). The nonnative, tropical sea slug *Elysia subornata* was also tested (Meinesz 2002). At 21°C, *E. subornata* fed on *C. taxifolia* at rates 2–11 times higher than the native ascoglossan species (Thibaut et al. 2001). Meinesz et al. (1995) projected that 1,000 *E. subornata*/m^2 were required to have a significant impact on the Mediterranean sea floor with 5,000 fronds/m^2. Unfortunately, a cold-resistant strain of this species could not be readily cultivated (Meinesz 2002) and, in 1997, the International Council for the Exploration of the Sea (ICES) stated that nonnative ascoglossans could not be introduced to Mediterranean Sea for biocontrol. The French Ministry of the Environment shared the same attitude based on (1) possible dietary switching by the herbivore, (2) competition with other important groups of herbivores in the event of eradication of the target food source, (3) introduction of pathogens, and (4) spread of the introduced herbivore itself.

15.5.7 Outreach, Signage, and Closures

The most cost-effective management strategy is to prevent the introduction of *C. taxifolia* into coastal waters. To ultimately be effective, outreach concerns need to be addressed at local, regional, national, and international levels (Hewitt 2003). Many countries have provided public and private funds for creating a wealth of outreach materials (animated videos, fact sheets, identification keys, lesson plans for educators, etc.) designed to change behaviors for every age and interest group that may contact *C. taxifolia*. The next goal is to get the information into the right hands. As aquarium releases are an important, intangible source of marine invasions, aquarium hobbyists and custom agents who inspect international aquarium shipments are prime outreach targets. However, both audiences are fluid with new customers entering the hobby every day and new individuals assigned to rapidly assess live aquarium shipments for illegal importation. For example, Zaleski and Walters (unpublished data) surveyed participants at MACNA 2006, the annual meeting of the Marine Aquarium Societies of North America in Houston, TX and found that 29% of the visitors had never heard of *C. taxifolia*, while an additional 25% had very limited understanding of the biology or global invasions of *C. taxifolia*.

In areas such as the Mediterranean and Australia that may never be able to eradicate *C. taxifolia*, outreach can slow additional spread. In infested Australian waters, signs are posted at all areas where boaters launch their craft and mandatory "wash-down" stations are provided (Fig. 15.5). NSW visibly marks and bans anchoring at

Fig. 15.5 Signage used in NSW, Australia to limit further dispersal of *Caulerpa taxifolia*

sites newly invaded by *C. taxifolia* (Glasby et al. 2005a). Additionally, when new infestations are found in NSW, NSW Fisheries establishes fishing closures to prevent hauling or mesh-netting in infested areas under the Fisheries Management Act of 1994.

15.5.8 Posteradication Surveys

Eradication only is successful when no living *Caulerpa* is present at a previously infested location. Although easy to describe, documentation of eradication can be difficult to obtain based on prevailing currents, water clarity, and epiphytism. In California, significant effort was devoted to surveying and developing defendable survey techniques once *C. taxifolia* was no longer located by trained divers. During both pre- and posteradication monitoring, diver teams followed prescribed parallel transect lines from GPS coordinates. The grid provided sufficient overlap to minimize missing *C. taxifolia* that was present (Anderson 2005). In Huntington Harbor, a full survey took 5 days. In Agua Hedionda Lagoon, a full survey of the lagoon involved swimming more than 500 miles and took 2–3 months (Anderson 2005). To ensure divers were searching effectively and not just missing plants, *C. taxifolia* mimics (plastic aquarium plants) were deployed in locations unknown to the divers (Anderson 2005). Survey efficacy varied substantially with water clarity and bottom type. The efficacy study was necessary to estimate how many surveys were necessary to find every last plant. After 24 months of monitoring with no sightings of live *C. taxifolia*, but continued sightings of mimics, California deemed eradication successful in July 2006.

In March 2007, California adopted a new survey protocol for persons applying for permits to disturb the benthos in areas where *C. taxifolia* was once found. Before commencing any permitted sediment-disturbing activity, a preconstruction *Caulerpa* survey of the area that covers at least 20% of the bottom that will be affected must be completed. The affected bottom includes the project footprint, areas where equipment were stored/moored, areas where vessel prop-wash could occur, and in-water disposal areas for sediment. The survey must be completed not earlier than 90 days prior and not later than 30 days prior. If any *C. taxifolia* is found, activity must be halted until the infestation was isolated, treated, and any risk eliminated. If work is to be undertaken in *Caulerpa*-infected waters, two surveys not less than 60 days apart must occur, one at the high intensity level (50% of bottom covered) and one at the eradication level (100% of bottom covered). If the project extends over 90 calendar days, additional surveys at the high-intensity level will be required. Additional surveys of dredged materials may also be required.

15.5.9 Modeling

Models allow managers to focus limited resources to survey only the most suitable sites for an invasive species, such as *C. taxifolia*. Glardon et al. (2008) have developed such a model for the genus *Caulerpa* in Florida waters. Glardon et al. (2008) conducted field surveys of 24 coastal areas around Florida in each of six zones chosen in a stratified manner and evaluated the association of potential indicators for the presence of *Caulerpa*. In total, 14 species of *Caulerpa*, but not *C. taxifolia*, were found at 31 of the 132 sites. Latitude, presence of seagrass beds, human population density, and proximity to marinas were simultaneously considered. A positive correlation between *Caulerpa* spp. presence and seagrass beds and proximity to marinas was documented while a negative correlation with latitude and human population density was also noted. The parameters in the logistic regression model assessing the association of *Caulerpa* occurrence with the measured variables then were used to predict current and future probabilities of *Caulerpa* spp. presence throughout the state. Percent correct for this model was 61.5% for presence and 98.1% for absence. While aquarium dumping provides an explanation for the positive correlation with marinas, the human population density results were surprising. This may be because, in Florida waters, high population densities enhance pollutant loads, freshwater inputs, and nutrient runoff, and these factors may decrease macroalgal growth.

A second type of useful model is one that accurately predicts the pace of an invasion once it has begun so that managers know how best to undertake eradication. Ruesink and Collado-Vides (2006) found that the model that best describes actual field distributions of *C. taxifolia* invokes local growth via rhizome expansion plus low levels of fragment dispersal and attachment (increases of 4–14-fold annually). The model goes on to suggest that the most effective plan for maximizing eradication is removal of established patches before summer and removal of fragments in the fall (Ruesink and Collado-Vides 2006). The times corresponded to just before maximum growth and just after maximum fragment production, respectively. Only a mixed strategy that combined 99% removal of all fragments and annual removal of 99% of established patches was predicted to entirely eliminate *C. taxifolia* (Ruesink and Collado-Vides 2006). This level of effort is only likely to be possible early in an invasion.

15.6 Reducing the Likelihood of Future Invasions Through Biosecurity Regulations

Marine algal invasions can transcend national boundaries, so the problem must be considered an international problem (Inderjit et al. 2006). To be successful, a global rapid response plan should be in place as well as immediate access to adequate funding. If biological invasions are treated the same way that governments respond to hurricanes, tornados, fires, floods, oil spills, or disease outbreaks, then rapid response should be possible. Anderson (2005) suggested that to be prepared to deal with an invasion, a drill must be performed to determine who will provide (1) biological experts, (2) ownership of the waterway, (3) knowledge of potentially successful eradication strategies, and (4) funding. Although many countries are concerned about future *C. taxifolia* invasions, currently only NZ appears ready to respond if an invasion were reported tomorrow.

In New Zealand, marine biological security is defined as protection of the marine environment from nonnative species. Biosecurity is a high profile topic, mainly because of the country's dependence on shipping (Hewitt et al. 2004). The NZ Marine Biosecurity Team was established in 1998 under the Biosecurity Act of 1993 with the dual goals of working on reducing knowledge gaps and establishing management frameworks. Active awareness campaigns by the government have led to a greater awareness of many nonnative species in the general population relative to other countries. Additionally, no discharge of unexchanged ballast water is permitted in NZ from any country unless exempted on the grounds of safety. Preborder and border management is likewise paramount to promote prevention, early detection, and rapid response (Wotton and Hewitt 2004). New Zealand is already on high alert, expecting *C. taxifolia* to arrive at any time. The decision making process in NZ for a *C. taxifolia* sighting follows a mostly universal rapid response protocol and would involve (1) confirming the genus species in NZ waters, (2) establishing the nature and magnitude of the incursion, and (3) risk analysis to determine the likelihood of an impact if the incursion was left untreated (insignificant, minor, moderate, major, catastrophic). Containment, management, or eradication would then be initiated if sustained, cost-effective action was possible and the organism posed an unacceptable risk. More than likely, any *C. taxifolia* in NZ waters would be considered an unacceptable risk. The actual level of response would then depend on the (1) potential impacts of the invasive organism on the environment, the economy, and people, (2) technical feasibility of response options, (3) ability to target the invasive species, (4) risks associated with treatment, (5) degree of public concern, and (6) likelihood of the organism being eradicated or managed. Monitoring and review of the response process itself completes this protocol.

In the USA, the US Department of the Interior was alerted to problems with *C. taxifolia* Mediterranean strain and the threat it posed to US coastal waters in 1998. The scientific community requested the Secretary of the Interior to be proactive and initiate action to prevent introduction of *C. taxifolia* into US waters. *Caulerpa taxifolia* was determined to pose a significant threat, so a comprehensive prevention plan was requested and the Mediterranean strain of *C. taxifolia* was banned from importation, entry, exportation, or movement in interstate commerce by the US Department of Agriculture (USDA) under the federal Noxious Weed Act (1999) and the federal Plant Protection Act (2000). Before the plan could be implemented and before importers/retailers knew about the ban, the invasive aquarium strain of *C. taxifolia* was discovered in California. The US was lucky that the invasion of *C. taxifolia* happened in southern California. Within 3 days of contract divers finding an unfamiliar macroalga in Agua Hediona Lagoon, specialists received and morphologically confirmed the unknown as *C. taxifolia*. DNA forensics shortly thereafter confirmed that the unknown alga was the invasive Mediterranean strain (Jousson et al. 2000). The Southern California *Caulerpa* Action Team (SCCAT) was created and included local, state, and federal agencies, university researchers, the San Diego and Santa Ana Regional Water Quality Control Boards, the local power company, and key local stakeholders (Anderson 2005). Treatment combining black tarpaulins and liquid chlorine began within days and rapidly was followed by new regulations at the state and county levels. California legislators banned nine species of *Caulerpa* after a public outcry by aquarists stated that a ban at the genus level would have dire effects on their industry. On 25 September 2001, California Fish and Game Code 2300 became law and prohibited the sale, possession, import, transport, transfer, release, or giving away of *C. taxifolia, C. cupressoides, C. mexicana, C. setularioides, C. floridana, C. ashmeadii, C. racemosa, C. verticillata*, and *C. scapelliformis*. All banned species were either known to be invasive in some location or were feathery look-alikes of *C. taxifolia*. Possession was permitted only for scientific research and any violators were subject to a civil penalty of not less than $500 and not more than $10,000 for each violation. Next, the city of San Diego, California adopted an ordinance that prohibited the sale and possession of all *Caulerpa* species within city limits. This comprehensive ban demonstrated the commitment of the city to protect its coastal resources. At the federal level in the USA, a National Management Plan was released in October 2005 for the genus *Caulerpa*. On a positive side, the plan identified research information gaps and how these gaps should be addressed. It also provided much-needed funding for research and outreach. The plan, however, did not have strong wording to create a federal ban at the genus or species level, so some states (Oregon, Massachusetts, South Carolina) were proactive and banned all strains of *C. taxifolia*. Dr. Susan Williams submitted a federal petition to increase the ban to the genus or species level. Williams correctly argued that current US regulations do not protect ecosystems from invasive Australian strains or other invasive strains not yet described. She also argued that the current regulation relies on expensive, time-consuming DNA technology. This petition has not yet been resolved.

In spite of the US regulations, Walters et al. (2006) found that of 60 eBay auctions, only four vendors provided information on interstate transport regulations for *C. taxifolia*. eBay stated that the sellers, not eBay, were responsible for knowing all federal and state regulations for all live plant products sold and that vendors stand to have their accounts suspended and forfeit all eBay fees on cancelled listings for breaking regulations. With internet commercial retailers, the ratio was similar with only 2 of 30 providing consumers with information on *Caulerpa* restrictions. Some large internet distributors now provide fact sheets promoting alternatives to release for all aquarium and water garden flora/fauna. The US Postal Service now prohibits mailing the aquarium strain of *C. taxifolia* by designating the alga as a "nonmailable plant pest." As with eBay, the burden of identification resides with the sender. Common names are used so often in the marine aquarium trade that the seller/buyer may not even know that they are distributing a banned species. From e-commerce purchases, Walters et al. (2006) learned that eBay, internet, and local retailers frequently do not identify products scientifically; only 14.1% used genus species names and sellers correctly identified algae only 10.6% of the time at the species level. None identified macroalgae by strain. Walters et al. (unpublished data) found that common names that may be the aquarium strain of *C. taxifolia* included feather *Caulerpa*, feather algae, aquacultured *Caulerpa* algae, *Caulerpa* algae, marine macroalgae, assorted *Caulerpa* species, coral reef algae, fern *Caulerpa*, and green *Caulerpa* tang heaven. Common names greatly reduce the ability of agencies to remove a species from the market.

Although the origin of *C. taxifolia* in NSW, Australia remains unclear, scientists hypothesize that it was associated with the aquarium industry (Creese et al. 2004). Australia's initial response was much less aggressive than in the USA, in part because *C. taxifolia* is native to Queensland. Eventually, *C. taxifolia* was declared a noxious weed in NSW and South Australia. Sandwiched between the two invaded states, uninvaded Victoria declared *C. taxifolia* a noxious weed in 2004. After a 90-day amnesty period, it was illegal to bring into Victoria, or take, hatch, keep, possess, sell, transport, put in any container, or release *C. taxifolia*. While total eradication of *C. taxifolia* from NSW waterways is unlikely, hopefully resource managers will be able to prevent further spread of *C. taxifolia*. Both Australia and NZ have established national systems of port baseline surveys using standardized methods (Hewitt and Martin 2001; Ruiz and Hewitt 2002).

The response of countries surrounding the Mediterranean documents how difficult it can be to create multinational regulations. In 1994, 10 years after the Monaco invasion, the Barcelona Appeal was a call by scientists to list the spread of *C. taxifolia* as a major threat to Mediterranean ecosystems. In 1996, Article 13 of the Barcelona Convention Protocol on Specially Protected Areas provided legislation regarding the introduction of nonnative species. In 1998, a delegation recommended that all affected countries establish regulations to limit the invasion. Several governments and regional entities have since banned selling, buying, using, and dumping this seaweed. Currently, however, there is no coordinated, multinational effort underway to eradicate *C. taxifolia* in the Mediterranean.

15.7 Problems with Other Species in the Genus *Caulerpa* and Final Concerns with *C. taxifolia*

As little as we know about controlling and managing *C. taxifolia*, we know even less about other species of *Caulerpa* that may cause ecological and economic problems. Included is *Caulerpa racemosa* var. *cylindracea* (Sonder) Verlaque, Huisman et Boudouresque, a southwestern Australian variety, that is expanding dramatically in the Mediterranean and surrounding Atlantic waters (Belsher et al. 2003; Verlaque et al. 2000, 2003, 2004; Ruitton et al. 2005; Piazzi and Ceccherelli 2006). In Mediterranean waters, *C. racemosa* has been spreading faster and to more locations than *C. taxifolia* (Ruitton et al. 2005). In some cases in the Mediterranean, *C. taxifolia* outcompeted *C. racemosa* (Ruitton et al. 2005), while in other locations the reverse was true (Piazzi and Ceccherelli 2002; Piazzi et al. 2003). After over 2 years of rapid growth, natural disasters in the form of repeated hurricanes removed nonnative *Caulerpa brachypus forma parvifolia* from coral reef areas in south Florida (Lapointe et al. 2006). Although native, blooms of *Caulerpa* also can alter ecosystems. Blooms of *C. verticillata* (south Florida, Lapointe et al. 2006), *C. prolifera* (west coast of central Florida, Stafford and Bell 2005), and *C. sertularioides* (Costa Rica, Fernandez and Cortes 2004) have been described in recent years. Preventing new invasions, locating and eradicating invaded populations while small, and managing large populations once eradication is deemed impossible continue to be issues of global concern with *C. taxifolia* and many additional species in this genus. At the present time, we are a long way from being successful. Much more research is needed to make us globally proactive.

References

Adolph S, Jung V, Rattke J, Pohnert G (2005) Wound closure in the invasive green alga *Caulerpa taxifolia* by enzymatic activation of a protein cross-linker. Angew Chem Int Ed 44:2806–2808

Agostini S, Desjobert J, Pergent G (1998) Distribution of phenolic compounds in the seagrass *Posidonia oceanica*. Phytochemistry 48:611–617

Amade P, Lemee R (1998) Chemical defense of the Mediterranean alga *Caulerpa taxifolia*: variations in caulerpenyne production. Aquat Toxicol 43:287–300

Anderson L (2005) California's reaction to *Caulerpa taxifolia*: a model for invasive species rapid response. Biol Invasions 7:1003–1016

Balata D, Piazzi L, Cinelli F (2004) A comparison among assemblages in areas invaded by *Caulerpa taxifolia* and *C. racemosa* on a subtidal Mediterranean rocky bottom. PSZN Mar Ecol 25:1–13

Bax N, Carlton J, Mathews-Amos A, Haedricj R, Howarth F, Purcell J, Rieser A, Gray A (2001) The control of biological invasions in the world's oceans. Conserv Biol 15:1234–1246

Belsher T, Meinesz A (1995) Deep-water dispersal of the tropical alga *Caulerpa taxifolia* introduced into the Mediterranean. Aquat Bot 51:163–169

Belsher T, Lunven M, Le Gall E, Caisey X, Dugornay O, Mingant C (2003) Observations concerning the expansion of *Caulerpa taxifolia* and *Caulerpa racemosa* in Rade d'Hyerees and Rade of Toulon (France). Oceanol Acta 26:161–166

Boudouresque C, Lemee R, Mari X, Meinesz A (1996) The invasive alga *Caulerpa taxifolia* is not a suitable diet for the sea urchin *Paracentrotus lividus*. Aquat Bot 53:245–250

Ceccherelli G, Cinelli F (1997) Short-term effects of nutrient enrichment of the sediment and interactions between the seagrass *Cymodocea nodosa* and the introduced green alga *Caulerpa taxifolia* in the Mediterranean Sea. J Exp Mar Biol Ecol 217:165–177

Ceccherelli G, Cinelli F (1998) Habitat effect on spatio-temporal variability in size and density of the introduced alga *Caulerpa taxifolia*. Mar Ecol Prog Ser 163:289–294

Ceccherelli G, Cinelli F (1999a) The role of vegetative fragmentation in dispersal of the invasive alga *Caulerpa taxifolia* in the Mediterranean. Mar Ecol Prog Ser 182:299–303

Ceccherelli G, Cinelli F (1999b) Effects of *Posidonia oceanica* canopy on *Caulerpa taxifolia* size in a north-western Mediterranean bay. J Exp Mar Biol Ecol 240:19–36

Ceccherelli G, Sechi N (2002) Nutrient availability in the sediment and the reciprocal effects between the native seagrass *Cymodocea nodosa* and the introduced rhizophytic alga *Caulerpa taxifolia*. Hydrobiologia 474:57–66

Ceccherelli G, Piazzi L, Balata D (2002) Spread of introduced *Caulerpa* species in macroalgal habitats. J Exp Mar Biol Ecol 280:1–11

Cheshire A, Westphalen G, Boxall V, Marsh R, Gilliland J, Collings G, Seddon S, Loo M (2002) *Caulerpa taxifolia* in the West Lakes and the Port River, South Australia: distribution, eradication options and consequences. SARDI Aquatic Sciences Publication Number RD02/0161. A Report to the PIRSA Fisheries Marine Habitat Program, 64 pp

Chisholm J, Jaubert J (1997) Photoautotrophic metabolism of *Caulerpa taxifolia* (Chlorophyta) in the NW Mediterranean. Mar Ecol Prog Ser 153:113–123

Chisholm J, Moulin P (2003) Stimulation of nitrogen fixation in refractory organic sediments by *Caulerpa taxifolia*. Limnol Oceanogr 48:787–794

Chisholm J, Marchioretti M, Jaubert J (2000) Effect of low water temperature on the metabolism and growth of a subtropical strain of *Caulerpa taxifolia* (Chlorophyta). Mar Ecol Prog Ser 201:189–198

Collado-Vides L, Robledo D (1999) Morphology and photosynthesis of *Caulerpa* (Chlorophyta) in relation to growth form. J Phycol 35:325–330

Collings G, Westphalen G, Cheshire A, Rowling K, Theil M (2004) *Caulerpa taxifolia* (Vahl) C. Agardh eradication efforts in West Lakes, South Australia. SARDI Aquatic Sciences. Report to PIRSA Marine Habitat Program. RD02/0161-8

Creese R, Davis A, Glasby T (2004) Eradicating and preventing the spread of the invasive alga *Caulerpa taxifolia* in NSW. NSW Fisheries Final Report Series No. 64, NSW, Sydney

Cuny P, Serve L, Jupin H, Bouderesque C (1995) Water soluble phenolic compounds of the marine phanerogam *Posidonia oceanica* in Mediterranean areas colonized by the introduced chlorophyte *Caulerpa taxifolia*. Aquat Bot 52:237–242

Cutignano A, Notti V, d'Ippolito G, Coll A, Cimino G, Fontana A (2004) Lipase-mediated production of defensive toxins in the marine mollusk *Oxynoe olivacea*. Org Biomol Chem 2:3167–3171

Davis A, Bekendorff K, Ward D (2005) Responses of common SE Australian herbivores to three suspected invasive *Caulerpa* spp. Mar Biol 146:859–868

DeHaro L, Treffot M, Jouglard J, Perringue C (1993) Trois cas d'intoxication de type ciguateresque après ingestion de Sparidae de Mediterranee. Ichthyol Physiol Acta 16:133–146

de Villele X, Verlaque M (1994) Incidence de l'algue introduite *Caulerpa taxifolia* sur le phytobenthos de Méditerranée occidentale. I. L'herbier de *Posidonia oceanica* (L.) Delile. In: Boudouresque C, Meinesz A, Gravez V (eds) First International Workshop on *Caulerpa taxifolia*, Marseille, France

de Villele X, Verlaque M (1995) Changes and degradation in a *Posidonia oceanica* bed invaded by the introduced tropical alga *Caulerpa taxifolia* in the North Western Mediterranean. Bot Mar 38:79–87

Dreher T, Grant B, Wetherbee R (1978) The wound response in the siphonous alga *Caulerpa simpliciuscula* C. Ag.: fine structure and cytology. Protoplasma 96:189–203

Dumay O, Pergent G, Pergent-Martini C, Amade P (2002) Variations in caulerpenyne contents in *Caulerpa taxifolia* and *Caulerpa racemosa*. J Chem Ecol 28:343–352

Fernandez C, Cortes J (2004) *Caulerpa sertularioides*. A green alga spreading aggressively over coral reef communities in Culebra Bay, North Pacific of Costa Rica. Coral Reefs 24:10

Francour P, Harmelin-Vivien M, Harmelin J, Duclerc J (1995) Impact of *Caulerpa taxifolia* colonization on the littoral ichthyfauna of north-western Mediterranean: preliminary results. Hydrobiologia 345–353

Galil B (2007) Loss or gain? Invasive aliens and biodiversity in the Mediterranean Sea. Mar Pollut Bull 55:314–322

Gianguzza P, Andaloro F, Riggio S (2007) Feeding strategy of the sacoglossan opisthobranch *Oxynoe olivacea* on the tropical green alga *Caulerpa taxifolia*. Hydrobiologia 580:255–257

Glardon C, Walters L, Olsen J, Stam W, Quintana-Ascencio P (2008) Predicting risks of invasion of *Caulerpa* species in Florida. Biol Invasions 10:1147–1157

Glasby T, Cresse R, Gibson P (2005a) Experimental use of salt to control the invasive marine alga *Caulerpa taxifolia* in NSW, Australia. Biol Conserv 122:573–580

Glasby T, Gibson P, Kay S (2005b) Tolerance of the invasive marine alga *Caulerpa taxifolia* to burial by sediment. Aquat Bot 82:71–81

Goddard R, Dawes C (1983) An ultrastructural and histochemical study of the wound response in the coenocytic green alga *Caulerpa ashmeadii*. Protoplasma 114:163–172

Gollan J, Wright J (2006) Limited grazing pressure by native herbivores on the invasive seaweed *Caulerpa taxifolia* in a temperate Australian estuary. Mar Freshw Res 57:685–694

Gribben P, Wright J (2006) Sublethal effects on reproduction in native fauna: are females more vulnerable to biological invasion? Oecologia 149:352–361

Guerriero A, Meinesz A, D'Ambossio M, Pietra F (1992) Isolation of toxic and potentially sesqui- and monoterpenes from the tropical green seaweed *Caulerpa taxifolia* which has invaded the region of Cap Martin and Monaco. Helv Chim Acta 75:689–695

Guiry M, Dhonncha N (2004) AlgaeBase, Version 2.1. www.algaebase.org

Harmelin-Vivien M, Francour P, Harmelin J (1999) Impact of *Caulerpa taxifolia* on Mediterranean fish assemblages: a six year study. In: Proceedings of the Workshop on Invasive *Caulerpa* in the Mediterranean. MAP Tech Rep Ser 125:127–138

Herren L, Walters L, Beach K (2006) Fragment generation, survival, and attachment of *Dictyota* spp. at Conch Reef in the Florida Keys, USA. Coral Reefs 25:287–295

Hewitt C (2003) Marine biosecurity issues in the world oceans: global activities and Australian directions. Ocean Yearb 17:193–212

Hewitt C, Martin R (2001) Revised protocols for baseline port security for introduced marine species – design considerations, sampling protocols and taxonomic sufficiency. Centre for Research on Introduced Marine Pests. Technical Report No. 22, CSIRO Marine Research, Hobart, Australia

Hewitt C, Willing J, Bauckham A, Cassidy A, Cox C, Jones L, Wotton D (2004) New Zealand biosecurity: delivering outcomes in a fluid environment. New Zealand J Mar Freshw Res 38:429–438

Inderjit (2005) Plant invasions: habitat invisibility and dominance of invasive plant species. Plant Soil 277:1–5

Inderjit, Chapman D, Ranelletti M, Kaushik S (2006) Invasive marine algae: an ecological perspective. Bot Rev 72:153–178

Ivesa L, Jaklin A, Devescovi M (2006) Vegetation patterns and spontaneous regression of *Caulerpa taxifolia* (Vahl) c. Agarrdh in Malinska (Northern Adriatic, Croatia). Aquat Bot 85:324–330

Jaubert J, Chisholm J, Ducrot D, Ripley H, Roy L, Passeron-Seitre G (1999) No deleterious alterations in *Posidonia* beds in the Bay of Menton (France) eight years after *Caulerpa taxifolia* colonization. J Phycol 35:1113–1119

Jousson O, Pawlowski J, Zaninetti L, Meinesz A, Boudouresque C (1998) Molecular evidence for the aquarium origin of the green alga *Caulerpa taxifolia* introduced to the Mediterranean Sea. Mar Ecol Prog Ser 172:275–280

Jousson O, Pawlowski J, Zaninetti L, Zechman W, Dini F, Di Guiseppe G, Woodfield R, Millar A, Meinesz A (2000) Invasive alga reaches California. Nature 408:157–158

Jung V, Pohnert G (2001) Rapid wound-activated transformation of the green algal defensive metabolite caulerpenyne. Tetrahedron 57:7169–7172

Kay S, Hoyle S (2001) Mail order, the internet and invasive aquatic weeds. J Aquat Plant Manage 39:88–91

Keppner S (2002) A prevention program for the Mediterranean strain of *Caulerpa taxifolia*. Submitted to the Aquatic Nuisance Species Task Force. US Fish and Wildlife Service, Amherst, NY

Komatsu T, Meinesz A, Buckles D (1997) Temperature and light responses of the alga *Caulerpa taxifolia* introduced into the Mediterranean Sea. Mar Ecol Prog Ser 146:145–153

Komatsu T, Ishikawa T, Yamaguchi N, Hori Y, Ohba H (2003) But next time?: unsuccessful establishment of the Mediterranean strain of the green seaweed *Caulerpa taxifolia* in the Sea of Japan. Biol Invasions 5:275–277

Lapointe B, Bedford B, Baumberger R (2006) Hurricanes Frances and Jeanne remove blooms of the invasive green alga *Caulerpa brachypus* forma *parvifolia* (Harvey) Cribb from coral reefs off northern Palm Beach County, Florida. Estuar Coast 29:966–971

Lemée R, Pesando D, Durand-Clément A, Dubreuil A, Meinesz A, Guerriero A, Pietra F (1993) Preliminary survey of toxicity of the green alga *Caulerpa taxifolia* introduced into the Mediterranean. J Appl Phycol 5:485–493

Levi F, Francour P (2004) Behavioral response of *Mullus surmuletus* to habitat modification by the invasive macroalga *Caulerpa taxifolia*. J Fish Biol 64:55–64

Longpierre S, Robert A, Levi F, Francour P (2005) How an invasive alga species (*Caulerpa taxifolia*) induces changes in foraging strategies of the benthivorous fish *Mullus surmuletus* in coastal Mediterranean ecosystems. Biodivers Conserv 14:365–376

Madl P, Yip M (2005) Literature review of the aquarium strain of *Caulerpa taxifolia*. http://www.sbg.ac.at/ipk/avstudio/pierofun/ct/caulerpa.htm. Accessed 1 February 2008

Magalhaes K, Pereira S, Guimaraes N, Amorin L (2003) Macroalga associated to *Halodule wrightii* beds on the Coast of Pernambuco, Northeastern Brazil. Gulf Mexico Sci 21:114

Meinesz A (1999) Killer algae: the true tale of a biological invasion. University of Chicago Press, Chicago, IL

Meinesz A (2002) Introduction for the international *Caulerpa taxifolia* conference. In: Williams E, Grosholz E (eds) International *Caulerpa taxifolia* Conference Proceedings. CA Sea Grant, University of California – San Diego, La Jolla, CA

Meinesz A, Hesse B (1991) Introduction et invasion de l'algue tropicale *Caulerpa taxifolia* en Mediterranee nord occidentale. Oceanol Acta 14:415–426

Meinesz A, de Vaugelas J, Hesse B, Mair X (1993) Spread of the introduced tropical green alga *Caulerpa taxifolia* in northern Mediterranean waters. J Appl Phycol 5:141–147

Meinesz A, Benichou L, Blachier J, Konatsu T, Lemee R, Molenaar H, Hari X (1995) Variations in the structure, morphology and biomass of *Caulerpa taxifolia* in the Mediterranean Sea. Bot Mar 38:499–508

Meinesz A, Belsher T, Thibault T, Antolic B, Mustapha K, Boudouresque C, Chiaverini D, Cinelli F, Cottalorda J, Djellouli A, El Abed A, Orestano C, Grau A, Ivesa L, Jaklin A, Langar H, Massuti-Pascual E, Peirano A, Tunesi L, DeVaugelas J, Zavodnik N, Zuljevic A (2001) The introduced green alga *Caulerpa taxifolia* continues to spread in the Mediterranean. Biol Invasions 3:201–210

Millar A (2002) The introduction of *Caulerpa taxifolia* in New South Wales, Australia. In: Williams E, Grosholz E (eds) International *Caulerpa taxifolia* Conference Proceedings. CA Sea Grant, University of California – San Diego, La Jolla, CA

Millar A (2004) New records of marine benthic algae from New South Wales, eastern Australia. Phycol Res 52:117–128

Montefalcone M, Morri C, Peirano A, Albertelli G, Bianchi C (2007) Substitution and phase shift within the *Posidonia oceanica* seagrass meadows of NW Mediterranean Sea. Estuar Coast Shelf Sci 75:63–71

Murphy N, Schaffelke B (2003) Use of amplified length polymorphism (AFLP) as a new tool to explore the invasive green alga *Caulerpa taxifolia* in Australia. Mar Ecol Prog Ser 307:307–310

O'Neill K, Schneider M, Glasby T, Redden A (2007) Lack of epifaunal response to the application of salt for managing the noxious green alga *Caulerpa taxifolia* in a coastal lake. Hydrobiologia 580:135–142

Padilla D, Williams S (2004) Beyond ballast water: aquarium and ornamental trades as sources of invasive species in aquatic ecosystems. Frontiers Ecol Environ 2:131–138

Paul V (2002) Seaweed chemical defenses on coral reefs. In: Paul V (ed) Ecological roles of marine natural products. Comstock, Ithaca, NY

Paul V, Fenical W (1986) Chemical defense in tropical green alga, order Caulerpales. Mar Ecol Prog Ser 34:157–169

Paul V, Van Alstyne K (1992) Activation of chemical defenses in the tropical green algae *Halimeda* sp. J Exp Mar Biol Ecol 160:191–203

Paul V, Littler M, Littler D, Fenical W (1987) Evidence for chemical defense in tropical green alga *Caulerpa ashmeadii* (Caulerpaceae: Chlorophyta): isolation of new bioactive sesquiterpenoids. J Chem Ecol 13:1171–1185

Paul V, Cruz-Rivera E, Thacker R (2001) Chemical mediation of macroalgal-herbivore interactions: ecological and evolutionary perspectives. McClintock J, Baker B (eds) Mar Chem Ecol CRC Press, Boca Raton, FL

Pedersen M, Borum J (1993) An annual nitrogen budget for a seagrass *Zostera marina* population. Mar Ecol Prog Ser 101:169–177

Pennings S, Paul V (1992) Effect of plant toughness, calcification and chemistry on herbivory by *Dolabella auricularia*. Ecology 73:1606–1619

Phillips J, Price I (2002) How different is Mediterranean *Caulerpa taxifolia* (Caulerpales: Chlorophyta) to other populations of the species? Mar Ecol Prog Ser 238:61–72

Piazzi L, Ceccherelli G (2002) Effects of competition between two introduced *Caulerpa*. Mar Ecol Prog Ser 225:189–195

Piazzi L, Ceccherelli G (2006) Persistence of biological invasion effects: recovery of macroalgal assemblages after removal of *Caulerpa racemosa* var. *cylindracea*. Estuar Coast Shelf Sci 68:655–661

Piazzi L, Balata D, Cecchi E, Cinelli F (2003) Co-occurrence of *Caulerpa taxifolia* and *C. racemosa* in the Mediterranean Sea: interspecific interactions and influence on native macroalgal assemblages. Cryptgamie: Algol 24:233–243

Pinnegar J, Polunin N (2000) Contributions of stable-isotope data to elucidating food webs of Mediterranean rocky littoral fishes. Oecologia 122:399–409

Raffaelli A, Pucci S, Pietra F (1997) Ionspray tandem mass spectrometry for sensitive, rapid determination of minor toxic sesquiterpenoids in the presence of major analogues of the foreign green seaweed *Caulerpa taxifolia* which is invading the northwestern Mediterranean. Anal Chem 34:179–182

Relini G, Relini M, Torchia G (1998) Fish biodiversity in a *Caulerpa taxifolia* meadow in the Ligurian Sea. Ital J Zool 65S:465–470

Relini G, Relini M, Torchia G (2000) The role of fishing gear in the spreading of allochtonous species: the case of *Caulerpa taxifolia* in the Ligurian Sea. ICES J Mar Sci 57:1421–1427

Rierra F, Pou S, Grau A, Delgado O, Weitzmann B, Ballesteros E (1994) Eradication of a population of the tropical green alga *Caulerpa taxifolia* in Cala d'Or (Mallorca, West Mediterranean): methods and results. In: Meinez A, Gravez V, Boudouresque CF (eds) Proc. First International Workshop in *Caulerpa taxifolia*. GIS Posidonie, Marseille, France, pp. 327–331

Ruesink J, Collado-Vides L (2006) Modeling the increase and control of *Caulerpa taxifolia*, an invasive marine macroalga. Biol Invasions 8:309–325

Ruitton S, Javel F, Culioli J, Meinesz A, Pergent G, Verlaque M (2005) First assessment of the *Caulerpa racemosa* (Caulerpales, Chlorophyta) invasion along the French Mediterranean coast. Mar Pollut Bull 50:1061–1068

Ruiz G, Hewitt C (2002) Towards understanding patterns of coastal marine invasions: a prospectus. In: Leppäkoski E, Gollasch S, Olenin S (eds) Invasive aquatic species of Europe: distribution, impact and management. Kluwer Academic, Dordrecht, The Netherlands

Ruiz-Carus R, Matheson R, Roberts D, Whitfield P (2006) The western pacific lionfish, *Pterois volitans* (Scorpaenidae), in Florida: evidence for reproduction and parasitism in the first exotic marine fish established in state waters. Biol Conserv 128:384–390

Sant N, Delghado O, Rodriguez-Prieto C, Ballesteros E (1996) The spreading of the seaweed *Caulerpa taxifolia* (Vahl) C. Agardh in the Mediterranean Sea: testing the boat transportation hypothesis. Bot Mar 39:427–430

Schaffelke B, Murphy N, Uthicke S (2002) Using genetic techniques to investigate the sources of the invasive alga *Caulerpa taxifolia* in three new locations in Australia. Mar Pollut Bull 44:204–210

Schaffelke B, Smith J, Hewitt C (2006) Introduced macroalgae – a growing concern. J Appl Phycol 18:529–541

Semmens B, Buhle E, Salomon A, Pettengill-Semmens C (2004) A hotspot for non-native marine fishes: evidence for the aquarium trade as an invasion pathway. Mar Ecol Prog Ser 266:239–244

Silva P (2003) Historical overview of the *Caulerpa*. Cryptogamie: Algol 24:33–50

Smith C, Walters L (1999) Vegetative fragmentation in three species of *Caulerpa* (Chlorophyta, Caulerpales): the importance of fragment origin, fragment length, and wound dimensions as predictors of success. PSZN Mar Ecol 20:307–319

Stafford N, Bell S (2005) Space competition between seagrass and *Caulerpa prolifera* (Forsskaal) Lamouroux following simulated disturbances in Lassing Park, FL. J Exp Mar Biol Ecol 333:49–57

Stam W, Olsen J, Zaleski S, Murray S, Brown K, Walters L (2006) A forensic and phylogenetic survey of *Caulerpa* species (Caulerpales, Chlorophyta) from the Florida coast, local aquarium shops and e-commerce: establishing a proactive baseline for early detection. J Phycol 42:1113–1124

Sureda A, Box A, Ensenat M, Alou E, Tauler P, Deudero S, Pons A (2006) Enzymatic antioxidant response of a labrid fish (*Coris julis*) liver to environmental caulerpenyne. Comp Biochem Physiol 144:191–196

Thibaut T (2001) Etude fonctionelle, controle et modelisation de l'invasion d'une algue introduite en Mediterranee: *Caulerpa taxifolia*. PhD, Universite de Paris VI, Paris, France

Thibaut T, Meinesz A (2000) Are the Mediterranean mollusks *Oxynoe olivacea* and *Lobiger serradifalci* suitable agents for a biological control against the invading tropical alga *Caulerpa taxifolia*. CR Acad Sci Ser III 323:1–12

Thibaut T, Meinesz A, Amado P, Charrier S, De Angelis K, Ierardi S, Mangialajo L, Melnick J, Vidal V (2001) *Elysia subornata* (Mollusca), a potential control agent of the alga *Caulerpa taxifolia* (Chlorophyta) in the Mediterranean Sea. J Mar Biol Assoc UK 81:497–504

Thibaut T, Meinesz A, Coquillard P (2004) Biomass seasonality of *Caulerpa taxifolia* in the Mediterranean Sea. Aquat Bot 80:291–297

Thresher R, Grewe P (2004) Feasibility study for genetic control of *Caulerpa* in SA and NSW. Summary of the Final Report for the Australian Government. Department of the Environment and Heritage, CSIRO Marine Research, Australia

Thresher R, Kuris A (2004) Options for managing invasive marine species. Biol Invasions 6:295–300

Uchimura M, Rival A, Nato A, Sandeaux R, Sandeaux J, Baccou J (2000) Potential use of Cu^{2+}, K^+ and Na^+ for the destruction of *Caulerpa taxifolia*: differential effects on photosynthetic parameters. J Appl Phycol 12:15–23

Verlaque M, Boudouresque C, Meinesz A, Gravez V (2000) The *Caulerpa racemosa* complex (Caulerpales, Ulvophyceae) in the Mediterranean Sea. Bot Mar 43:49–68

Verlaque M, Durand C, Huisman J, Boudouresque C, Parco Y (2003) On the identity and origin of the Mediterranean invasive *Caulerpa racemosa* (Caulerpales, Chlorophta). Eur J Phycol 38:325–339

Verlaque M, Afonso-Carrillo J, Gil-Rodriguez M, Durand C, Boudouresque C, Le Parco Y (2004) Blitzkrieg in a marine invasion: *Caulerpa racemosa* var. *cylindracea* (Bryopsidales, Chlorophyta) reaches the Canary Islands (NE Atlantic). Biol Invasions 6:269–281

Vroom P, Smith C (2001) The challenge of siphonous green algae. Am Sci 89:525–531

Walters L (2003) Understanding marine bioinvasions: classroom experiments with macroalgae. J Mar Ed Assoc 18:7–10

Walters L, Smith C (1994) Rapid rhizoid production in *Halimeda discoidea* Decaisne (Chlorophyta, Caulerpales) fragments: a mechanism for survival after separation from adult thalli. J Exp Mar Biol Ecol 175:105–120

Walters L, Brown K, Stam W, Olsen J (2006) E-Commerce and *Caulerpa*, unregulated dispersal of invasive species. Frontiers Ecol Environ 4:75–79

West E (2003) The role of anthropogenic activity in the fragmentation and spread of the invasive alga, *Caulerpa taxifolia*. Thesis, University of Wollongong, NSW, Australia

West E, West R (2007) Growth and survival of the invasive alga, *Caulerpa taxifolia*, in different salinities and temperatures: implications for coastal lake management. Hydrobiologia 577:87–94

West E, Barnes P, Wright J, Davis A (2007) Anchors aweigh: fragment generation of invasive *Caulerpa taxifolia* by boat anchors and its resistance to desiccation. Aquat Bot 87:196–202

Westphalen G, Rowling K (2005) *Caulerpa taxifolia* surveys of the North Haven coast. A report for PIRSA Biosecurity. SARDI Aquatic Sciences Publication No. RD02/0161-16. SARDI Aquatic Sciences, Adelaide, Australia

Whitfield P, Gardner T, Vives S, Gilligan M, Courtenay W, Ray G, Hare J (2002) Biological invasion of the Indo-Pacific lionfish *Pterois volitans* along the Atlantic coast of North America. Mar Ecol Prog Ser 235:289–297

Wiedenmann J, Baumstark A, Pillen T, Meinesz A, Vogel W (2001) DNA fingerprints of *Caulerpa taxifolia* provide evidence for the introduction of an aquarium strain into the Mediterranean Sea and its close relationship to an Australia population. Mar Biol 138:229–234

Williams S (1984) Decomposition of the tropical macroalga *Caulerpa cupressoides*: field and laboratory studies. J Exp Mar Biol Ecol 80:109–124

Williams S (1990) Experimental studies of Caribbean seagrass bed development. Ecol Monogr 60:449–469

Williams S, Grosholz E (2002) Preliminary reports from the *Caulerpa taxifolia* invasion in southern California. Mar Ecol Prog Ser 223:307–310

Williams S, Schroeder S (2004) Eradication of the invasive seaweed *Caulerpa taxifolia* by chlorine bleach. Mar Ecol Prog Ser 272:69–76

Wotton D, Hewitt C (2004) Marine biosecurity post-border management: developing incursion response systems for New Zealand. NZ J Mar Freshw Res 38:553–559

Wright J (2005) Differences between native and invasive *Caulerpa taxifolia*: a link between asexual fragmentation and abundance in invasive populations. Mar Biol 147:559–569

Wright J, Davis A (2006) Demographic feedback between clonal growth and fragmentation in an invasive seaweed. Ecology 87:1744–1754

York P, Booth D, Glasby T, Pease B (2006) Fish assemblages in habitat dominated by *Caulerpa taxifolia* and native seagrasses in south-eastern Australia. Mar Ecol Prog Ser 312:223–234

Zaleski S, Murray S (2006) Taxonomic diversity, geographic distribution and commercial availability of aquarium-traded species of *Caulerpa* (Chlorophyta, Caulerpaceae) in southern California. Mar Ecol Prog Ser 314:97–108

Zuljevic A, Antolic B (2000) Synchronous release of male gametes of *Caulerpa taxifolia* (Caulerpales, Chlorophyta) in the Mediterranean Sea. Phycology 39:157–159

Zuljevic A, Meinesz A (2002) Appearance and eradication of *Caulerpa taxifolia* in Croatia. In: Williams E, Grosholz E (eds) International *Caulerpa taxifolia* Conference Proceedings. CA Sea Grant, University of California – San Diego, La Jolla, CA

Zuljevic A, Thibaut T, Elloukal H, Meinesz A (2001) Sea slug disperses the invasive *Caulerpa taxifolia*. J Mar Biol Assoc UK 81:343–344

Chapter 16
Approach of the European and Mediterranean Plant Protection Organization to the Evaluation and Management of Risks Presented by Invasive Alien Plants

Sarah Brunel, Françoise Petter, Eladio Fernandez-Galiano, and Ian Smith

Abstract Invasive alien plants may be introduced intentionally with trade (80% of current invasive alien plants in Europe were introduced as ornamental or agricultural plants) or unintentionally (as contaminants of grain, seeds, soil, machinery etc., or with travellers). Preventing the introduction of invasive alien plants is considered more cost-effective, from both environmental and economic points of view, than managing them after introduction. Pest risk analysis (PRA) standards have been developed by the International Plant Protection Convention (IPPC) and the European and Mediterranean Plant Protection Organization (EPPO) to allow assessment of the phytosanitary risk presented by invasive alien plants, and the development of appropriate measures to prevent their introduction and spread. These measures may in turn have an impact on international trade, and the obligations arising from trade agreements have also to be taken into account when phytosanitary measures are established. PRA basically consists in a framework for organizing biological and other scientific and economic information to assess risk. This leads to the identification of management options to reduce the risk to an acceptable level. Within the EPPO context, the results of these PRAs are translating into recommendations for countries to implement their national regulations. This article gives an overview of the international framework for regulation of invasive alien plants under the IPPC. It then presents the approach followed by EPPO for the evaluation and management of risks presented by such plants, as well as its application.

Keywords European and Mediterranean Plant Protection Organization • Pest risk analysis

S. Brunel(✉), F. Petter, E. Fernandez-Galiano, and I. Smith
European and Mediterranean Plant Protection Organization (EPPO), 1 rue Le Nôtre, 75016 Paris, France 33-1-42-24-89-43,
brunel@eppo.fr

16.1 Introduction

Modern methods of travel, trade and communication have allowed an enormous increase in the movement of people, commodities and conveyances over the past century and this is still accelerating. This has resulted in a higher risk of introduction and spread of organisms harmful to plants and plant products, including invasive alien species. Trade in agricultural products provides a clear economic benefit but prevention of introduction of pests with trade is recognized as an important target for countries. In this respect, the International Plant Protection Convention (IPPC) (FAO 1997) has been the major agreement for countries that trade in agricultural, horticultural and forestry products. The Conference of the Parties to the Convention on Biological Diversity (CBD, http://www.cbd.int/default.shtml. Accessed on 1 February 2008) has responsibility for global policies on invasive alien species, but has recognized the role of the IPPC in this sector. In the framework of the IPPC, the European and Mediterranean Plant Protection Organization (EPPO) has recently developed a work programme specifically addressing invasive alien plants, as part of its ongoing programme on quarantine pests.

Invasive alien plants may be introduced intentionally with trade (80% of current invasive alien plants in Europe were introduced as ornamental or agricultural plants; Hulme 2007) or unintentionally (as contaminants of grain, seeds, soil, machinery, etc., or with travellers). Preventing the introduction of invasive alien plants is considered more cost-effective, from both environmental and economic points of view, than managing them after introduction. Pest risk analysis (PRA) standards have been developed by IPPC and EPPO to allow assessment of the phytosanitary risk presented by invasive alien plants, and the development of appropriate measures to prevent their introduction and spread. These measures may in turn have an impact on international trade, and the obligations arising from trade agreements have also to be taken into account when phytosanitary measures are established.

This article gives an overview of the international framework for regulation of invasive alien plants under the IPPC. It then presents the approach followed by EPPO for the evaluation and management of risks presented by such plants. Terms used are defined in the *Glossary of phytosanitary terms* (IPPC 2007c).

16.2 International Context

16.2.1 World Trade Organization (WTO)

The WTO was established in January 1995 and deals with the rules of trade between nations at a global or near-global level. It is a negotiation forum for 150 member countries. It results from the 1986 to 1994 negotiations called the Uruguay Round and earlier negotiations under the General Agreement on Tariffs and Trade (GATT).

The GATT (1994) is WTO's core agreement with respect to trade in commodities, and its objective is to limit tariff and non-tariff barriers to trade. Its two main requirements are that (1) imported commodities should not be treated less favourably than equivalent domestic commodities (the "national treatment" obligation), (2) there should not be discrimination for imported commodities between countries where the same conditions prevail. Nevertheless, article XX of the GATT states that "nothing in the Agreement shall prevent the adoption or enforcement by any contracting party of measures necessary to protect human, animal or plant life or health, provided that measures are not applied in a manner which would constitute a means of arbitrary or unjustifiable discrimination between countries where the same conditions prevail, or a disguised restriction on international trade".

The Agreement on the Application of Sanitary and Phytosanitary Measures (SPS Agreement) (WTO 1994) elaborates rules for the application of the provision of GATT related to the use of sanitary or phytosanitary measures, in particular article XX. It defines the basic rights and obligations of members to protect animal and plant life or health from risks arising from the entry, establishment or spread of pests, where such measures may directly or indirectly affect international trade. Consequently this agreement covers phytosanitary regulations established to prevent the introduction of invasive alien plants. Preventive measures have to comply with a set of principles such as "harmonization", "equivalence", "assessment of risk", "transparency", etc. The agreement also provides for a dispute settlement mechanism so that in case of dispute between countries, the two contracting parties should consult bilaterally with the aim of resolving the problem. In the SPS agreement, the IPPC is recognized as the relevant international standard-setting organization for the elaboration of international standards ensuring that phytosanitary measures are not used as unjustified barriers to trade.

16.2.2 IPPC

The IPPC is an international treaty to which 165 governments currently adhere (as of September 2007). Its objectives are to secure action to prevent the spread and introduction of pests of plants and plant products, and to promote appropriate measures for their control. It came into force in 1952. It is governed by the Commission on Phytosanitary Measures (CPM), which adopts International Standards on Phytosanitary Measures (ISPMs). The IPPC Secretariat coordinates the activities of the Convention and is hosted by the Food and Agriculture Organization of the United Nations (FAO). See IPPC website at https://www.ippc.int/IPP/En/default.jsp.

The IPPC is implemented at a national level by phytosanitary authorities called National Plant Protection Organizations (NPPOs), usually within the Ministry of Agriculture. NPPOs carry out the important task of preventing the introduction and spread of quarantine pests. An efficient infrastructure (such as border controls, national surveillance programmes, technical and scientific institutions, as well as export-oriented certification programmes) has been established to achieve the tasks of phytosanitary authorities (Lopian 2005).

As explained before, IPPC is recognized as the standard-setting organization for phytosanitary measures and is developing ISPMs. So far, 29 ISPMs have been adopted, of which 3 are of particular interest for risk analysis:

- ISPM no.1 *Phytosanitary principles for the protection of plants and the application of phytosanitary measures in international trade.* (IPPC 2007a)
- ISPM no. 2 *Framework for pest risk analysis.* (IPPC 2007b)
- ISPM no. 11 *Pest risk analysis for quarantine pests including analysis of environmental risks and living modified organisms.* (IPPC 2007d)

At one time, the IPPC was interpreted as referring mainly to the protection of cultivated plants, but in 1999 the CPM recognized that it always had a wider scope, extending to wild plants and the environment. Major changes were made to two ISPMs in consequence. Firstly, a supplement (no. 2) was added to the *Glossary of phytosanitary terms*, providing "*Guidelines on the understanding of potential economic importance and related terms including reference to environmental considerations*". This made it clear that "potential economic importance" (as referred to in the IPPC definition of a quarantine pest) can include environmental concerns. Thus, the scope of the IPPC covers the protection not only of cultivated plants in agriculture (including horticulture and forestry), but also of uncultivated/unmanaged plants, wild flora, habitats and ecosystems. Secondly, extensive changes were made to ISPM no. 11 on *Pest risk analysis for quarantine pests.* This standard describes the integrated processes to be used for the assessment of risks presented by plant pests, as well as the selection of risk management options. The concerns for the environment originally concerned only the side effects on the environment of pests mainly affecting cultivated plants. This was now extended to any organisms having harmful effects on plants in the environment, whether or not they affect cultivated plants. The analysis of risks to the environment and biological diversity, including risks affecting uncultivated/unmanaged plants, wild flora, habitats and ecosystems contained in the PRA area, was set out in greater detail and, most importantly for the present purpose, invasive alien plants were recognized as an important hazard for the environment. As a result, invasive alien plants can now be the subject of PRA under the IPPC.

16.2.3 CBD

In June 1992, the United Conference on Environment and Development (UNCED) known as the "Earth Summit" was held in Rio de Janeiro. One of the main results of this summit was the signature of the CBD, which aims at the conservation and sustainable use of biological diversity and the fair and equitable sharing of benefits arising out of the utilization of genetic resources. To date, it has been signed by 150 governments, including those of all the European countries. More information is available at http://www.cbd.int/default.shtml. In its Article 8(h), the CBD asks its members "to prevent the introduction of, control or eradicate those alien species which threaten ecosystems, habitats or species" as far as possible and when appropriate.

In 2002, at the sixth meeting of the CBD Conference of the Parties in The Hague, "*Guiding Principles for the prevention of introduction and mitigation of impacts of alien species that threaten ecosystems, habitats or species*" were adopted (CBD 2002). This text provides further advice to members on Article 8(h) of the Convention. More recently, the eighth CBD Conference of the Parties held in Brazil in 2006 encouraged members to work at a regional level and to ensure close inter-agency cooperation at the national and regional levels among the various sectors (Ministries of Environment and of Agriculture, traders), as well as sharing information necessary for risk analysis (COP 8 2006).

16.2.4 Cooperation Between the IPPC and the CBD

Since activities of the CBD in relation to invasive alien species correspond to a certain degree with those of the IPPC for those invasive alien species that are harmful to plants, cooperation between the CBD and the IPPC has been established since 2004. This avoids overlap and duplication of work between the two Conventions. The respective Secretariats participate in each other's meetings. A Memorandum of Understanding has been established between CBD and IPPC and the revision of ISPMs no. 5 and no. 11 (see previous paragraph) was accordingly done in consultation.

The relationship between the CBD guiding principles on invasive alien species and the IPPC and its ISPMs has been described by Schrader and Unger (2003) and Lopian (2005), and will be the subject of a new supplement to the *Glossary of phytosanitary terms*, whose purpose is to give an interpretation of the terminology of the Convention on Biological Diversity in relation to the *Glossary of phytosanitary terms*. Essentially, the CBD defines an "alien" as a "species … introduced outside its natural … distribution" and an invasive alien species as "an alien species whose introduction and/or spread threatens biological diversity" (annex footnote 57, CBD 2002). The *Glossary of phytosanitary terms* defines a quarantine pest as "a pest of potential economic importance to the area endangered thereby and not yet present there, or present but not widely distributed and being officially controlled". Evidently, the two definitions cover similar ground. The main differences are that, unless "biodiversity" is taken in a very wide sense to include agro-ecosystems, a quarantine pest does not necessarily threaten biodiversity and may only affect agriculture (Lopian 2005). On the other hand, according to the CBD, an invasive alien species has already been introduced. If it has also spread to the point that it is widely distributed, it can no longer be considered as a quarantine pest. Thus, ISPM no. 11 on "*Pest risk analysis for quarantine pests including analysis of environmental risks and living modified organisms*" applies to invasive alien plants that have been introduced but are not widely distributed. It also applies to potentially invasive plants that have not yet been introduced.

16.3 European Regional Context

16.3.1 EPPO

To promote regional cooperation, the IPPC includes provisions for the establishment of Regional Plant Protection Organizations (RPPOs) functioning as coordinating bodies in the areas they cover. EPPO is the RPPO for Europe and the Mediterranean area, and establishes regional standards on phytosanitary measures. It was created in 1951, and in 2008 it has 49 member countries, including all members of the European Union, Russia and several other countries of the Commonwealth of Independent States, and Mediterranean countries in North Africa and the Near East. EPPO's members are represented by their NPPOs, i.e. the official services that are responsible for plant protection in each country (usually part of the Ministry of Agriculture). One of EPPO's main priorities is to prevent the introduction of dangerous pests from other parts of the world, and to limit their spread within the region should they be introduced. EPPO is also conducting regional PRA activities for the European and Mediterranean region. More information on EPPO's activities is available at www.eppo.org.

16.3.2 Bern Convention

The Convention on the Conservation of European Wildlife and Natural Habitats (Bern 1979), generally known as "the Bern Convention" is a nature conservation treaty, which deals with a wide array of aspects concerning the conservation of natural heritage in Europe. It counts at present 44 Contracting Parties, including the 27 Member States of the European Union, the European Community and four African states. It is administered by the Council of Europe, based in Strasbourg. It implements the CBD within its region and has a threefold objective: to conserve wild flora and fauna and their natural habitats, to promote co-operation between states in the field of conservation of biological diversity and, in particular, to protect endangered and vulnerable species and endangered natural habitats.

The Bern Convention requires Contracting Parties "to strictly control the introduction of non-native species" (Article 11 par. b). The Convention coordinates action of European Ministries of the Environment in matters related to the conservation of biological diversity. It started activities on invasive alien species in 1984 with the launch of a general recommendation to the member states of the Council of Europe, followed by the establishment of a group of experts on invasive alien species. Specific recommendations were then adopted, as for instance on the control of *Caulerpa taxifolia* (an invasive alga in the Mediterranean). In 2002, the Convention adopted a European Strategy on invasive alien species, with the aim of providing guidance to countries in drawing up and implementing their national strategies (Genovesi and Shine 2002). The Strategy identifies priorities and key actions in this field and includes precise proposals on (1) awareness and information issues concerning invasive alien species, (2) the need to strengthen national and regional capacities, (3) prevention of new introductions and early warning systems for new arrivals, (4) reduction of the adverse impacts of invasive alien species on biological diversity, (5) measures required to recover species and natural habitats affected by invasive alien species.

16.3.3 Cooperation Between EPPO and the Bern Convention

The Bern Convention and EPPO have established a partnership on the topic of invasive alien species, and work closely together on invasive alien plants at the regional scale, as recommended by the CBD (COP 8 Decision VIII/27). As IPPC-related activities are in most countries under the responsibility of the Ministry of Agriculture, while CBD matters are under the responsibility of the Ministry of the Environment, this partnership allows a concrete partnership to be established between the Plant Health and Biodiversity Conservation sectors.

16.4 EPPO Regional Approach to the Evaluation and Management of Risks Presented by Invasive Alien Plants

16.4.1 PRA Systems in Place Within EPPO

The EPPO Convention lays down that one of the aims of EPPO is "to pursue and develop, by cooperation between the Member Governments, the protection of plants and plant products against pests and the prevention of their international spread and especially their introduction into endangered areas". EPPO Council has consequently decided to draw up lists of pests, which present an unacceptable risk, and whose regulation is relevant for the whole of, or large parts of, the EPPO region. The first list is of A1 pests, not present in the EPPO region. The second list is of A2 pests, present in the EPPO region but not widely distributed (i.e. absent from or not widely distributed in certain countries, where they are therefore subject to official control). The first lists were approved in 1975. In 2007, they contained 298 quarantine pests recommended for regulation (available on the EPPO website).

Addition of a pest to the A1 or A2 list may be proposed by a member government, or result from the appearance of the pest on the EPPO Alert List (a pest warning system managed by the Secretariat). In either case, the proposal has been since the mid 1990s subject to PRA following the standards of the IPPC and EPPO. Originally, this PRA was usually put forward by the proposing member, commissioned from an expert or prepared by the Secretariat. Since 2006, however, in addition to the process mentioned earlier, PRAs on specific pests are performed by Expert Working Groups (EWG), following EPPO PM 5/3 *Decision-support scheme for quarantine pests*. Expert Working Groups have already been organized for several plant pests (*Phytophthora lateralis*, *Iris yellow spot virus, Megaplatypus mutatus and Tetranychus evansi*), which are not plants. EWG will also be organized on plants from 2008 on.

The output of a PRA takes the form of a general recommendation to countries, with measures proposed for each organism concerned, distinguishing different levels of risks for different parts of the EPPO region if necessary (Smith 2005). This recommendation has then to be adopted by consensus by the EPPO Members, after appropriate consultation. Members decide individually whether the reported risks concern them, and select appropriate measures if they do. The EPPO Convention creates no greater obligation on members than that they should "endeavour to implement" EPPO recommendations. However, there is a general policy of "regional solidarity", by which Members do take phytosanitary measures against A1 pests (unless the risk of establishment on their territory is very low) and do select their measures from those recommended.

The PRA documents are freely available on the EPPO website as recommended in Decision VIII/27 of the CBD Conference of the Parties held in 2006 in Brazil (CBD 2006). EPPO organizes periodic training sessions on PRA for staff of the NPPOs of EPPO countries.

16.4.2 Initiation of an EPPO Work Programme on Invasive Alien Plants

In 2002, the EPPO Council recognized that invasive alien species that have an effect on plants are quarantine pests under the IPPC (and therefore should be evaluated following ISPM no. 11), and that NPPOs should consider their responsibility for the management of invasive alien plants (which are considered quarantine pests under the IPPC), in cooperation with the environmental authorities. As a consequence, EPPO initiated a work programme on invasive alien species (Schrader 2004) and a Panel on invasive alien species was created to help the EPPO member countries to achieve this aim. This Panel now has experts from 18 countries of the EPPO region.

The Panel started its work by assembling a preliminary list of approximately 500 invasive alien plants in the EPPO region from the scientific and technical literature, from web sites and from official contacts in EPPO member countries (by questionnaire). Technical evaluation of this list led to the first achievement of the Panel: a list of 40 terrestrial or aquatic invasive alien plants identified as posing an important threat to plant health, environment and biodiversity in the EPPO region.

The prioritization of these species was done by expert judgment based on the following factors:

- Whether the plant is considered invasive or potentially invasive by several EPPO countries
- Whether the plant is absent or still containable by appropriate measures in several EPPO countries
- Whether the plant has potential for further spread and damage into significant areas where it is absent
- Whether the plant is reported to be actively spreading or becoming more damaging in its current distribution area

EPPO strongly recommends countries endangered by these species to take measures to prevent their introduction and spread or to manage unwanted populations (for example by publicity, restrictions on sale or planting, control campaigns). The species mentioned on this list may in fact be quite widely distributed, and the EPPO recommendations concerning them are intended to be applied nationally (by the NPPO, or more probably by some other national or subnational authority).

This list, with additional information on the individual plants, is available on the EPPO website. It is open to revision and extension, and the Panel is further developing a prioritization process to take more species into account and to determine priorities for PRA.

Besides this, the EPPO Reporting Service (the EPPO monthly web-based phytosanitary newsletter) has been extended to include many items on invasive alien species, for example reports from individual countries concerning the species to which they give priority, and information on pathways for the introduction of invasive alien plants such as aquatic plants (EPPO RSE 2007/016) or bird seed (EPPO RSE 2007/123).

16.4.3 PRA of Invasive Alien Plants

The EPPO Panel's work as described earlier was not based on formal PRA, and concentrated on species already present within the region and already recognized to be invasive. The aim was to develop activities and to reach a rapid consensus on priorities for the EPPO region. The next phase of the work programme was to apply the EPPO PRA system to invasive alien plants, in the same way as it is used for other plant pests, i.e. to place certain invasive alien plants in the EPPO A1 or A2 lists, and to recommend measures against them.

16.4.3.1 A1 List

The first question that arose was whether to place invasive alien plants on the EPPO A1 list. It should be recalled that A1 pests are not present in the EPPO region (i.e. have not been introduced), so that they do not fit the CBD definition of having already been introduced. In practice, the EPPO Panel ignored this distinction, and considered that, in principle at least, EPPO could conduct PRA for plants not present in the EPPO region, which could be considered potentially invasive. The difficulty was, however, that there is a great number of plant species, which may be introduced into Europe and there they become invasive alien plants. Furthermore, the horticultural industry introduces new plant species into cultivation in Europe every year. Performing a PRA is a time-consuming and laborious process, and it is difficult to make confident predictions of the behaviour of alien species in Europe. Accordingly, although a preventive CBD approach would be ideal, it has not appeared feasible within EPPO to conduct PRAs on potentially invasive plants for addition to an A1 list. In particular, European countries do not currently regulate the import of non-European plants as such, except as pathways for plant pests or for quite other reasons (e.g. regulations on illegal drugs). The possibilities of reaching international agreement on a list of plants to be internationally regulated, and of undertaking the work programme to establish such a list, seem remote. For this reason, the main focus of attention for PRA for plants has been on the A2 list, i.e. invasive alien plants that are already present in Europe.

16.4.3.2 A2 List

Adding invasive alien plants to the A2 list implies performing PRAs on alien species, which already have a limited distribution in the EPPO region and which have shown invasive behaviour in Europe and/or elsewhere in the world. It also implies that international phytosanitary measures are appropriate concerning the movement of these species between countries, which is not always the case, since national measures may be more appropriate. A2 candidates are more likely to be very recent arrivals, present in very few countries, and could include species that have only just been introduced into cultivation in Europe and have not established in the wild. In these cases, international measures are especially appropriate.

16.4.4 The EPPO A2 List and the EPPO List of Invasive Alien Plants

As noted earlier, EPPO has established a list of "invasive alien plants identified as posing an important threat to plant health, environment and biodiversity in the EPPO region". All these listed species do not necessarily qualify as potential A2 pests, subject to international regulation, because of the following:

- They may be widely distributed (and so not fit the definition of a quarantine pest)
- It may not be possible to apply national measures equivalent to those required internationally, so that non-discrimination cannot be assured
- It may not be relevant to apply measures related to international movement

In fact, the EPPO Panel has undertaken to operate as an Expert Working Group (see earlier) to perform PRAs on alien plants for the EPPO A2 list. Five species have been subjected to PRA and are now recommended for regulation to the 49 EPPO countries (*Crassula helmsii, Hydrocotyle ranunculoides, Lysichiton americanus, Pueraria lobata, Solanum elaeagnifolium)*. As of September 2007, further PRAs are in preparation for *Heracleum sosnowskyi, H. persicum, Polygonum perfoliatum* and *Eichhornia crassipes*. On the basis of information gathered by the EPPO Secretariat, it appears that all these species have a limited distribution within the EPPO region and their entry into other countries of the region could be prevented. A process is now being developed to identify further candidates on the basis of simple transparent criteria.

16.4.5 *Practical Application*

So far, of the five species recommended for regulation by EPPO, only *Hydrocotyle ranunculoides* is regulated: its possession and trade are prohibited in The Netherlands. These preventive measures are implemented in a single country, and may be compromised if efforts by neighbouring countries are inadequate (Burgiel et al. 2006). Indeed, there are no international measures established for this plant and, other than in an extreme emergency, the EU phytosanitary system would not allow an individual country to put in place such measures unilaterally (though non EU countries could do so). So, at a practical level, it must be recognized that the recommendations made by EPPO on invasive alien plants are fairly recent, and time will be needed before national (or EU) regulations are implemented. In addition, NPPOs may also have to consult with national environmental authorities in evaluating the risk to their territory and in determining the measures to be established (Smith 2005). It is possible to regulate invasive alien plants under the IPPC, and EPPO has taken the first steps in creating a situation in which the European countries (and the EU) can do so.

16.5 Application of the EPPO Decision-Support Scheme on PRA to Invasive Alien Plants

As outlined earlier, EPPO has developed a scheme for PRA of quarantine pests, and has also started to perform PRAs on invasive alien plants. Since PRA is a technical analysis providing a basis for administrative and legislative decisions, it is important that it should be done transparently according to accepted standards. Thus, EPPO has adapted and extended its decision-support scheme so that it can be used for all sorts of plant pests, including invasive alien plants. This scheme therefore provides an example of how a PRA scheme developed in the framework of the IPPC can be used to assess invasive alien plants (Schrader 2004).

The EPPO scheme originally took the form of two separate standards: pest risk assessment (PM 5/3 adopted in 1997) and pest risk management (PM 5/4 adopted in 2000). More recently, these have been merged into a single revised EPPO Standard PM 5/3: *Decision support scheme for quarantine pests*, compatible with ISPM no. 11 *Pest risk analysis for quarantine pests including analysis of environmental risks and living modified organisms*. The scheme provides detailed instructions for the successive stages of PRA: initiation, pest categorization, probability of introduction, assessment of potential economic consequences and pest risk management. Basically, it is a framework for organizing biological and other scientific and economic information, and using it to assess risk. This leads to the identification of management options to reduce the risk to an acceptable level. PRAs can be very short and simple, or very long and complex. There is no fixed criterion for the quantity of information needed. The evaluation does not necessarily have to be quantitative and it can include qualitative considerations, as long as it is scientifically sound (Burgiel et al. 2006). Expert judgement may be used in answering the questions.

The successive stages of the scheme are reviewed here, with particular reference to its use for invasive alien plants. The scheme follows the sequences presented in Appendices 1, 2 and 3.

16.5.1 Initiation

Initiation aims to identify the pests or pathway to be considered for risk analysis in relation to the identified PRA area. The EPPO scheme is primarily concerned with the assessment of individual pests, since this is the basis on which European countries formulate their phytosanitary regulations. So, European countries do PRAs for pests, and thus for individual invasive alien species if appropriate.

However, ISPM no. 11 also provides for PRA of a pathway. Countries that prohibit the import of most plants and plant products frequently have to consider whether a new trade can be opened for a previously prohibited plant. The PRA then concerns all the pests that might be carried by this new pathway. Such PRAs are not normally done in Europe (though the EPPO Standard follows ISPM no. 11 in allowing the possibility). For invasive alien plants, the evaluation of a pathway such as internationally traded birdseed could be relevant for EPPO, and could be the initiation point for PRAs of new candidate plants.

In doing PRAs for individual pests, it is important to establish that their identity is clear. The pests should as far as possible be well and accurately documented before the PRA starts. The information generally needed is listed in EPPO Standard PM5/1(1) *Check-list of information required for pest-risk analysis* (PRA) (EPPO 1998), though it needs revision to cover invasive alien plants. While using the scheme, the user should specify all details that appear relevant to the replies to individual question, indicating the source of the information (Schrader 2005).

Although the EPPO scheme specifies many possible initiation points for PRAs, most are not relevant for invasive alien plants. From experience so far, there are broadly two initiation points for invasive alien plants. PRAs may be appropriate for the following:

- Plants that have been (or are proposed to be) intentionally introduced for ornament, and that have, or might in future, escaped from plantations to invade and threaten unmanaged ecosystems (i.e. semi-natural or natural habitats). According to Hodkinson and Thompson (1997), these species tend to be spreading perennials with transient seed banks. Such species represent about 80% of invasive alien plants (Hulme 2007). With respect to the PRAs performed so far by EPPO, *Crassula helmsii, Hydrocotyle ranunculoides, Lysichiton americanus, Pueraria lobata, H. sosnowskyi, H. persicum*, and *Eichhornia crassipes* fall into this category.
- Plants that are unintentionally introduced as contaminants associated with international movement of various commodities and articles, including soil and vehicles. According to Hodkinson and Thompson (1997), these plant species are often small and fast growing, but their most unifying characteristic is the production of numerous, small, persistent seeds. Grain and seeds for planting are important commodities likely to act as a pathway for unintentional introduction of such plants. Because such plants are originally associated with the agricultural or managed plants or plant products that are traded, they are also likely to be a greater threat to agriculture and cultivated ecosystems (as weeds) than to uncultivated ecosystems. With respect to the PRAs performed so far by EPPO, *Solanum elaeagnifolium* and *Polygonum perfoliatum* fall into this category.

16.5.2 Pest Risk Assessment

16.5.2.1 Pest Categorization

A rapid qualitative assessment is first made, with little information, to determine whether the organism meets the criteria of the definition of a quarantine pest (see paragraph *Cooperation between the IPPC and the CBD*) and could therefore be regulated in international trade. The main aim of this step is to avoid conducting a full PRA in a case that can immediately be seen not to require one.

If the pest categorization step leads to a positive answer, the main PRA starts. It is essentially composed of a series of questions, made in terms of "likeliness" for qualitative questions (very unlikely, unlikely, moderately likely, likely, very likely), and an estimate for quantitative questions (very few, few, moderate number, many, very many).

16.5.2.2 Probability of Introduction

Introduction, as defined by the *Glossary of phytosanitary terms* is the entry of a pest resulting in its establishment. Entry and establishment are separate processes and need to be evaluated separately. It may be noted that, in CBD terminology, introduction does not include establishment, and is thus effectively entry. This text follows IPPC terminology.

16.5.2.3 Probability of Entry

For quarantine pests other than invasive alien plants, there may be many alternative pathways of entry to be considered. For any of these to be regulated in international trade, the PRA should show that other relevant pathways have been considered. Each has to be considered in turn.

For invasive alien plants, the possibilities of entry are in practice more limited. Following Burgiel et al. (2006) and Genovesi (2007), pathways of entry of invasive plants can be categorized as follows:

Intentional entry		Unintentional entry
Direct entry into the environment	Entry into a containment facility or in a controlled environment	• The alien species unintentionally enters as a contaminant of a specific commodity: plant products such as plants for planting, seeds, grain, and soil and packaging
• For ornament in landscaping (the most frequent case)	• In botanical and private gardens	
	• In greenhouses	
• For agriculture	• In aquarium and horticultural pond trade	
• For forestry	• For research	• The alien species unintentionally enters with movements of people or of machinery

For invasive alien plants, the pathway most often assessed is intentional import for ornamental purposes (including aquatic plants). In this case, entry is certain and does not need to be considered as a variable. The assessor can go directly to the probability of establishment (in particular the probability of establishment in non-intended habitats).

Nevertheless, species introduced for ornament may also be introduced as contaminants. For instance, seeds of *Heracleum* spp. may contaminate soil and growing media.

In the cases that EPPO has considered so far, intentional entry was the only pathway evaluated for *Crassula helmsii, Hydrocotyle ranunculoides, Lysichiton americanus* and *Pueraria lobata*. In the other case (*Solanum elaeagnifolium*), unintentional entry by several pathways was considered: contaminant of plants for planting, soil/growing media, used machinery, grain, seeds for planting.

A plant associated with a pathway is assessed first for the probability that it should enter, then for its survival during transport and its probability of transfer to a suitable habitat. Thus, a plant that occurs in nurseries in the exporting country is likely to be carried by plants for planting in growing media moving in international trade, is likely to survive especially if it is in the form of seeds rather than young plants, and is likely to escape as a weed in the nursery of destination. A plant that contaminates grain at harvest is likely to survive as seeds in the grain, but relatively unlikely to reach a suitable habitat if the grain is processed in the usual way.

16.5.2.4 Probability of Establishment

Whatever the type of pest, an organism that enters does not necessarily establish. Many exotic plants enter intentionally or unintentionally, but few escape. Of those that do, many are only reported as casual and then disappear since they cannot maintain sustainable populations. Only a small fraction can establish in the wild, and it is this probability that has to be assessed.

The first parameter necessary for the establishment of the plant is the presence of suitable habitats. These are listed and their number and distribution are assessed to determine whether the invasive plant will find adequate environment to establish. A plant like *Pueraria lobata*, for example, which colonizes disturbed habitats such as roadsides, fallows and edges of forests, has numerous potential habitats.

The second parameter is the suitability of the environment. The similarity of climatic conditions in the PRA area and in the current area of distribution of the species is considered. When possible, a climatic prediction analysis can be performed with softwares such as CLIMEX indicating different levels of risk. Full details of the software can be found on the Hearne website (http://www.hearne.com.au/products/climex/) and in the CLIMEX User's Guide (Sutherst et al. 2004). For instance, in the case of *Solanum elaeagnifolium*, Mediterranean countries are considered more at risk than temperate countries, and northern countries are not at risk. Other relevant environmental factors are abiotic factors such as soil type, and biotic factors such as competition and natural enemies.

The reaction of an introduced plant to current management practices and possible control measures will affect the probability of establishment, together with various other characteristics of the plant such as reproductive strategy, genetic diversity, and adaptability.

16.5.2.5 Probability of Spread

A plant that can rapidly spread after establishment presents a much greater risk. An assessment is made of the risk of natural spread, including movement by wind or water dispersal, transport by vectors such as insects or birds, natural migration, rhizome growth, combined with the presence of natural barriers and the quantity of pest to be dispersed, and also of the risk of spread by human assistance, through movement of soil, irrigation waters, footwear, used machinery, etc. The possibility of containing the plant is also considered, since herbicide treatments may easily contain a plant even if it has established.

16.5.2.6 Potential Economic Consequences (Including Environmental Impacts)

In the case of introduced plants, establishment and spread do not necessarily imply that there is a negative impact. Introduced species may even increase biological diversity (the Mediterranean flora contains about 20% of exotic species). So it is necessary to evaluate further whether there are potential negative economic impacts (including environmental and social impacts). Any such effects are documented and evaluated for the current area of distribution of the plant, and estimated for the PRA area. This may be done in monetary terms, especially for control costs. For example, in the EPPO PRA for *Crassula helmsii*: "one recent estimate puts the cost of control of *C. helmsii* at between 1.45 and 3 million euros based on the treatment of 500 sites over a period of 2–3 years in the British Isles" (Leach and Dawson 1999).

For invasive alien plants, it is particularly important to evaluate environmental impacts such as reduction of keystone species; reduction of species that are major components of ecosystems, and of endangered species; significant reduction, displacement or elimination of other species; indirect effects on plant communities (species richness, biodiversity); significant change in ecological processes, and the structure, stability of an ecosystem (including further effects on plant species), etc. are evaluated. For example, in the assessment of the environmental impact of *Crassula helmsii*, part of the information provided in the PRA is: "[…]. The rare starfruit *Damasonium alisma*, one of the rarest plants in UK, is thought to be threatened by *C. helmsii* (Watson 2001). Moreover, Leach and Dawson (1999) state that in an artificially managed lake (Priors Down Lake, Stalbridge, Dorset), evidence suggests changes in floral dominance, *C. helmsii* excluding *Ludwigia palustris* and *Galium debile* (Dawson and Warman 1987) […]."

Invasive alien plants may also have social impacts, which can be taken into account as they would be for any other kind of pest. For example, these social impacts could include damaging the livelihood of a proportion of the human population and affecting human activities (e.g. water quality, recreational uses, tourism, animal grazing, hunting and fishing). Some of these effects, such as those on human or animal health, the water table or tourism, might have also to be considered, as appropriate, by other agencies/authorities. Information provided for *Crassula helmsii* was: "The mats formed by the plant choke ponds and drainage ditches. Strongly invaded waters lose their attractiveness for recreation and flooding may be caused. The mats can be dangerous to pets, livestock and children who mistake them for dry land".

Whether for entry, establishment or economic effects, the areas and degree of uncertainty should be noted. They ensure transparency of the process (according to the SPS Agreement principle of transparency) and may orientate additional research to complete the PRA or give it more accuracy.

The overall conclusion of the pest risk assessment is to decide whether the pest qualifies as a quarantine pest, on the basis of the answers given. If so, PRA continues with the selection of risk management options, provided the risk identified is considered unacceptable.

16.5.3 Pest Risk Management

This part of the analysis identifies measures to prevent entry, establishment or spread of the pest. It explores options that can be implemented: (1) at origin or in the exporting country, (2) at the point of entry or (3) within the importing country or invaded area. The options are structured so that, as far as possible, the least stringent options are considered before the most expensive/disruptive ones, and are consistent with the SPS-Agreement and Plant Health principles (described in ISPM no. 1).

The methods whereby risk management options are selected for invasive alien plants differ according to whether the introduction is intentional or unintentional, whether the organism is absent or already present in the PRA area and the type of entry pathway. Different measures will apply for these different categories.

If the invasive alien plant is to be intentionally imported, the possible measures will generally be either to prohibit import (e.g. in the case of *Pueraria lobata*) or to take action only within the importing country. An EPPO Standard PM/3 67 on *Guidelines for the management of invasive alien plants or potentially invasive alien plants which are intended for import or have been intentionally imported* has been adopted in 2006 (http://www.blackwell-synergy.com/doi/abs/10.1111/j.1365 2338.2006.01031.x. Accessed on 1 February 2008). These measures can be used either nationally or within specified endangered areas and include the following:

- Publicity (existing regulations and lists of invasive or potentially invasive plants, information about threats and pathways should be publicized to raise awareness among all the persons concerned, e.g. horticultural industry, botanical gardens, gardeners)
- Labelling or marking of plants explaining the risks and appropriate actions/uses
- Surveillance
- Control plan
- Restrictions or codes of conduct on sale
- Restrictions or codes of conduct on holding
- Restrictions or codes of conduct on movement (e.g. prevention of movement to specified areas)
- Restrictions or codes of conduct for importers (including notification before import, limitation of quantities)
- Import restricted to specified non-invasive cultivars or clones
- Restrictions or code of conduct on planting (including authorization to plant in intended habitats, prohibition of planting in unintended habitats, required growing conditions for plants).

If the invasive alien plant is likely to be unintentionally introduced as a contaminant, classical plant health measures are appropriate, including prohibition of certain consignments, detection in consignments, removal from consignments, exclusion from consignments or prevention of natural spread. Pre-entry measures are preferred to post-entry measures since they are considered more efficient in preventing introduction. For some invasive plants, it will be possible to prevent the contamination of the pathway by treatment of the crop or consignment, or by other phytosanitary procedures, in the exporting country, under the responsibility of the NPPO. For example, the crop can be treated with herbicides, or grown in a specified way, or the consignment can be cleaned. Consignments can be required to originate in a crop free from the invasive plant, or in place of production, or area free from that plant (according to the capacity of the plant for local spread).

If entry with travellers and their luggage is a significant pathway, possible measures are inspection, publicity to enhance public awareness of pest risks, fines or incentives. For example, EPPO recommends its members to promote public awareness of pest risks due to the unintentional movement of seeds or rhizomes of *Solanum elaeagnifolium* with travellers. Contaminated machinery or means of transport may be cleaned or disinfected.

Finally, measures applied when the commodity has entered the country may also be envisaged, such as prevention of establishment by limiting the use of the consignment, or import under special licence/permit and specified restrictions.

16.6 Other Relevant EPPO Standards

Although preventive measures are considered the most effective tool to tackle the problem of invasive alien plants, conducting PRA on many individual species is likely to take time, and other approaches may be taken.

16.6.1 National Regulatory Control Systems

National measures such as monitoring, eradication, containment and/or control may be implemented by countries. EPPO provides such information with Standards in the series PM 9 "National regulatory control systems". So far, drafts are being prepared for *Ambrosia artemisiifolia* and *Heracleum* spp.

16.6.2 Codes of Conduct

Codes of conduct for plant producers, sellers and users may be an effective tool for the management of invasive alien plants, if regulation is too complex and costly. Partnerships with the nursery industry and elaboration of codes of conduct have already been undertaken within the EPPO region (United Kingdom), and have given fruitful results and a better understanding of the problem. Such initiative is undertaken by EPPO in partnership with the Bern Convention at the European and Mediterranean level. Such codes should provide technical information to professionals in order to allow them to manage the problem themselves.

16.7 Further Improvements

Despite these advances, a recent study predicts that the number of plant pests establishing in Europe will increase significantly in the next 10 years, based on current trends (Waage et al. 2005). PRA must therefore be made even more effective. As noted earlier, performing PRA on individual species takes time and it is important that the use of the international and regional standards for PRA is enhanced at the national level. Better coordination and synergy is needed between relevant bodies at the national level (Ministries of Agriculture and Environment, traders, producers). EPPO plans to provide basic training on PRA and to improve its information systems for PRA, while EPPO countries plan to operate more effective international systems for PRA. At the present time, management of invasive alien plants in Europe remains a national or even a sub-national concern, but the systems exist that will allow the European countries to agree on common policies for preventing the introduction and spread of invasive alien plants in the framework of the IPPC.

16.8 PRATIQUE: A Project Within the Seventh European Union Framework Programme

The EPPO Decision-support scheme is widely used by EPPO countries for their internal purposes, but is confronted by the fact that the application of phytosanitary measures in 27 of those countries requires decisions at the EU level. PRA at the EU level is still under development. The data required to make accurate analyses of the risks throughout the EU are often lacking. The existing systems in the EU respond slowly to new developments, and are very complex to operate with full participation of the member states. PRATIQUE (Enhancements of PRA Techniques), a project within the seventh framework programme of the European Union has the objective to develop more efficient risk analysis techniques for pests and pathogens of phytosanitary concern.

Between 2008 and 2011, a consortium of 15 bodies will work in order to do the following:

- Provide data sets valid for PRAs concerning the whole of the EU, with appropriate information on trade, on new pests, etc.
- Conduct multi-disciplinary research to enhance the techniques used in PRA for the assessment of impacts, standardizing and summarizing risks, pathway analysis, etc.
- Ensure that the PRA scheme is fit for its purpose and user-friendly.

16.9 Conclusion

Experience in the EPPO region as well as in other parts of the world shows the essential and successful role of PRA, in the IPPC framework, as a basis for phytosanitary import regulations.

EPPO has made its Decision-support scheme for PRA evolve so that invasive alien plants can bc assessed. Consequently appropriate tools exist in the IPPC framework to address risks presented by invasive alien plants. These tools now need to be promoted and used by countries, and collaboration should be established between the different sectors involved.

References

Burgiel S, Foote G, Orellana M, Perrault A (2006) Invasive alien species and trade: integrating prevention measures and international trade rules. Defenders of wildlife. Center for International Environmental Law. The Nature Conservancy. 54 p. http://www.cleantrade.net. Accessed on 1 February 2008

CBD (2002) Sixth conference of the parties. The Hagues, the Netherlands, 7–19 April 2002: Decision VI/23: Alien species that threaten ecosystems, habitats or species to which is annexed guiding principles for the prevention, introduction and mitigation of impacts of alien species that threaten ecosystems, habitats or species. Available at www.biodiv.org

CBD (2006) Alien species that threaten ecosystems, habitats or species (Article 8(h)): Further consideration of gaps and inconsistencies in the international regulatory framework. (COP 8, Decision VIII/27). http://www.cbd.int/decisions/cop-08.shtml?m = COP-08&id = 11041&lg = 0. Accessed on 1 February 2008

Dawson FH, Warman EA (1987) *Crassula helmsii* (T. Kirk) cockayne: Is it an aggressive alien aquatic plant in Britain? Biological Conservation 42, 247–272.

EPPO (1997) Decision-support scheme for quarantine pests. EPPO Standard PM 5/3 (2). Available at www.eppo.org. Accessed on 1 February 2008

EPPO (1998) Pest risk analysis: check-list of information required for pest risk analysis (PRA). EPPO standard PM 5/1(1). Available at www.eppo.org. Accessed on 1 February 2008

EPPO Reporting Service no. 1 (2007) Pathway analysis: aquatic plants imported in France. 2007/016. http://archives.eppo.org/EPPOReporting/2007/Rse-0701.pdf. Accessed on 1 February 2008

EPPO Reporting Service no. 6 (2007) Pathway analysis: alien plants introduced through the bird seed pathway. 2007/123. http://archives.eppo.org/EPPOReporting/2007/Rse-0706.pdf. Accessed on 1 February 2008

Genovesi P (2007) Assessment of existing lists of invasive alien species for Europe, with particular focus on species entering Europe through trade, and proposed responses. Convention on the Conservation of European wildlife and natural habitats. T-PVS/Inf (2007) 2. 37 p.

Genovesi P, Shine C (2002) European Strategy on invasive alien species. Convention on the Conservation of European wildlife and natural habitats. T-PVS (2003) 7 revised. 50 p. http://www.coe.int/t/e/Cultural_Co-operation/Environment/Nature_and_biological_diversity/Nature_protection/sc23_tpvs07erev.pdf?L = E. Accessed on 1 February 2008

Hulme PE (2007) Biological invasions in Europe: Drivers, pressures, states, impacts and responses. In: Hester R, Harrison RM (eds) Biodiversity under threat, Issues in Environmental Science and Technology, Vol. 25, Royal Society of Chemistry, Cambridge, pp. 55–79

FAO (1997) International Plant Protection Convention (new revised text). FAO, Rome, Italy

IPPC (2007a) Principles of plant quarantine as related to international trade. ISPM no. 1 in International Standards for Phytosanitary Measures, pp. 17–26. IPPC Secretariat, FAO, Rome (IT). https://www.ippc.int/IPP/En/default.jsp. Accessed on 1 February 2008

IPPC (2007b) Guidelines for pest risk analysis ISPM no. 2 in international standards for phytosanitary measures, pp. 27–43. IPPC Secretariat, FAO, Rome, Italy. https://www.ippc.int/IPP/En/default.jsp. Accessed on 1 February 2008

IPPC (2007c) Glossary of phytosanitary terms. ISPM no. 5 in international standards for phytosanitary measures, pp. 63–85. IPPC Secretariat, FAO, Rome, Italy. https://www.ippc.int/IPP/En/default.jsp Accessed on 1 February 2008

IPPC (2007d) Pest risk analysis for quarantine pests including analysis of environmental risks and living modified organisms. ISPM no. 11 in international standards for phytosanitary measures, pp. 135–160. IPPC Secretariat, FAO, Rome, Italy. https://www.ippc.int/IPP/En/default.jsp. Accessed on 1 February 2008

Leach J, Dawson H (1999) *Crassula helmsii* in the British Isles – An unwelcome invader. British Wildlife 10(4), 234–239

Lopian R (2005) The International Plant Protection Convention and invasive alien species. Proceedings of the workshop on invasive alien species and the International Plant Protection Convention, Braunschweig, Germany, 22–26 September 2003. FAO, Rome, Italy. pp. 6–16. http://www.fao.org/docrep/008/y5968e/y5968e00.htm Accessed on 1 February 2008

Hodkinson DJ, Thompson K (1997) Plant dispersal: The role of man. Journal of Applied Ecology 34, 1484–1496

Schrader G (2004) A new working programme on invasive alien species started by a multinational European organisation dedicated to protecting plants. Weed Technology 18, 1342–1348.

Schrader G (2005) Adaptation of regional pest risk assessment to the revised ISPM 11. In: IPPC Secretariat – Identification of risks and management of invasive alien species using the IPPC framework. Proceedings of the workshop on invasive alien species and the International Plant Protection Convention, Braunschweig, Germany, 22–26 September 2003. FAO, Rome, Italy. pp. 110–113. http://www.fao.org/docrep/008/y5968e/y5968e00.htm. Accessed on 1 February 2008

Schrader G, Unger JG (2003) Plant quarantine as a measure against invasive alien species: The framework of the International Plant Protection Convention and the plant health regulations in the European Union. Biological Invasions 5(4), 357–364

Shine C (2005) Overview of the management of invasive alien species from the environmental perspective. In IPPC Secretariat – Identification of risks and management of invasive alien species using the IPPC framework. Proceedings of the workshop on invasive alien species and the International Plant Protection Convention, Braunschweig, Germany, 22–26 September 2003. FAO, Rome, Italy. pp. 20–34. http://www.fao.org/docrep/008/y5968e/y5968e00.htm Accessed on 1 February 2008

Smith IM (2005) EPPO's regional approach to invasive alien species. In: IPPC Secretariat – Identification of risks and management of invasive alien species using the IPPC framework. Proceedings of the workshop on invasive alien species and the International Plant Protection Convention, Braunschweig, Germany, 22–26 September 2003. FAO, Rome, Italy. pp. 45–48.http://www.fao.org/docrep/008/y5968e/y5968e00.htm. Accessed on 1 February 2008

Sutherst GW, Maywald GF, Bottomley W, Bourne A (2004) CLIMEX v2. User's guide. Hearne Scientific Software, Melbourne, Australia

Waage JK, Fraser RW, Mumford JD, Cook DC, Wilby A (2005) A new agenda for biosecurity. Horizon Scanning Programme, Department for Environment, Food and Rural Affairs, UK. 198 pp.

Watson WRC (2001) An unwelcome acquatic invader! Worcestershire Record, issue 10. http://www.wbrc.org.uk/WorcRecd/Issue10/invader.htm [accessed in October 2008]

World Trade Organization – WTO (1994) Agreement on the application of sanitary and phytosanitary measures. In: Agreement establishing the World Trade Organization: Annex 1A: Multilateral agreements on trade in goods. Geneva, Switzerland. Available at www.wto.org. Accessed on 1 February 2008

Appendix 1
Initiation of the EPPO Decision-Support Scheme for PRA

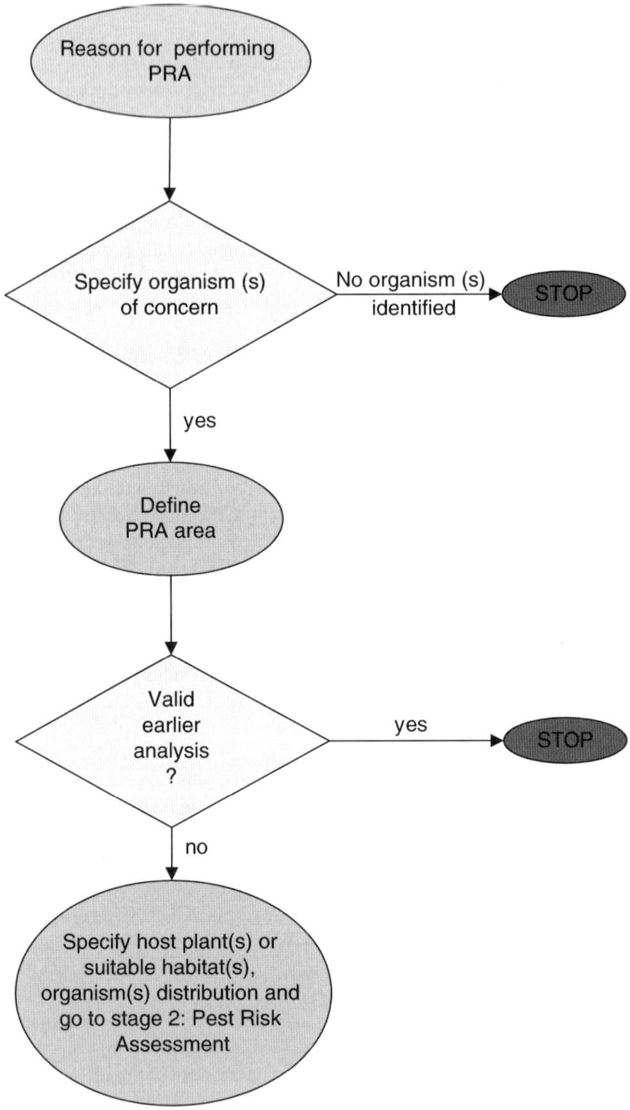

Appendix 2
Decision-Support Scheme For Quarantine Pests. Stage 2: Pest Risk Assessment

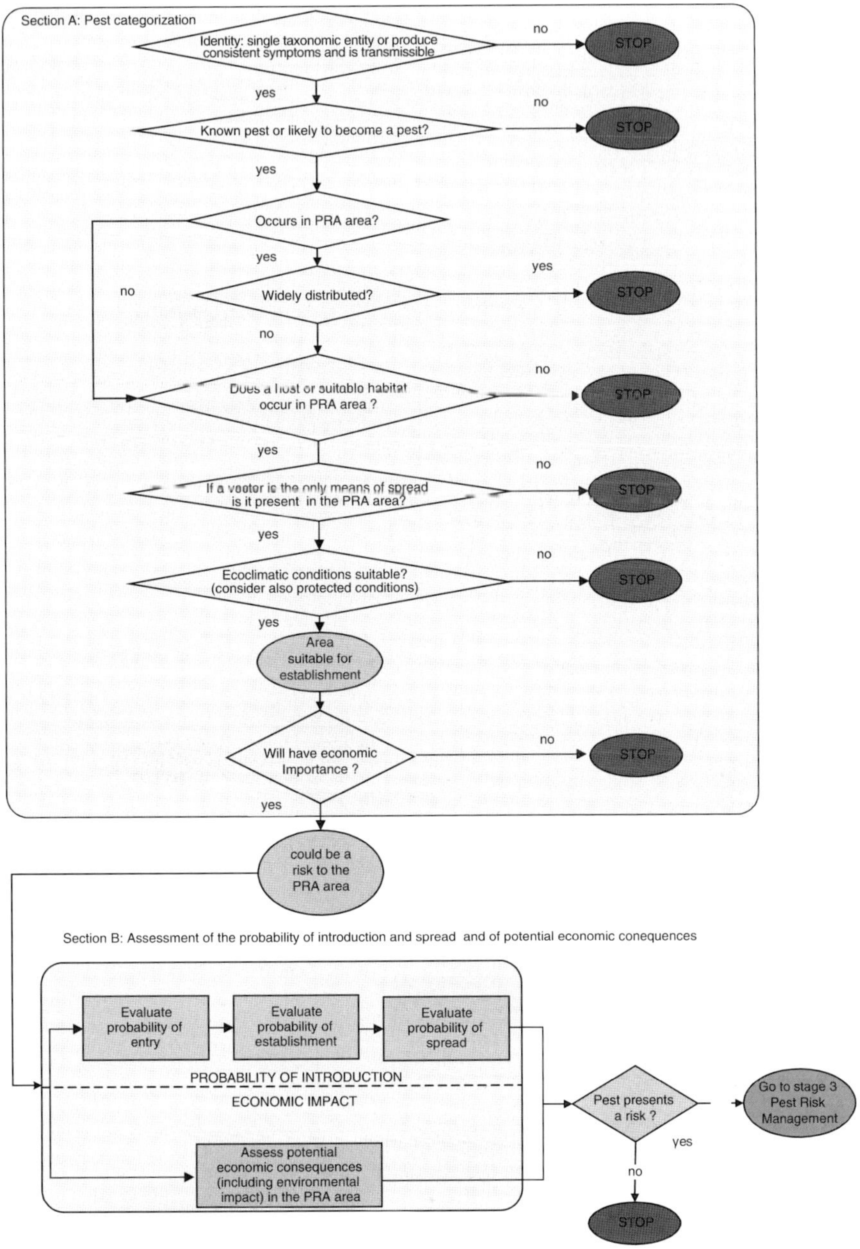

Appendix 3
Decision-Support Scheme for Quarantine Pests. Stage 3: Pest Risk Management

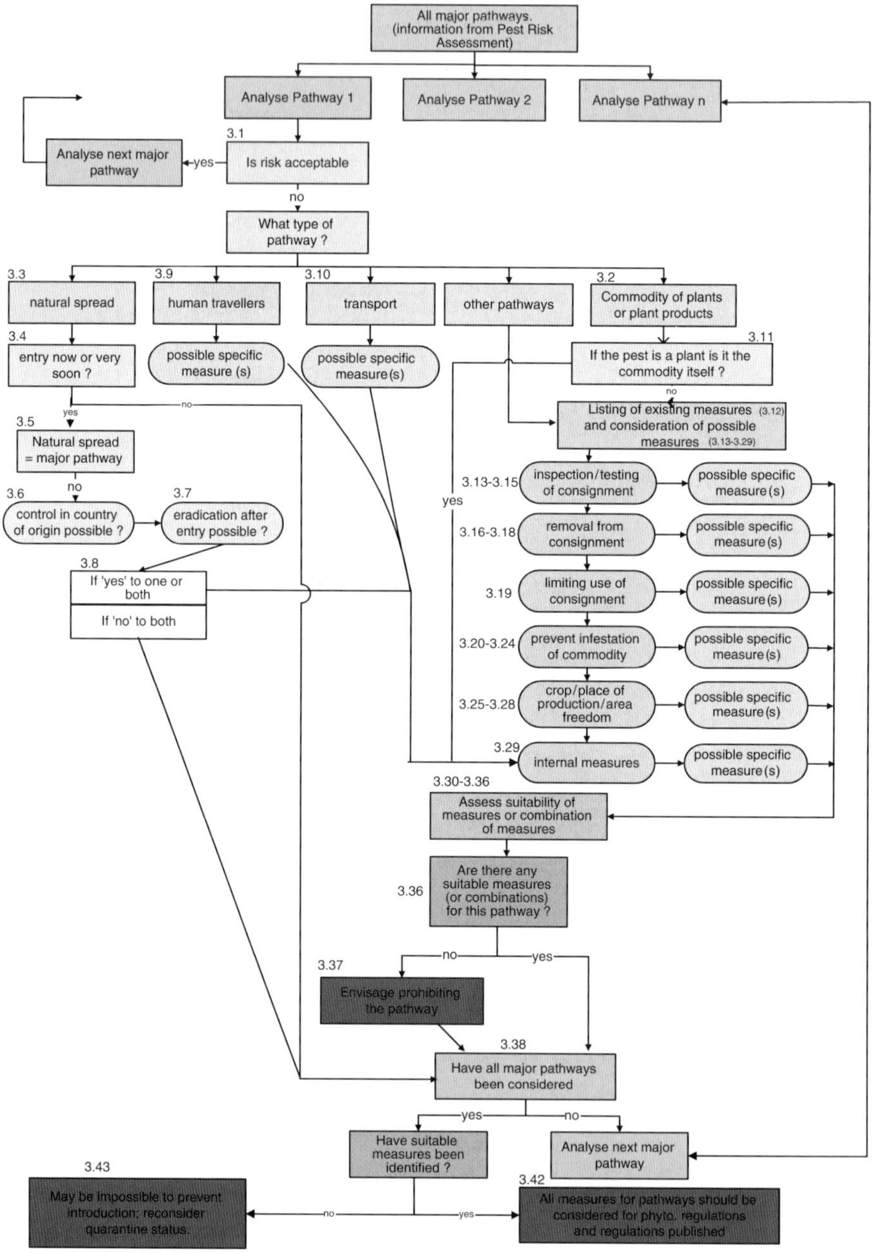

In the framework of the IPPC, PRAs are initiated by importing countries in order to develop appropriate phytosanitary measures to prevent the introduction and spread of quarantine pests, and to justify these measures to trading partners. The measures usually concern the unintentional movement of pests with traded commodities. Under the IPPC, exporting countries should, if requested, provide adequate information in support of the PRAs of importing countries. This model does fit invasive alien plants in some circumstances.

However, many potentially invasive plants are intentionally imported, as such, for agricultural, horticultural or other purposes. CBD Guiding Principle 10 on intentional introduction states that the "the burden of proof that a proposed introduction is unlikely to threaten biological diversity should be with the proposer of the introduction or be assigned as appropriate by the recipient State". As already explained, there is in Europe no general measure restricting the import of plants from other continents. Exporters and importers agree on what is traded. The IPPC framework makes no provision for PRA to be conducted by exporters or importers, so in fact only the NPPO of the importing country can in practice perform PRAs for invasive alien plants, and besides has the systems in place to do so.

When is a PRA initiated? The EPPO scheme provides many possible scenarios appropriate for other plant pests. For invasive alien plants, the situation is relatively simple: an established infestation may exist or be discovered in the PRA area, a plant may be reported to be an invasive alien in some other part of the world, or a new plant may be intentionally imported.

Chapter 17
Implementing Science-Based Invasive Plant Management

Steven R. Radosevich, Timothy Prather, Claudio M. Ghersa, and Larry Lass

Abstract Invasive nonnative plants impact landscapes worldwide through changes in the structure, composition, and succession pathways of plant communities. Whether and where new and existing exotic plant populations will expand influences decisions and willingness of land managers to expend resources on proactive management. Thus, land managers generally prefer to contain existing patches of invasive plants rather than find and eradicate new ones. Although preventive strategies are an effective way to limit plant invasions, they are difficult to achieve because adequate descriptions of biological and environmental characteristics are often lacking, and predictive models of invasive plant expansion have been elusive. Studies of invasive plant management usually focus on tools to control weeds or competition in simplified natural production systems. It is difficult to determine with such studies how practices to control nonnative invasive plants influence naturally occurring control mechanisms. This paper examines the theories and practices of integrated pest management for exotic invasive plant containment. We also explore the development and use of invasive plant expansion models. Both approaches are needed to manage weeds in increasingly complex agricultural and natural production systems.

Keywords Invasive species management • Approaches • Models • Framework

17.1 Introduction

An increasingly global economy, worldwide transport of biological commodities, and opportunities for transworld travel have all promoted the introduction and subsequent colonization of exotic plants in many parts of the world (Fig. 17.1). For example, Rejmánek (2000) indicates that over 21% of the 22,000 vascular plants found in North America are nonnative or exotic. If this magnitude of plant introductions continues at

S.R. Radosevich (✉), T. Prather, C.M. Ghersa, and L. Lass
Oregon State University, Corvallis, OR, USA
Steve.radosevich@oregonstate.edu

Inderjit (ed.), *Management of Invasive Weeds*,

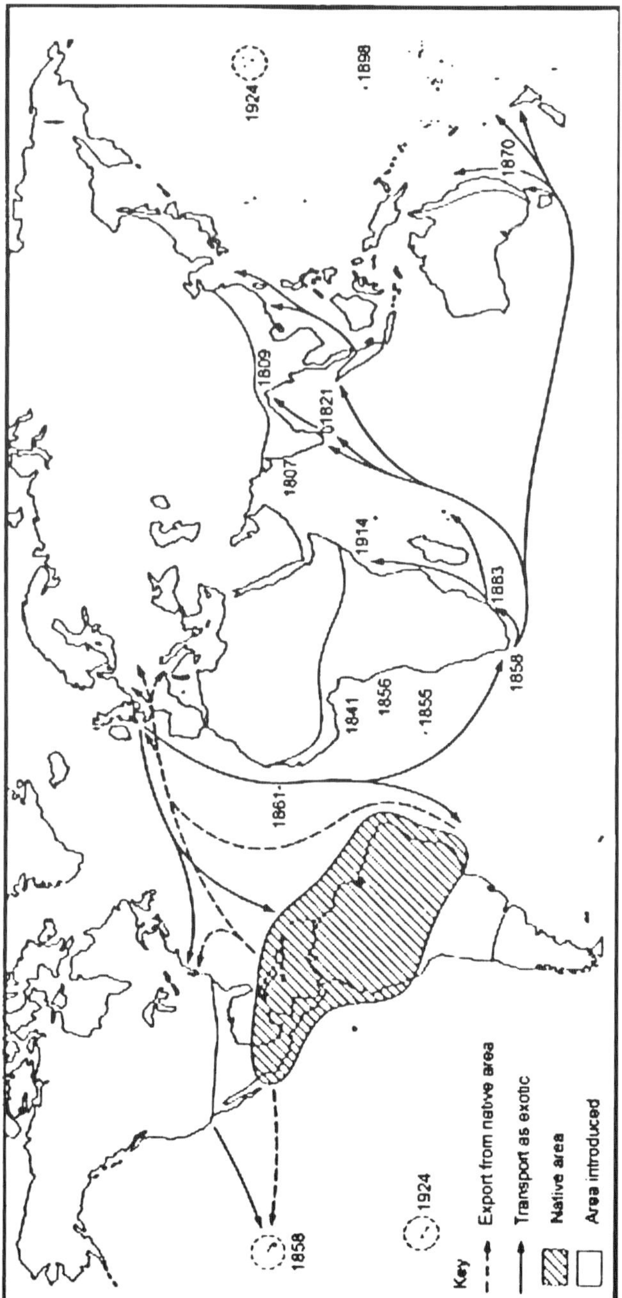

Fig. 17.1 Some invaders, such as the shrub lantana (*Lantana camara*), have been introduced repeatedly in new ranges as a result of global human colonization and commerce. As the array of estimated years indicates, lantana was introduced throughout the nineteenth and beginning of the twentieth centuries in many subtropical and tropical areas. In each new range it has become highly destructive, both in agricultural and natural communities (Cronk and Fuller 1995 from Mack et al. 2000)

its current pace, the earth's flora could eventually homogenize to only a few highly successful species (Luken and Thieret 1997; Ewel et al. 1999; McNeely 1999).

Plant invasion is generally divided into a biological component, or the capacity of a plant to spread beyond the site of introduction (*invasiveness*) and an environmental component, which is the susceptibility of a habitat to the colonization and establishment of individuals from species not currently part of the local community (*invasibility*) (Rosonsweig 2001 ; Davis et al. 2005; Radosevich et al. 2007). Environmental differences among habitats and communities contributing to invasibility are often easier to identify than the biological traits associated with invasiveness (Lonsdale 1999; Reichard 1997), although certain habitats, such as those of mature forests and dense grassland, tend to have relatively few exotic plant species (Richardson et al. 1994; Harrison 1999; Parks et al. 2005). Knowing the susceptibility of different habitats and plant communities to invasion can help design programs to manage invasive plants or protect native habitats.

17.2 Basis for Management Decisions About Invasive Plants

Land managers use a variety of ways to manage invasive plants. Weeds are invasive plants that increase in habitat range because of the expansion of human activities for production (Radosevich et al. 2007). These activities alter the structure and function of many plant communities around the world. Decisions about weed control are influenced by the biology of the species, existing technology, and social considerations that are often represented by people affected by management procedures (Radosevich and Ghersa 1992). These three factors are linked through six fundamental scientific disciplines (Fig. 17.2), which collectively generate a base of

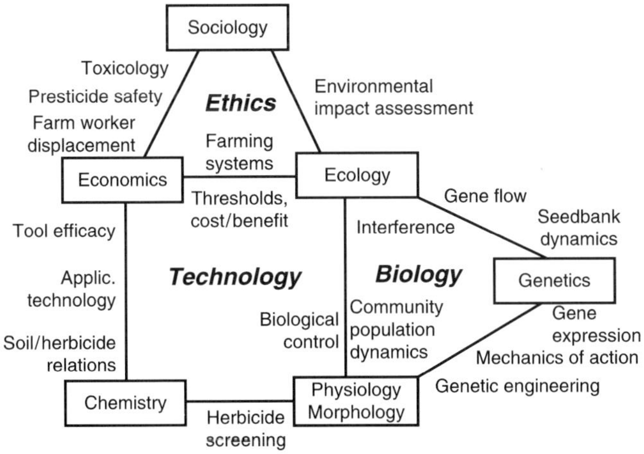

Fig. 17.2 A diagram depicting the interrelationships of six fundamental disciplines in weed science. Major areas of activity are weed technology, weed biology, and the ethics of weed control (reproduced from Radosevich and Ghersa 1992)

empirical information to develop or modify weed control procedures, and also to justify invasive plant control. In contrast to agricultural systems, a major focus of weed management in natural production systems is to assess the risk that new plant species will become invasive in order to prevent their spread. Byers et al. (2002) identify four levels of *risk assessment* associated with the biological stages of exotic species invasion:

- Arrival (risk associated with entry pathways)
- Establishment (risk of forming viable, reproductive populations)
- Spread (risk of expanding the range or extent)
- Impact (risk of having a measurable effect on existing species or communities)

Thus, management of weeds in natural systems focuses primarily on detection and eradication of potentially invasive plants that are not yet widespread. The assessment of invasive plants, however, has proven more difficult than simply finding, counting, and controlling them (Auld et al. 1987; Hobbs and Humphries 1995; Leung et al. 2002; Pitafi and Roumasset 2005).

17.3 The Question of Whether, Not How

Management of invasive plants is a general strategy that encompasses prevention, eradication, and control. *Prevention* of species invasion involves procedures that inhibit or delay establishment of weeds in areas that are not already inhabited by them. These practices restrict the introduction, propagation, and spread of weeds on a local or regional level. Quarantines, surveys, and monitoring are the first steps in prevention of invasive species. *Eradication* is the elimination of a plant species from a field, area, or entire region. It requires the complete removal of seed and vegetative parts of a species. Eradication is usually attempted only in small areas or those with high-value crops or land use because of the difficulty and high costs associated with the practice. Eradication of small weed patches, however, is a low-cost tactic for management of invasive plant species, especially when compared with control costs of species with much broader distribution (Rejmánek 2000; Radosevich et al. 2007). *Control* practices reduce or suppress weeds in a defined area but do not necessarily result in the elimination of any particular species. Similar to control, *containment* is often a goal of management of invasive plants, where the infestation is held to a defined geographic area and not allowed to spread. This strategy involves habitat manipulation through plant community restoration, often at landscape and regional scales.

Finnoff et al. (2005) developed models to examine the economics of invasive species. They identified allocations of manager time and capital to prevent *versus* control invasive species in order to achieve an acceptable risk (Leung et al. 2002). Figure 17.3 is an analysis of managers who are risk neutral (RN), mildly risk-adverse (RA1), moderately risk-adverse (RN2), and highly risk-adverse (RN3) at four monetary discount rates. These results indicate that managers select activities that seem least risky, which means less prevention and more control (Finnoff et al.

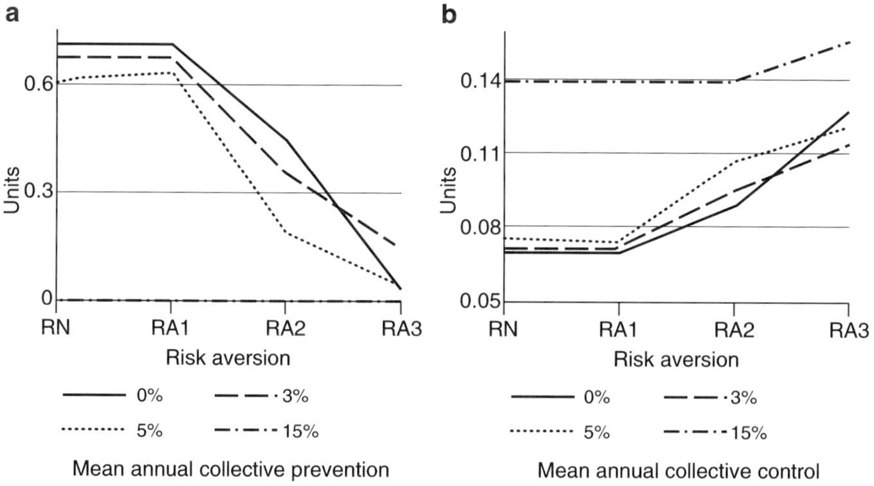

Fig. 17.3 The impact of risk aversion over four monetary discount rates. For both figures the horizontal axes are increasing levels of risk aversion by managers. Units of collective prevention in the left figure are the average number of prevention events that take place on an annual basis, whereas units of collective control in the right figure are the average number of control events (e.g., molluscicide applications) on an annual basis (from Finnoff et al. 2005)

2005). Control is intuitively more attractive to managers because it negatively impacts existing invaders from the ecosystem, whereas prevention is perceived to only eliminate the chance of invasion. Finnoff et al. also indicate that prevention and control are substitutes for each other, and that delays in the implementation of control increase the probability of invasions occurring.

17.3.1 Tools

Many tools used to control, contain, or eradicate invasive plants are available to land managers of natural resource production systems. In general, the methods used to reduce the abundance or vigor of unwanted vegetation vary only slightly according to habitat, i.e., whether they are used in agriculture, natural resource production systems such as rangeland or forest plantations, or natural ecosystems. Excellent discussions of the methods and tools used for weed control are provided by Aldrich (1984), Ross and Lembi (1985, 1999), Radosevich et al. (1997), Muyt (2001), and Coombs et al. (2004).

17.3.2 Establishing Priorities

Hobbs and Humphries (1995) suggest an approach to set management priorities based on land value and the degree of site disturbance (risk of invasion) for areas

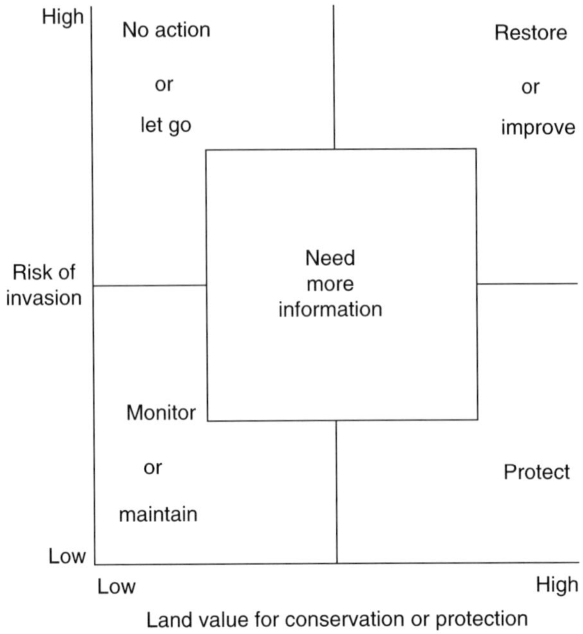

Fig. 17.4 Assessment of management priorities for a region based on the relative value of different sites for conservation and/or production, and their relative degree of risk of invasion (modified from Hobbs and Humphries 1995 in Radosevich et al. 2007)

occupied by invasive plants. Specific approaches to prioritize species of invasive plants for management are discussed later. The approach proposed by Hobbs and Humphries (Fig. 17.4) depicts four categories of management based on the characteristics of the site. These are as follows:

- Sites of high value that are relatively undisturbed, i.e., the risk of invasion is low (prevent/protect)
- Locations of high value that are subject to greater levels of disturbance (risk) and, hence, are more susceptible to invasion (protect and improve)
- Sites of low value that are subject to low levels of disturbance (monitor)
- Sites of low value that are subject to high levels of disturbance ("let go")

Unfortunately, the prevailing trend is one of transition from the bottom right (protect) to the top left (let go) of Fig. 17.4 as plant communities come into higher risk of invasion by exotic plants from continued environmental degradation (Hobbs and Humphries 1995).

17.4 Approaches for Management of Invasive Plants

Rejmánek (2000) reviews the approaches used to achieve prevention, eradication, or control of invasive plants. These approaches are as follows: *stochastic*, *empirical-taxon specific*, *evaluation of biological characters*, *evaluation of habitat compatibility*, and *experimentation*.

The *stochastic* approach focuses on initial plant population size and number and timing of introductions as factors that increase the probability of invasion success. For example, a significant correlation can usually be demonstrated between the total number of known localities of an invasive plant species and the years since the first observation of the species was recorded (Rejmánek 2000). An *empirical-taxon specific* management approach is based on information about the invasiveness of a species elsewhere. Knowing the experiences of others helps land managers make decisions about control and/or eradication of invasive plants. Williams et al. (2000) indicate that 80% of the exotic weed species in New Zealand are also described as invasive outside that country. These two approaches form the basis for a commonly used management tactic of many land management agencies in the USA, i.e., early detection and rapid response (EDRR). The shortcoming of the approaches, however, is that they tell little about the real impact of the species or the environmental and biological factors necessary to manage them (Mashadi and Radosevich 2003).

Rappoort (1991) *evaluated the biological characters* of invasive plants and reported that 10% of the estimated 260,000 vascular plant species on earth are potential invaders. Only 15% of these potential invaders have actually invaded an area outside their native range. The traits that make some plants more invasive than others have been extensively studied since Baker (1965) conceived the concept of an "ideal weed." Statistical tools such as discriminate analysis, multiple logistic regression, and classification and regression trees can be used to assess biological characters responsible for invasiveness (Endress et al. 2007). However, adaptive change induced by selection plays a central role in plant speciation and in molding traits of weeds and invasive plants. During biological invasions, significant genetic change can occur in species that are no longer limited by their native environment (Gray et al. 1986). It is unlikely, therefore, that studies of the shared attributes of successful weeds or invasive plants will provide the adequate information to assess biological characters responsible for invasiveness. Studies also should focus on information regarding the genetic changes that plants undergo during colonization and the factors controlling these evolutionary changes. Adaptability of a species may be more important than its tolerance or plasticity to environmental change. There are many examples of differentiation among plant populations that occur across short spatial distances and over relatively short time periods. The importance of evolutionary adaptation is well known for both agricultural weeds (Dekker 2003) and invasive plants (Ellstrand and Schierenbeck 2000), making a strong case for natural and human-induced selection (Radosevich et al. 2007). Unfortunately, few studies evaluate invasiveness by combining the stochastic approach with particular traits that control the adaptability of a species, i.e., successive introductions and the probability of generating adaptive variability through hybridization (Gray et al. 1986).

The fourth approach indicated by Rejmánek (2000) for management of invasive plants is the *evaluation of habitat compatibility* (invasibility) to determine whether a particular species can invade a particular habitat type. This approach assumes that climate is the overriding factor that determines the suitability of a site for an invasive species (Woodward 1987; Panetta and Mitchell 1991; Mack 1996). Several models such as GARP and Bioclim now predict the potential new range of an invasive species by identifying regions that are climatically similar to the species' native range (Sutherst and Maywald 1985; Peterson and Vieglais 2001; Peterson 2003; Thuiller et al. 2005). This approach is most powerful when combined with analysis of other factors, such as soil type, that can influence invasive plant distribution. Finally, *experiments* can be conducted to test empirically the predictions made from the four approaches described earlier (Rejmánek 2000). However, the time lag that is inherent in most invasive episodes (Kowarik 1995) usually makes such experiments unappealing for land managers (Radosevich et al. 2005).

17.4.1 Need for Surveys and Monitoring

Rew et al. (2005) indicate that, historically, surveys have been linked to invasive plant management, while monitoring may not occur until after management activity has happened, if at all. Rew et al. also indicate that prior information should be used to develop any survey scheme in order to maximize finding invasive plants. For example, since invasive plants are most often introduced by humans, the vectors of human transport such as roads and rails should be examined. Although some exceptions exist, occupancy by invasive plants usually declines as plant cover increases such that a gradient of decreasing occupancy from areas of low to high cover should be expected. Rew et al. (2005) offer an approach (Fig. 17.5) where monitoring is used to improve the reliability of management tactics in areas inhabited by invasive plants. They suggest that land managers (1) develop monitoring plans, (2) select methods that will quickly meet monitoring objectives, and (3) clearly link monitoring output to management decisions. Rew and Pokorny (2006) describe both on-the-ground and remote sensing survey and monitoring approaches used in western North America to assess invasive plant occurrence.

17.4.2 Risk Assessment Models

Although several models of range expansion from source populations have been developed for invasive plants, this approach has not been incorporated widely into management decisions. Only recently have simulation models of invasion been developed that incorporate management options and outcomes (Goslee et al. 2006; Kriticos et al. 2003). Endress (in Radosevich et al. 2005) provides an example of how to construct a risk assessment model based on the susceptibility of native plant communities to invasion, disturbance history of sites, and proximity to current infestations. Although models of risk assessment can be valuable tools for land managers, they require good information on species biology, site characteristics, and reliable position coordinates for an area, watershed, or region. Heger and Trepl (2003) and Prather (2006, in press) also describe computer programs that predict

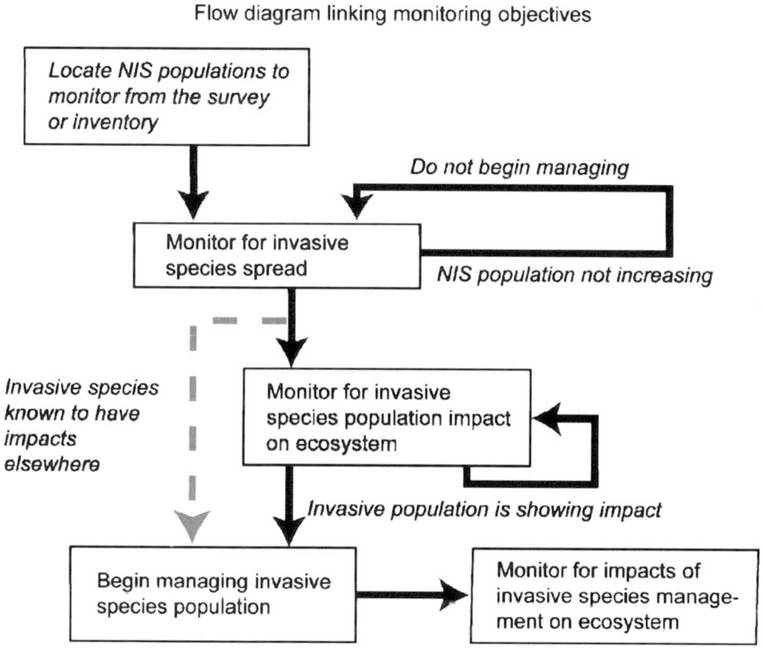

Fig. 17.5 Flow diagram linking monitoring and land management objectives for exotic invasive plants (NIS in the figure indicates nonnative invasive species; from Maxwell 2005)

the occurrence and spread of invasive plants using a community or habitat approach.

Prather and Lass (in press) are developing a spatially explicit plant movement model for yellow starthistle (Fig. 17.6) and sulfur cinquefoil. They combined biological information on species dispersal and population dynamics with a network movement model that included both slope and aspect of currently occupied and potential sites to predict the probability of occurrence of each species using the following equation:

$$\ln \frac{Ps_{ij}}{1 - Ps_{ij}} = \beta_0 + \beta_1 u + \beta_2 u^2 + \beta_3 v + \beta_4 v^2 + \beta_2 \ln(u^2 + v^2),$$

where

Ps_{ij} is the proportion of YST present at the ith and jth classification levels of aspect and slope, u and v are polar transformations of slope and aspect given by: $u =$ slope$_i$cos(aspect$_j$), v = slope$_i$ × sin(aspect$_j$), and $\beta1$–$\beta5$ are regression coefficients.

The model (Fig. 17.6) corresponded closely to observed values. It predicted that 92% of existing area had yellow starthistle infestation and located another 39% of the area that was susceptible to invasion by the species. The likelihood of yellow starthistle occurrence was affected by both slope (31 ± 8°) and aspect (237 ± 89°), which indirectly provided information on the plant's ability to reproduce and disperse seed. The approach also was used to develop a likelihood of survival function for plant communities subject to invasion by the two invasive plant species and to predict the movement of either species across a landscape over time (Fig. 17.6).

17.4.3 Thresholds and Succession

Thresholds are fundamental to integrated pest management (IPM) in agriculture and other natural resource production systems because cost-effective weed man-

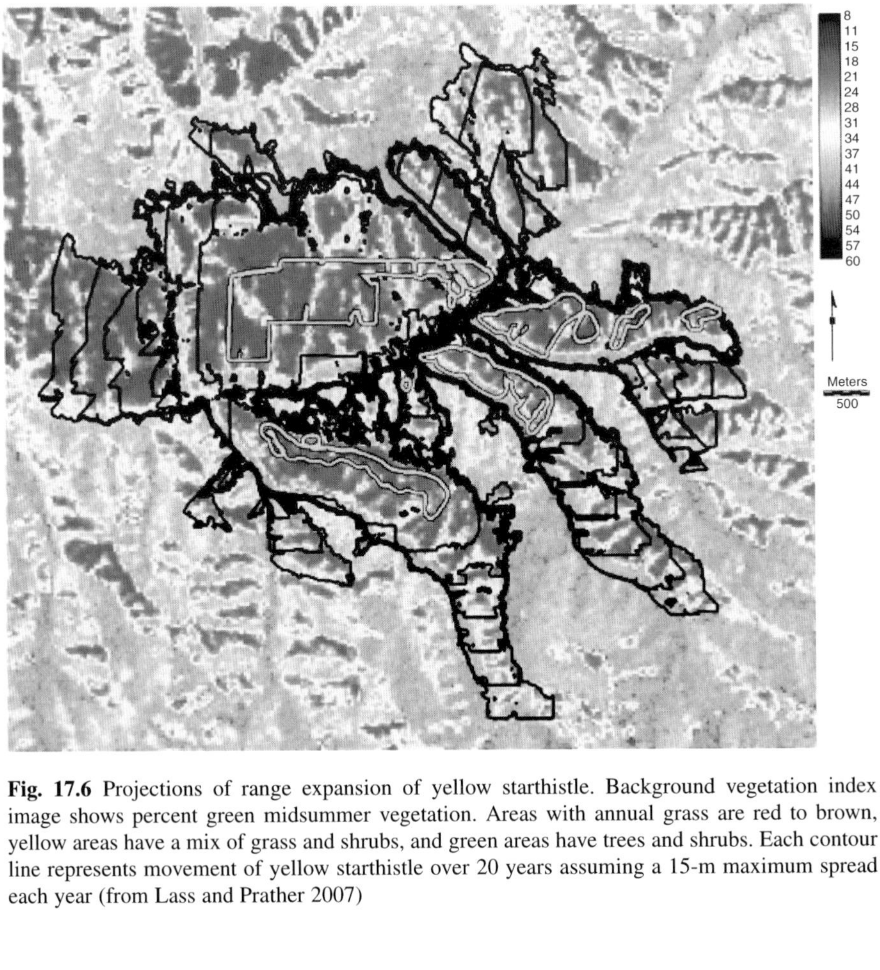

Fig. 17.6 Projections of range expansion of yellow starthistle. Background vegetation index image shows percent green midsummer vegetation. Areas with annual grass are red to brown, yellow areas have a mix of grass and shrubs, and green areas have trees and shrubs. Each contour line represents movement of yellow starthistle over 20 years assuming a 15-m maximum spread each year (from Lass and Prather 2007)

agement requires that an assessment of both possible and real damage be made prior to the introduction of weed control tactics (Norris et al. 2003; Radosevich et al. 2007). The most common thresholds applied to weeds are those that relate to damage, economics, and action. A *damage* threshold describes the plant population at which negative impact to a crop is detected. It is usually expressed as plant density or biomass per area. An *economic* threshold is the weed density or damage level at which control measures should be taken to prevent further economic injury from being incurred. The economic threshold is also called the *economic injury level* (EIL) (Norris et al. 2003), which implies that the costs of control should be less than the loss that would have occurred had nothing been done. An *action* threshold is the weed population level at which some intervention is needed to preclude further damage. A method to construct thresholds is discussed in Radosevich et al. (2007).

In addition, the traditional views of succession are now being challenged by scientists who believe that the process is more climate- and disturbance-driven than driven by competition (Westoby et al. 1989; Briske et al. 2003). These scientists argue that many equilibrium states probably exist among plant communities during succession and that transitions among these states occur when an *ecological threshold* is crossed (Kimmins 1997). According to Briske et al. (2003), ecological thresholds separate multiple equilibrium states and can be distinguished by changes in community structure and composition. Many exotic invasive plant species are believed to disrupt native ecosystem function by their presence (Vitousek et al. 1996; Hobbs and Huenneke 1992; Quigley and Arbelbide 1997; Sheley and Petroff 1999; Harrod 2001).

The possible impacts of exotic invasive plant species on ecological thresholds of natural production systems are only now being considered by land managers. The introduction of IPM concepts also has been slow to emerge in natural production systems, although the threshold most easily recognized by land managers is the recent directive for EDRR. In this case, control tactics are employed to eradicate the patch once a new exotic species that is deemed to be harmful is found in a new area. This *action* implies that any damage by the weed to the plant community is too much. According to Hobbs and Humphries (1995), this type of threshold should be employed in a natural production system only when a resource or area is extremely valuable and the risk to it by presence of invasive plant is great. It must be recognized, however, that eradication of a few small patches or isolated plants may be an extremely cost-effective form of weed management. In late stages of succession, if disturbance to the natural system is severe or if the presence of exotic plants is ubiquitous, the economic threshold (EIL) would be a better measure of the cost-effectiveness for restoration than EDRR. In this case, the damage to a plant community by various levels of invasive plant species and the costs of restoration would be compared with the long-term gain in ecosystem function from the action. In most well-established natural plant communities, species diversity, complexity, and coexistence are the rule, rather than direct competition (Vandermeer 1989). Thus, it is possible in many cases that the threshold for action is higher than 0. It also seems possible that both economic and ecological thresholds could be lower than anticipated if the long-term benefit of creating and maintaining a self-perpetuating natural production system, which is relatively free of weeds, is high.

17.5 Framework to Implement Science-Based Management of Invasive Plants

Preventing, reducing, or eliminating undesirable impacts of invasive plants is a challenge facing land managers around the world. Because of the potential seriousness of the invasive plant problem, there is also a need to develop models and other elements of a research program that can facilitate relationships between scientists and land managers. Radosevich et al. (2005) suggest a framework in which empirical experiments, risk assessments, and projections of invasive plant species introduction and spread across susceptible landscapes can help land managers conduct management activities (Fig. 17.7). The approach incorporates habitat-level (Werner and Soule 1976; Zouhar 2003) and species-level (Sheley and Petroff 1999; DiTomaso 2000) experiments on age structure, population dynamics, competitive ability, dispersal, disturbance, and herbivory into a landscape-level model (Neubert and Caswell 2000). When these activities are combined with a GIS-based risk assessment, e.g., Fig. 17.6, it is possible to project expansion of the species over time. This approach is helpful to land managers because the consequences of management or no action can be determined and policies derived from it can be justified. Results of the approach (Fig. 17.7) also provide land managers guidance in integrating management tools (e.g., herbicides, fire, and native plant seeding) with information on plant invasiveness and community invasibility.

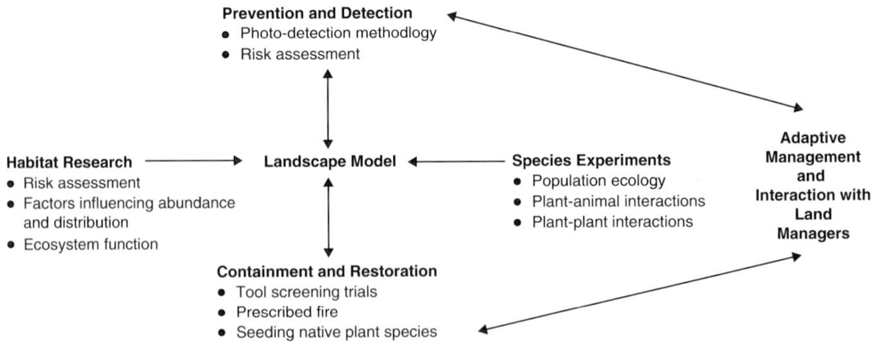

Fig. 17.7 Framework for implementing science-based management of invasive plants (from Radosevich et al. 2005)

References

Aldrich RJ (1984) Weed-Crop Ecology: Principles in Weed Management. Breton, North Scituate, MA.

Auld BA, Menz KM, Tisdell CA (1987) Weed Control Economics. Academic Press, London

Baker HG (1965) Characteristics and modes of origin of weeds. pp. 147–168. In Baker HG, Stebbins GL (eds) The Genetics of Colonizing Species. Academic Press, New York

Briske DD, Fuhlendorf SD, Smeins FE (2003) Vegetation dynamics on rangelands: A critique of the current paradigms. J Appl Ecol 40: 601–614

Byers JE, Reichard S, Randall JM et al (2002) Directing research to reduce the impacts of nonindigenous species. Conserv Biol 6: 630–640

Coombs EM, Clark JK, Piper GL et al (2004) Biological Control of Invasive Plants in the United States. Oregon State University Press, Corvallis, OR

Davis MA, Thompson K, Grime JP (2005) Invasibility: The local mechanism driving community assembly and species diversity. Ecography 28: 696–704

Dekker J (2003) Evolutionary biology of the foxtail (*Setaria*) species-group. pp. 65–114. In Inderjit K (ed) Weed Biology and Management. Kluwer Academic, Dordrecht, Netherlands

DiTomaso JM (2000) Invasive weeds in rangeland: Species, impacts and management. Weed Sci 48: 255–265

Ellstrand NC, Schierenbeck KA (2000) Hybridization as a stimulus for invasiveness in plants? Proc Natl Acad Sci USA 97: 7043–7050

Endress BA, Naylor BJ, Parks CG et al (2007) Factors influencing the abundance, distribution and dominance of the invasive plant *Potentilla recta*. Rangeland Ecol Manage 60: 218–224

Ewel JJ, Dowd DO, Bergelson J et al (1999) Deliberate introduction of species: Research needs. Bioscience 49: 619–630

Finnoff D, Shogren JF, Leung Bet al (2005) Risk and nonindigenous species management. Rev Agric Econ 27: 475–482

Gray AJ, Mack RN, Harper JL et al (1986) Do invading species have definable genetic characteristics. Philos Trans R Soc Lond 314: 655–674

Goslee SC, Peters DPC, Beck KG (2006) Spatial prediction of invasion success across heterogeneous landscapes using an individual-based model. Biol Invas 8: 193–200

Harrison S (1999) Native and alien species diversity at the local and regional scales in a grazed California grassland. Oecologia 121: 99–106

Harrod JR (2001) The effects of invasive and noxious plants on land management in eastern Oregon and Washington. Northwest Sci 75: 85–90

Heger T, Trepl L (2003) Predicting biological invasions. Biol Invas 5: 313–321

Hobbs RJ, Huenneke LF (1992) Disturbance, diversity, and invasion: Implications for conservation. Conserv Biol 6: 324–337

Hobbs RJ, Humphries SE (1995) An integrated approach to the ecology and management of plant invasions. Conserv Biol 9: 761–770

Kimmins JP (1997) Forest Ecology: A Foundation for Sustainable Management. Prentice-Hall, Upper Saddle River, NJ

Kowarik I (1995) Time lags in biological invasions with regard to the success and failure of alien species. pp. 15–38. In Pyšek P, Prach K, Rejmánek M, Wade M (eds) Plant Invasions: General Aspects and Special Problems. SPB Academic, Amsterdam, Netherlands

Kriticos DJ, Brown JR, Maywald G et al (2003) SPANDX: A process-based population dynamics model to explore management and climate change impacts on an invasive alien plant, *Acacia nilotica*. Ecol Model 163: 187–208

Leung B, Lodge DM, Finnoff D et al (2002) An ounce of prevention or a pound of cure: Bioeconomic risk analysis of invasive species. Proc R Soc Lond Series B Biol Sci 269: 2407–2413

Lonsdale WM (1999) Global patterns of plant invasions and the concept of invasibility. Ecology 80: 1522–1536

Luken JO, Thieret JW (1997) Assessment and Management of Plant Invasions. Springer, New York, 324 p
Mack RN (1996) Predicting the identity and fate of plant invaders: Emergent and emerging approaches. Biol Conserv 78: 107–121
Mashadi HR, Radosevich SR (2003) Invasive plants. pp. 1–28. In Inderjit (ed) Weed Biology and Management. Kluwer Academic, The Netherlands.
McNeely JA (1999) The great reshuffling: How alien species help feed the global economy. pp. 11–31. In Sandlund OT, Schei PJ, Viken Å (eds) Invasive Species and Biodiversity Management based on a selection of papers presented at the Norway/UN Conference on Alien Species, Trondheim, Norway. Population and Community Biology Series, Vol. 24, Kluwer Academic, Dordrecht, Netherlands
Muyt A (2001) Bush Invaders of Southeast Australia. Richardson RG and Richardson FJ, Meredith, VIC
Neubert MG, Caswell H (2000) Demography and dispersal: Calculation and sensitivity analysis of invasion speed for structured populations. Ecology 81: 1613–1628
Norris RF, Caswell-Chen EP, Kogan M (2003) Concepts in Integrated Pest Management. Pearson Education, Upper Saddle River, NJ
Panetta FD, Mitchell ND (1991) Homoclime analysis and the prediction of weediness. Weed Res 31: 273–284
Parks CG, Radosevich SR, Endress BA et al (2005a) Natural and land-use history of the northwest mountain regions (USA) in relation to patterns of plant invasions. Perspect Plant Ecol Syst 7: 137–158
Peterson AT (2003) Predicting the geography of species' invasions via ecological niche modeling. Q Rev Biol 78: 419–433
Peterson AT, Vieglais DA (2001) Predicting species invasions using ecological niche modeling: New approaches from bioinformatics attack a pressing problem. BioScience 51: 363–371
Pitafi BA, Roumasset JA (2005) Proceedings of Northeastern Agricultural Resource Economics Association. Annapolis, MD
Prather T (2006) How can you incorporate risk assessment into invasive plant management? In McFadzen M (Coordinator) Understanding and Assessing Plant Invasions: A Framework for Prioritizing Management Strategies. Online workshop, Center for Invasive Plant Management, Bozeman, MT. Available at http://www.weedcenter.org/education/syllabus_07.htm
Quigley TM, Arbelbide SJ (1997) An Assessment of Ecosystem Components in the Interior Columbia Basin and Portions of the Klamath and Great Basins. General Technical Report PNW-GTR-405. USDA Forest Service, Pacific Northwest Research Station, Portland, OR
Radosevich SR, Ghersa CM (1992) Weeds, crops, and herbicides: A modern day "neckriddle." Weed Technol 6: 788–795.
Radosevich SR, Holt JS, Ghersa CM (1997) Weed Ecology: Implications for Management, 2nd Edn. Wiley, New York
Radosevich SR, Endress BA, Parks CG (2005) Defining a regional approach for invasive plant research and management. pp.141–166. In Inderjit K (ed) Ecological and Agricultural Aspects of Invasive Plants. Birkhäuser-Verlag, Basel, The Netherlands
Radosevich SR, Holt JS, Ghersa CM (2007) Ecology of Weeds and Invasive Plants: Relationship to Agriculture and Natural Resource Management, 3rd edn. Wiley, Hoboken, NJ
Rappoort EH (1991) Tropical versus temperate weeds: A glance into the present and future. pp. 441–51. In Ramakrishnan PS (ed) Ecology of Biological Invasion in the Tropics. International Scientific Publications, New Delhi
Reichard SE (1997) Prevention of invasive plants introduction on national and local levels. pp. 215–240. In Luken JO, Thieret JW (eds) Assessment and Management of Plant invasions. Springer, New York
Rejmánek M (2000) Invasive plants: Approaches and predictions. Aust Ecol 25: 497–506
Rew LJ, Pokorny ML (eds) (2006) Inventory and Survey Methods for Nonindigenous Plant Species. Montana State University Extension Service, Bozeman, MT

Rew LJ, Maxwell BD, Aspinall R (2005) Predicting the occurrence of nonindigenous species using environmental and remotely sensed data. Weed Sci 53: 236–241

Richardson DM, Williams PA, Hobbs RJ (1994) Pine invasions in the southern hemisphere: Determinants of spread and invadability. J Biogeogr 21: 511–527

Rosonzweig ML (2001) What does the introduction of exotic species do to diversity. Evol Ecol Res 3: 361–367

Ross MA, Lembi CA (1985) Applied Weed Science. Burgess, Minneapolis, MN

Ross MA, Lembi CA (1999) Applied Weed Science, 2nd edn. Burgess, Minneapolis, MN

Sheley R, Petroff J (1999) Biology and Management of Noxious Rangeland Weeds. Oregon State University Press, Corvallis, OR

Sutherst RW, Maywald GF (1985) A computerized system for matching climates in ecology. Agric Ecosyst Environ 13: 281–299

Thuiller W, Richardson DM, Pyšek P et al (2005) Niche-based modeling as a tool for predicting the risk of alien plant invasions at a global scale. Glob Change Biol 11: 2234–2250

Vandermeer J (1989) The Ecology of Intercropping. Cambridge University Press, Cambridge, UK

Vitousek PM, D'Antonio CM, Loope LL et al (1996) Biological invasions as global environmental change. Am Sci 84: 648–678

Werner PA, Soule JD (1976) The biology of Canadian weeds. 18. *Potentilla recta* L., *P. norvegica* L., and *P. argena* L. Can J Plant Sci 56: 591–603

Westoby M, Walker BH, Noy-Meir I (1989) Opportunistic management for rangelands not at equilibrium. J Range Manage 42: 266–274

Williams PA, Nicol E, Newfield M (2000) Assessing the risk to indigenous biota of new plant taxa not yet in New Zealand. In Virture JG, Groves RH (eds) First International Workshop on Weed Risk Assessment, Adelaide, Australia, Commonwealth Scientific and Industrial Research Organization.

Woodward FI (1987) Climate and Plant Distribution. Cambridge University Press, New York

Zouhar K (2003) *Potentilla recta*. In Fire Effects Information System. Online. USDA Forest Service, Rocky Mountain Research Station, Fire Sciences Laboratory. Available at http://www.fs.fed.us/database/feis/

Index